Differential Equations and Their Applications

Analysis from a Physicist's Viewpoint

Differential Equations and Their Applications

Analysis from a Physicist's Viewpoint

Noboru Nakanishi

Kyoto University, Japan

Kenji Seto

Hokkai-Gakuen University, Japan

World Scientific

NEW JERSEY · LONDON · SINGAPORE · BEIJING · SHANGHAI · HONG KONG · TAIPEI · CHENNAI · TOKYO

Published by

World Scientific Publishing Co. Pte. Ltd.

5 Toh Tuck Link, Singapore 596224

USA office: 27 Warren Street, Suite 401-402, Hackensack, NJ 07601

UK office: 57 Shelton Street, Covent Garden, London WC2H 9HE

Library of Congress Cataloging-in-Publication Data
Names: Nakanishi, Noboru, 1932–   author. | Seto, Kenji, 1940–   author.
Title: Differential equations and their applications : analysis from a physicist's viewpoint /
    Noboru Nakanishi, Kyoto University, Japan, Kenji Seto, Hokkai-Gakuen University, Japan.
Description: New Jersey : World Scientific, [2022] | Includes index.
Identifiers: LCCN 2022006321 | ISBN 9789811247453 (hardcover) |
    ISBN 9789811247460 (ebook for institutions) | ISBN 9789811247477 (ebook for individuals)
Subjects: LCSH: Differential equations.
Classification: LCC QC20.7.D5 N346 2022 | DDC 530.15/535--dc23/eng20220421
LC record available at https://lccn.loc.gov/2022006321

British Library Cataloguing-in-Publication Data
A catalogue record for this book is available from the British Library.

For any available supplementary material, please visit
https://www.worldscientific.com/worldscibooks/10.1142/12564#t=suppl

Preface

This book consists of two parts. Part I, written by Noboru Nakanishi, presents the theory of differential equations. Part II, written by Kenji Seto, discusses various concrete problems which can be solved using differential equations. At the end of some of the sections, the authors have included some additional comments and materials as a sidebar that may be of interest to the reader. These are presented as boxed columns.

Nakanishi and Seto would like to express their sincere thanks and deep appreciation to Dr K. K. Phua and Ms. Nargiza Amirova for the opportunity to publish this book. Seto would like to especially thank Nargiza for her kind assistance with her skillful Japanese language to facilitate communication with World Scientific Publishing Company. Seto would also like to express his gratitude to Ms. Lai-Fun Kwong and Ms. Ying-Oi Chiew for their assistance and patience at the production stage of this book.

Thanks are also due to Professor Taichi Kugo,[1] Professor Toshiharu Kawai,[2] and Professor Tetsuro Sakuma[3] for supporting the preparation of this book.

The picture on the book cover is designed by Seto. It is a superposition of two images of Fig. 7.8 in PART II, one rotated 45 degrees clockwise, the other 45 degrees anticlockwise.

---

[1]Professor Emeritus of Kyoto University.
[2]Professor Emeritus of Osaka City University.
[3]Professor Emeritus of Hokkaido University.

Contents

PART I
Theories

Prologue to PART I

Let me introduce myself. I am Noboru Nakanishi, the author of PART I of this book. Although I present here the theory of differential equations, I am not a mathematician but a mathematical physicist, that is, a user of mathematics. For six decades, I have worked in quantum field theory (the theory of elementary particles and quantum gravity).

In PART I, I present the theoretical aspect of differential equations. I neither intend to write a mathematical textbook nor a rigorous advanced mathematics book. My readers, I suppose, are people who are not specialists of mathematics but fond of mathematical sciences, especially, theoretical physics. What I wish to present is not a completed majestic palace but a building under construction.

I agree to the assertion that mathematics must be rigorous. But I believe that the more important thing is the mathematical essence. We should be free from the insistence in the description of "standard" mathematics.

Some mathematicians who are interested in theoretical physics wish to make physical theories mathematically rigorous. Such attempts are important, but one must not forget that there is a pitfall. In order to construct a mathematical theory, it is unavoidable to set up a set of axioms — its initial postulates. But the physical evidences are not sufficient to determine those axioms. It is, therefore, necessary to introduce some axioms artificially on the basis of mathematical convenience. There is no assurance that such axioms do not bring non-physical consequences to the theory.

Eighteenth century mathematics was very comfortable. Many important results were discovered alongside the development of theoretical physics, with little regard for mathematical rigor. For example, the concept of "infinitesimal" was freely used. But it often brought about inconsistency. Then the so-called "(ϵ, δ)-definition of limit" was introduced to make its magnitude clear in terms of ordinary positive numbers. The new aspect of analysis became the set-theoretical description, which employs $\forall$, $\exists$, $\emptyset$, etc. rather than a sequence of equations. Even today, physicists are fond of doing analysis the 18th century way. Physicists have invented concise ways of avoiding inconsistency, based on the belief that *nature contains no inconsistency.*

$$***************************$$

PART I consists of three chapters.

Chapter 1 presents the introduction to the theory of analysis. The contents are essentially elementary, but include somewhat advanced concepts. The description is not necessarily rigorous; its preference is mainly devoted to the understanding of the mathematical essence. The concept of the function and the definitions of differentiation and integration are critically explained. The theory of the analytic function is presented briefly. Various formulas concerning Euler's gamma function and beta function are proved. At the end of Chapter 1, Schwartz's distribution is introduced, and its connection with the boundary value of the analytic function is discussed.

Chapter 2 presents the theory of differential equations. Various methods for solving ordinary differential equations are explained. The second-order linear differential equation is solved by means of power-series expansion. The special functions such as Gauss' hypergeometric function are discussed. Then the boundary value problem of the second-order linear differential equation is considered as the eigenvalue problem.

Chapter 3 presents three topics: differential operator calculus, non-integer order derivative, and the method for solving linear differential equations with constant operator coefficients. The last one is my original contribution.

PART I was originally published as a Japanese book entitled「微分方程式 —物理的発想の解析学—」(Differential Equation — Analysis from the physicist's viewpoint —) by Maruzen Publishing in 2016. I translated this book into English with various modifications; in particular, the boxed comments have been changed almost completely.

****

I am very grateful to my coauthor, Kenji Seto for the TeX setting of my manuscript. Thanks are also due to him and to Professor Toshiharu Kawai[1] for their invaluable comments.

---

[1]Professor Emeritus of Osaka City University.

Chapter 1

Introduction to the Theory of Analysis

This chapter is an introduction to the theory of analysis. Rather than presenting the rigorous mathematics, the essential points of mathematics are discussed critically from the standpoint of the outsider.

1.1 Elementary Facts on Functions

In this section, we summarize some elementary, fundamental facts on functions. The reader who is familiar with elementary mathematics may skip this section.

1.1.1 *Polynomials*

The simplest example of a function is a polynomial. A **polynomial** is a sum over positive powers of a certain variable, say x, multiplied by a constant coefficient, plus a constant term, that is, a polynomial $P(x)$ is

$$P(x) \equiv a_0 x^n + a_1 x^{n-1} + \cdots + a_{n-1}x + a_n \equiv \sum_{k=0}^{n} a_k x^{n-k}. \qquad (1.1.1)$$

If $a_0 \neq 0$, n is called the degree of $P(x)$. The expression written at the rightmost of (1.1.1) is the compact expression written by using the summation symbol $\sum$, which is often used. Here we must understand $x^0 \equiv 1$ even if $x = 0$.

The polynomial on several variables is similarly defined.

1.1.2 *Algebraic functions*

A function which can be written as a ratio of two polynomials is called a **rational function**. A negative power of x, namely, x^{-n} ($n > 0$) is identified with the n-th power of the inverse of x, namely, $1/x^n$. Furthermore, a fractional power of x can be identified with an integral power of a root of x in such a way that $x^{m/n} = (\sqrt[n]{x})^m$. In that way, for any fractional number m/n, $x^{m/n}$ becomes sensible so as to be consistent with the rules of exponentials: $x^p x^q = x^{p+q}$, $(x^p)^q = x^{pq}$. As for the power of any fractional number, or more generally a complex, it is necessary to introduce the general consideration concerning the exponential function. The imaginary power is defined by means of the Euler formula stated in §1.7.

A polynomial in both x and y, denoted by $P(x, y)$, can be regarded as a polynomial in y with the coefficients being polynomials in x. The solution to the algebraic equation $P(x, y) = 0$ with respect to y is called the **algebraic function** of x. If the degree of $P(x, y)$ with respect to y is 5 or more, the algebraic function is not necessarily expressed in terms of rational powers.

Any function which is not an algebraic function is called a **transcendental function**. Therefore, exponential functions and trigonometric functions are transcendental functions.

In general, given a function $F(x, y)$ of two variables x and y, the solution to the equation $F(x, y) = 0$ with respect to y defines a function of x. Such a representation of a function is called an **implicit function**. An algebraic function is an example of an implicit function.

1.1.3 *Exponential functions and logarithmic functions*

As stated above, if the exponent of x is a rational number, then the rules of the exponent are satisfied. Using the continuity with respect to the exponent, we can define the power of a real number so as to be consistent with the rules of the exponent. Thus, we can define a function of the exponent; regarding it as the function of the exponent only, we call it the **exponential function**. In order to define the exponential function, it is necessary to use some limiting procedure. There are several ways of defining it, the most ordinary way is

$$\lim_{n \to \infty} \left(1 + \frac{x}{n}\right)^n \equiv e^x \equiv \exp\ x. \tag{1.1.2}$$

Here, the value of (1.1.2) at $x = 1$, denoted by e, is a special constant called the **base of natural logarithm** or the **Napier constant**; it is an irrational number $e = 2.71828\cdots$.

Given a function $y = f(x)$, if x can be regarded as a function of y, it is called the **inverse function** of $y = f(x)$, and (generically) it is denoted by $x = f^{-1}(y)$. The inverse function of the exponential function is called the **logarithmic function**. That is, when $y = e^x$, we have $x = \log y$. Here, $\log$ (sometimes, it is also denoted by $\ln$) means the **natural logarithm**, but we simply call it the logarithm. Since $e^{\log a} = a$, we have $a^x = e^{x\log a}$. The inverse function of $y = a^x$ is often denoted by $x = \log_a y$, which is nothing but $x = \log y/\log a$.

In the above consideration, we have always assumed $a > 0$ tacitly. If one wishes to extend to the case of $a < 0$ correctly, one must extend everything to the case of a complex number (stated in §1.7).

1.1.4 *Trigonometric functions*

Originally, trigonometric functions were defined by the ratios of two sides as a function of a non-rectangular angle θ (in the units of radian) in a rectangular triangle. But in this definition, θ is restricted to the range $0 < \theta < \pi/2$.[1] The more reasonable definition of a **trigonometric function** is as follows. In the Cartesian coordinate system on a plane, the equation for the unit circle is given by $x^2 + y^2 = 1$. Let θ be the direction angle; the cosine function, the sine function and the tangent function of θ are defined by $x = \cos\theta$, $y = \sin\theta$ and $y/x = \tan\theta = \sin\theta/\cos\theta$, respectively. Hence the identity $\cos^2\theta + \sin^2\theta = 1$ holds. From the above definition of trigonometric functions, we see that they are periodic functions; the period of $\sin\theta$ and $\cos\theta$ is 2π, while that of $\tan\theta$ is π. The zero points of $\sin\theta$ are $\theta = n\pi$, n being any integer. There is a relation $\cos\theta = \sin(\pi/2 \pm \theta)$. The addition theorem for sine and cosine are

$$\begin{aligned}\sin(\theta + \phi) &= \sin\theta\cos\phi + \cos\theta\sin\phi, \\ \cos(\theta + \phi) &= \cos\theta\cos\phi - \sin\theta\sin\phi.\end{aligned} \tag{1.1.3}$$

They are of course proved in elementary geometry. But later, we see that (1.1.3) is easily derivable from the addition theorem for the exponential function.

---

[1] As is well known, π denotes the ratio of the circumference of a circle to its diameter ($\pi = 3.14159\cdots$).

The inverse functions of the trigonometric functions are called the **inverse trigonometric functions**. The inverse function of sin is denoted by arcsin or $\sin^{-1}$; the others are similar. Unless the range of θ is restricted, the inverse trigonometric functions do not have a unique value because of the periodicity of the trigonometric functions. While the inverse function of $y = x^2$ has two values, the inverse trigonometric functions have *infinitely many values*.

The functions which can be composed of algebraic functions, exponential functions and trigonometric functions are called **elementary functions**. There are many important functions which are not elementary functions.

1.1.5 *Differential formulas for basic elementary functions*

We present some fundamental formulas of differentiation for concrete functions without proof:

$$\frac{d}{dx}x^n = nx^{n-1} \qquad (1.1.4)$$

(this formula holds even for n non-integer),

$$\frac{d}{dx}e^x = e^x, \quad \frac{d}{dx}\log x = \frac{1}{x}, \qquad (1.1.5)$$

$$\frac{d}{dx}\sin x = \cos x, \quad \frac{d}{dx}\cos x = -\sin x. \qquad (1.1.6)$$

1.2 What is a Function?

1.2.1 *Definition of a function*

In the previous section, we have discussed the function without defining it. That is, we have tacitly assumed that the concept of a function exists *a priori*. There is one (or several) variable, say x, which can take various values freely, and there is another variable, say y, whose values are determined by the values of x automatically. The variables x and y are called the **independent variable** and the **dependent variable**, respectively. If there exists a formula by which y is calculable from x, then y is a function of x. This was the "definition" of the function in the 18th century. But what is the formula by which y is expressed in terms of x? It is quite obscure. If one is not allowed to take any limiting procedure, almost all important transcendental functions are not included in the concept of a function. But once one allows a limiting procedure, there arise extraordinarily queer "functions". Contrarily, in the 19th century, assuming a defiant attitude, one defined the function as a mapping: given a non-empty set X of numbers, or more generally, a set of elements and another set Y of numbers; if for any $x \in X$, only one $y \in Y$ corresponds to x, then y is called the **function** of x. That is, essentially, a mapping $f : x \mapsto y$ is nothing but a function $y = f(x)$. This is, however, not what we naively imagine as a function. Indeed, for example, if x takes integral numbers only, we want to call y a sequence rather than a function, except in number theory. We expect that x can take continuous values rather than discrete ones. In mathematical language, we should consider the correspondence between topological spaces, but we refrain from indulgence into higher mathematics. Here, we emphasize the importance of considering the behavior of y in a neighborhood of x.

The above-mentioned definition of a function is too broad, but it is also too narrow in the sense that it restricts the function to one-valuedness. We wish to allow many-valuedness otherwise most algebraic functions and inverse trigonometric functions are not accepted as functions. That is, we must set up some artificial range where x can take values. But, then, the inverse function of the inverse function does not coincide with the original function. The function, which we tacitly expect to be the function, namely, the so-called "heavenly-given functions" are found to be the analytic functions, as is seen later. Unfortunately, however, in order to construct the theory of the analytic function, we must construct the theory of the real function which requires the introduction of artificial techniques.

1.2.2 *Continuous functions*

The functions which we actually need in physics are at most those which can be expressed in terms of a finite number (in any finite range) of continuous functions. We do not consider such functions as discontinuous everywhere (for example, $f(x) = 0$ for x rational, $f(x) = 1$ for x irrational).

The **continuity** means that any very small change of the independent variables induces only a very small change in the dependent variable. More precisely, the definition of continuity is as follows. (For simplicity, we state it in the one-variable case.) Inside a certain interval, we consider two points $x = a$ and $x = a + h$. If $|f(a + h) - f(a)|$ can always be made arbitrarily small when $|h|$ is sufficiently small, then $f(x)$ is continuous at $x = a$. We write the above statement as

$$\lim_{h \to 0} f(a + h) = f(a). \tag{1.2.1}$$

In the many-variable case, we have only to extend h to a vector. It is important to note that a function continuous with respect to each variable is *not* necessarily continuous. For example, consider a function $f(x, y) \equiv xy/(x^2 + y^2)$ for $(x, y) \neq (0, 0)$ with $f(0, 0) = 0$. It is continuous with respect to each of x and y. But it is discontinuous at $(x, y) = (0, 0)$, because if it is expressed in the polar coordinates for $r > 0$, we have $f(x, y) = \cos \theta \sin \theta$, θ being arbitrary. In this connection, it is interesting to note that for the corresponding property of the "heavenly-given function", we have the Hartogs theorem, which states, "A many-variable function analytic with respect to each variable is always analytic as the many-variable function."

Since it is cumbersome to substitute a generic value $x = a$ into $f(x)$, we often use x to represent an arbitrary point. Furthermore, in order to indicate that h is a small quantity automatically, we hereafter replace h by Δx. Then the continuity of $f(x)$ at x is written as $\lim_{\Delta x \to 0} f(x + \Delta x) = f(x)$. In this description, we suppose that x is again a variable belonging to an interval I. If $f(x)$ is continuous at every point $x \in I$, $f(x)$ is said to be continuous in I.

If the upper bound on $|\Delta x|$ exists independently of x throughout in I, $f(x)$ is said to be **uniformly continuous** in I. The uniform continuity is an important mathematical concept, which assures the commutability of the order of two limiting processes. But, rather, the need of the uniform continuity might be regarded as a defect of the pointwise definition of the function. Later (in §1.9), we see a reason for taking such a viewpoint.

1.3 Definition of a Differential

As is well known, Isaac Newton introduced the notion of differentiation in order to define the velocity and the acceleration of a point particle at a particular instant of time t. If one draws the motion of the particle as a graph in the Cartesian coordinate system, the direction of the tangent line at t is the **differential coefficient**. On the other hand, independently, Gottfried Wilhelm Leibniz considered the zero limit of the ratio of a small change Δy of a function $y = f(x)$ to a small change Δx of the independent variable x, that is, $\lim_{\Delta x \to 0} \Delta y / \Delta x$ defines the **differential quotient** of $y = f(x)$. Although the differential coefficient and the differential quotient are essentially the same, these two names of the derivative unintendedly symbolize the two "faces" of its characteristics. All elementary functions and other useful functions are differentiable (except for some special points), that is, the limit

$$\lim_{\Delta x \to 0} \frac{\Delta y}{\Delta x} \equiv \lim_{\Delta x \to 0} \frac{f(x + \Delta x) - f(x)}{\Delta x} \tag{1.3.1}$$

exists. If a function is differentiable, then it is continuous at the same point; but the converse is not true.

Regarding (1.3.1) as a function of x, we call it the **derivative** of $y = f(x)$. It is denoted in various ways: y', $f'(x)$, dy/dx, $(d/dx)f(x)$. When the independent variable is time t, it is still customary to use the dot notation like $\dot{y}$ as used by Newton.

1.3.1 *Derivatives and the quotients of differentials*

It is due to Leibniz that we denote the derivative by dy/dx. It is a very convenient notation, but it is also quite misleading, because it is not the ratio of dy to dx as is seen in (1.3.1). In spite of this fact, it is usually treated as if it were a quotient. This way of writing is quite confusing for beginners. In a textbook, which I studied by myself at a pre-teen age, it was stated in the following way: "We define the differential dy by $dy \equiv (dy/dx)\Delta x$. Setting $y = x$ in particular, we find $dx = (dx/dx)\Delta x = \Delta x$. Substituting $\Delta x = dx$ into the original formula, we find $dy = (dy/dx)dx$." I wondered whether it was all right to substitute the formula obtained in a particular case into the formula in the general case; is this a trick? By the above reasoning, the differential quotient has really become the quotient of differentials! Originally, both Δx and Δy should be small but *non-zero* numbers. On the contrary, the differentials dx and dy are special symbols, expressing the **infinitesimal** which cannot be regarded as concrete numbers. Given a formula, all terms must have the same degree with respect to the differentials, while no such restriction exists for Δx and Δy. Then the following question may arise: "What is the infinitesimal?" It is difficult to answer. It may be said to be something like a state, just like infinity.[2] If one wishes to treat the infinitesimals and the infinity as some quantities, one must study the theory of non-standard analysis.

Differentiating the derivative of $y = f(x)$ again, we obtain the second-order derivative, which is denoted by y'', $f''(x)$, d^2y/dx^2, $(d^2/dx^2)f(x)$.

---

[2]In the theory of the exterior differential form, dx_k $(k = 1, 2, \ldots)$ is regarded as a "Grassmann number", which is totally anticommutative.

Here dx^2 means $(dx)^2$; it is different from d^2x. That is,

$$\frac{d^2y}{dx^2} \equiv \left(\frac{d}{dx}\right)^2 y. \qquad (1.3.2)$$

If x is an independent variable, we have $d^2x = 0$, of course. But if x is a **dependent variable**, then we have $d^2x \neq 0$. To make sure, we present an explicit example. Consider a function $y = x^2$. Differentiating it twice with respect to x, we have $d^2y/dx^2 = 2$. On the other hand, if $x = t^2$ where t is an independent variable, then we have $y = t^4$; and therefore, $dy = 4t^3 dt$ and $d^2y = 12t^2 (dt)^2$. Since $dx = 2tdt$, we have

$$\frac{d^2y}{(dx)^2} = \frac{12t^2 (dt)^2}{(2tdt)^2} = 3. \qquad (1.3.3)$$

It is not equal to 2. The reason for the disagreement is as follows: while $d^2x = 0$ in the first calculation, $d^2x = 2(dt)^2$ in the second calculation. Thus the formulas written in terms of differentials may *not* necessarily be rewritten in terms of derivatives, unless we restrict ourselves to consider the *first*-order derivatives only.

1.3.2 *Partial derivatives and total differentials*

In order to clarify the above confusing situation, it is convenient to introduce the notion of partial derivative. Given a function of two or more variables, a derivative with respect to a particular one variable, keeping all other variables constant, is called the **partial derivative**. More concretely, consider a function $y = f(x, u)$. The partial derivative with respect to x is the derivative in which y is regarded as a function of x only, namely, u is fixed. It is denoted by $\partial y/\partial x$. Suppose, for a moment, that u is absent, that is, the set of all other variables is empty. In this setting, we may write

$$dy = \frac{\partial y}{\partial x} dx, \qquad (1.3.4)$$

though contrary to custom. The reason for preferring to write (1.3.4) is that the differentials dx, dy, etc. are definable regardless of whether x, y, etc. are independent variables. On the contrary, there are *no* ∂x and *no* ∂y separately, because the partial derivative is meaningful *only if* the set of the other independent variables is indicated.[3] That is, the symbol ∂ is

---

[3]In thermodynamics, the three physical variables (pressure P, volume V and temperature T) are constrained by the equation of state. Hence the partial derivative must be specified by the other independent variable chosen (e.g. denoted by $(\partial S/\partial T)_P$).

used only in the partial derivative. If one employs the notation of (1.3.4), one can easily avoid the misunderstanding explained above.

In (1.3.4), if dx and dy are replaced by $X - x$ and $Y - y$, respectively, then it becomes the equation for the tangent line at the point (x, y) in the X-Y coordinate system. We extend this consideration to the many-variable case (for simplicity, we describe the two-variable case); then we encounter the concept of the **total differential**.

Suppose that $z = f(x, y)$ is the equation for a surface S in the X-Y-Z coordinate system. If S has the tangent plane at (x, y), then there exists the total differential

$$dz = \frac{\partial f}{\partial x} dx + \frac{\partial f}{\partial y} dy, \tag{1.3.5}$$

and if dx, dy and dz are replaced by $X - x, Y - y$ and $Z - z$, respectively, then (1.3.5) becomes the equation for the tangent plane of S at (x, y). It is more natural to rewrite (1.3.5) into

$$df = \left(dx \frac{\partial}{\partial x} + dy \frac{\partial}{\partial y} \right) f. \tag{1.3.6}$$

As long as f is totally differentiable, f is arbitrary; therefore, we may *formally* remove f from (1.3.6) to obtain

$$d = dx \frac{\partial}{\partial x} + dy \frac{\partial}{\partial y}. \tag{1.3.7}$$

The right-hand side of (1.3.7) can be regarded as the inner product of two vectors (dx, dy) and $(\partial/\partial x, \partial/\partial y)$.

Comments on the Definition of dx and dy

It seems that the following way of defining the differentials dx and dy is adopted in many mathematical textbooks:

(1) One defines dy by

$$dy = \frac{dy}{dx} \cdot \Delta x.$$

(2) In particular, when $y = x$ one has

$$dx = \frac{dx}{dx} \cdot \Delta x = \Delta x.$$

(3) One substitutes $\Delta x = dx$ in (1), and obtains

$$dy = \frac{dy}{dx} dx.$$

The drawback of the above reasoning is the confusion of the dependent variable and the independent one. While the x used in "dx" is an independent variable in (1) and in (3), it is a dependent one in (2). In the text, we have shown, by means of a concrete example, that if this distinction is not clearly made, we encounter an inconsistency in the second-order differential. Indeed, differentiating $dy = y'dx$, we should have

$$d^2y = d(y'dx) = d^2y + y'd^2x$$

owing to the Leibniz rule. Therefore, we must have $d^2x \equiv 0$, but this is true *only when x is an independent variable*.

****

It should be noted that the result, $dx = \Delta x$, obtained in (2) is completely unnecessary. Furthermore, it induces an unwelcome question why one is not allowed to write the integration by $\int \cdots \Delta x$ instead of $\int \cdots dx$. For the dependent variable, one has $dy \neq \Delta y$ generically. But one never needs to use a formula involving both Δy and dy simultaneously.

The only reason for introducing the curious assumption (1) is to avoid introducing the concept of "infinitesimal" in the framework of standard analysis. (In non-standard analysis, infinitesimal and infinity are treated just like a number.) But the introduction of a new concept is a technique widely used in standard mathematics. As long as no inconsistency is encountered, there is no reason for not allowing the introduction of the concept of "infinitesimal".

****

In my opinion, the widely-accepted procedures (1)–(3) are quite unnatural. We should adopt a more natural procedure in which the concept of "infinitesimal" is explicitly used.

Since it is important to discriminate clearly the differential of the independent variable from that of the dependent one, we denote the former by $\tilde{d}x$ for clarity. It is an infinitesimal quantity. For the dependent variable $y = f(x)$, we define

$$dy \equiv \frac{dy}{dx}\tilde{d}x.$$

Hence for $f(x) \equiv x$, we have

$$dx = \tilde{d}x,$$

that is, the first-order differential of the independent variable is equal to that of the dependent one. As for the higher-order differentials, as shown above, we have $\tilde{d}^2 x = 0$ and therefore

$$\tilde{d}^n x = 0, \quad \text{for} \quad n \geq 2.$$

Thus, it is possible to express any formula without using $\tilde{d}$.

1.4 Leibniz Rule

1.4.1 *Fundamental properties of differentiation*

Differentiation is a **linear operation**, that is, the differentiation of a linear combination of several functions is equal to the same linear combination of their derivatives. If we want to write this property as a formula, we have the following (the two-function case is sufficient to reproduce the general case):

$$\frac{d}{dx}\left(af(x) + bg(x)\right) = a\frac{df(x)}{dx} + b\frac{dg(x)}{dx}, \tag{1.4.1}$$

where a and b are arbitrary constants and $f(x)$ and $g(x)$ are arbitrary functions.

As for the product of two functions, we have a very important formula,

$$\frac{d}{dx}\left(f(x)g(x)\right) = \frac{df(x)}{dx}g(x) + f(x)\frac{dg(x)}{dx}, \tag{1.4.2}$$

which is called the **Leibniz rule**. Furthermore, there is a rule characteristic to the differentiation. The differentiation of a function is expressed in the following simple form:

$$\frac{d}{dx}f\left(g(x)\right) = \frac{dg(x)}{dx}\frac{df\left(g(x)\right)}{dg(x)}. \tag{1.4.3}$$

This formula is formally derivable by the reduction of the differential $dg(x)$ on the right-hand side. (Such a formal calculation is always true as long as the first-order differentials only are involved.) This formula is very strong. For example, since $(d/dx)(1/x) = -1/x^2$, by setting $f(x) = 1/x$ in (1.4.3), we have

$$\frac{d}{dx}\left(\frac{1}{g(x)}\right) = -\frac{1}{[g(x)]^2}\frac{dg(x)}{dx}. \tag{1.4.4}$$

From (1.4.2) and (1.4.4), we obtain the differentiation formula for the quotient of two functions,

$$\frac{d}{dx}\left(\frac{f(x)}{g(x)}\right) = \frac{\dfrac{df(x)}{dx}g(x) - f(x)\dfrac{dg(x)}{dx}}{[g(x)]^2}. \tag{1.4.5}$$

As for the implicit function defined by $f(x, y) = 0$, we have the following formula. Setting $z = 0$ in the formula for the total differential (1.3.5), we have

$$0 = \frac{\partial f}{\partial x}dx + \frac{\partial f}{\partial y}dy; \tag{1.4.6}$$

and therefore

$$\frac{dy}{dx} = -\frac{\partial f/\partial x}{\partial f/\partial y}. \tag{1.4.7}$$

1.4.2 *Higher-order derivatives*

If we know the derivatives of some fundamental functions, we can calculate the derivative of any function obtainable by combining them, because we have the general formulas, (1.4.1), (1.4.2), (1.4.3), (1.4.5) and (1.4.7) for the differentiation. This property of differentiation is in sharp contrast with integration which is discussed in the next section.

Now, we can calculate the derivative of the derivative of a function $y = f(x)$, and its further derivative, ..., and the n-th-order derivative, which is denoted by $y^{(n)}$, $f^{(n)}(x)$, $d^n y/dx^n$ and $(d/dx)^n f(x)$.

By using the Leibniz rule (1.4.2) repeatedly, we obtain

$$\frac{d^2}{dx^2}\big(f(x)g(x)\big) = \frac{d^2 f(x)}{dx^2}g(x) + 2\frac{df(x)}{dx}\frac{dg(x)}{dx} + f(x)\frac{d^2 g(x)}{dx^2},$$

$$\frac{d^3}{dx^3}\big(f(x)g(x)\big) = \frac{d^3 f(x)}{dx^3}g(x) + 3\frac{d^2 f(x)}{dx^2}\frac{dg(x)}{dx}$$

$$+ 3\frac{df(x)}{dx}\frac{d^2 g(x)}{dx^2} + f(x)\frac{d^3 g(x)}{dx^3}, \ldots\ldots\ldots \tag{1.4.8}$$

The general formula is written as

$$\frac{d^n}{dx^n}\big(f(x)g(x)\big) = \sum_{k=0}^{n} {}_nC_k \frac{d^{n-k}f(x)}{dx^{n-k}} \frac{d^k g(x)}{dx^k}, \qquad (1.4.9)$$

where ${}_nC_k \equiv n!/\big(k!(n-k)!\big)$ is the number of combinations, namely, the coefficient of the binomial expansion. The reason why there arise the coefficients of the binomial expansion can be understood by the following consideration. We rewrite the Leibniz rule (1.4.2) into

$$\left(\frac{\partial}{\partial x} + \frac{\partial}{\partial x'}\right)\big(f(x)g(x')\big)\bigg|_{x'=x} = \left(\frac{\partial f(x)}{\partial x}g(x') + f(x)\frac{\partial g(x')}{\partial x'}\right)\bigg|_{x'=x}.$$
$$(1.4.10)$$

Then the n-th-order derivative is evidently written as

$$\left(\frac{\partial}{\partial x} + \frac{\partial}{\partial x'}\right)^n \big(f(x)g(x')\big)\bigg|_{x'=x}. \qquad (1.4.11)$$

Then the theorem of binomial expansion works to reproduce (1.4.9).

1.4.3 *Linear operators*

Although $\partial/\partial x$ is not a quantity but an operation, it is often convenient to regard it as if it were a special kind of quantity. In general, the operation which is regarded as a "quantity" is called the **operator**. If it is linear, then it is called a **linear operator**. The simplest example of a linear operator is the operation defined by multiplying a certain number.

If a linear operator satisfies the Leibniz rule, then it is called a **derivation**. There are derivations other than differentiation. Such an example is constructed as follows: consider two linear operators A and B, they are *not* commutative, that is, $AB \neq BA$ (suppose matrix, for example). Their difference is called the **commutator**, which is denoted by using a pair of square brackets:

$$[A, B] \equiv AB - BA. \qquad (1.4.12)$$

Let A be a fixed operator, while X be a variable operator. We consider the commutator $[A, X]$; then $[A, \cdot]$ can be regarded as an operator acting on X. Since

$$[A, \ XY] = AXY - XYA = AXY - XAY + XAY - XYA$$
$$= [A, X]Y + X[A, Y], \qquad (1.4.13)$$

we see that the operator $[A, \cdot]$ satisfies the Leibniz rule; that is, the commutator operation is a derivation. This fact is very important in quantum mechanics because its fundamental equation, the Heisenberg equation, has the form $i(\partial/\partial t)X = [H, X]$.

The operator defined by the multiplication by x is also denoted by x. The operator x and the quantity x are not the same concept, because the former is a mapping $f(x) \mapsto xf(x)$. The multiplication operator x and the differential operator $\partial/\partial x$ are not commutative. Indeed, from the Leibniz rule, we have

$$\frac{\partial}{\partial x}\big(xf(x)\big) = f(x) + x\frac{\partial}{\partial x}f(x). \qquad (1.4.14)$$

We rewrite (1.4.14) as

$$\left[\frac{\partial}{\partial x},\ x\right] f(x) = f(x). \qquad (1.4.15)$$

Since (1.4.15) is independent of what function $f(x)$ is, we omit writing it; then we see that a relation between two operators,

$$\left[\frac{\partial}{\partial x},\ x\right] = 1 \qquad (1.4.16)$$

holds. In general, if the commutator between two operators are expressed by some known operator, such a relation is called a **commutation relation**. The commutation relation (1.4.16) is an important relation in the Schrödinger equation of quantum mechanics, because the momentum operator is expressed by $-i\partial/\partial x$ there.

1.5 Definitions of Integration

Unlike differentiation, integration has two clearly different definitions. It is important to recognize this fact. The first definition of integration is the inverse operation of differentiation. This is the fundamental theorem of the theory of analysis. The second definition of integration defines an area by expressing a function as a graph and then regarding it as the limit of a histogram. This is a useful tool for calculating the area whose boundary is curved.

1.5.1 *Definition of integration as the inverse operation of differentiation*

Firstly, we consider the definition of integration as the inverse operation of differentiation. We consider two functions $F(x)$ and $f(x)$. If $f(x)$ equals the derivative of $F(x)$ (i.e., $f(x) = F'(x)$), $F(x)$ is called the **primitive function** of $f(x)$ or the **indefinite integral** of $f(x)$. Looking from the side of $F(x)$, we call $f(x)$ the **integrand**.

We express $F(x)$ in terms of $f(x)$ in the following way:

$$F(x) = \int f(x)dx. \tag{1.5.1}$$

Since the derivative of a constant is zero and since any other differentiable function cannot have the derivative of identically zero, the linearity of the integration implies that $F(x)$ is undetermined up to an additional constant. This constant is called the **integration constant**, and is usually denoted by C. It is important to note that C is not a particular number but an indefinite constant.

Most mathematicians like to write the integration as in (1.5.1). But most physicists prefer instead to write the integration as

$$\int dx f(x). \tag{1.5.2}$$

In what way we write it is, of course, a matter of taste, but we think that (1.5.2) is more reasonable than (1.5.1), because the symbols $\int$ and dx express one and only one operation. Indeed, even mathematicians do not write differentiation as $df(x)(dx)^{-1}$. Furthermore, from a practical point of view, (1.5.2) is more convenient than (1.5.1). In physics, in contrast with mathematics, one often encounters multiple integrations. As is seen later, the information concerning the range of integration is written in the position of the symbol $\int$. Hence if one applies (1.5.1) to a multiple integral, the correspondence between the integration variable and the range becomes very confusing. Such confusion will not occur if one uses (1.5.2).

1.5.2 *Integration formulas*

It is straightforward to rewrite the elementary differential formulas (1.1.4), (1.1.5) and (1.1.6) into integration ones:

$$\int dx\, x^n = \frac{x^{n+1}}{n+1} + C \quad (n \neq -1); \tag{1.5.3}$$

$$\int dx\, e^x = e^x + C, \quad \int \frac{dx}{x} = \log x + C; \tag{1.5.4}$$

$$\int dx\, \cos x = \sin x + C, \quad \int dx\, \sin x = -\cos x + C. \tag{1.5.5}$$

As stated in §1.4, the general rules for the derivative are the linearity, the Leibniz rule and the formula for a function of function. As for the linearity there is no problem; the integration is a linear operation. The third rule (1.4.3) is also easily translated into the integration formula with a change of the integration variable, that is, setting $x = g(u)$, we have

$$\int dx \, f(x) = \int du \, \frac{dg(u)}{du} f\big(g(u)\big). \qquad (1.5.6)$$

This formula is called **integration by substitution**, and is often used in the practical calculation of integrals.[4]

Note the fact that $\log x$ is not real for $x < 0$ in the second formula of (1.5.4). Setting $x = -x'$ ($x' > 0$), we can make $\log x'$ real, while using (1.5.6) we see that the expression for the integral in the left-hand side is formally the same as the original one. Thus we obtain

$$\int \frac{dx}{x} = \log |x| + C. \qquad (1.5.7)$$

Of course, in (1.5.7) C cannot always be real.

Now, the problematic one is the Leibniz rule (1.4.2). In this formula, while only one term exists on the left-hand side, there are two terms on the right-hand side. Hence when we rewrite it into the integration formula, an integral cannot be expressed in the form in which the symbol $\int$ is absent. More concretely, we have

$$\int dx \, \frac{df(x)}{dx} g(x) = f(x)g(x) - \int dx \, f(x) \frac{dg(x)}{dx}. \qquad (1.5.8)$$

That is, its right-hand side still contains an integral. Therefore, (1.5.8) is called the formula of **integration by parts**. It is a characteristic property of integration that a product of two functions is generally difficult to be integrated out even if both functions are elementary. For example, such integrals as

$$\int dx \frac{\log x}{1 - x}, \quad \int dx \, e^{-x^2} = \frac{1}{2} \int dx' \frac{e^{-x'}}{\sqrt{x'}},$$

$$\int dx \, \frac{dx}{\sqrt{(1 - x^2)(1 - k^2 x^2)}}, \quad (0 < k < 1) \qquad (1.5.9)$$

are known to be non-elementary functions. They are called the **di-logarithm integral**, the **Gaussian integral** and the **elliptic integral**,

---

[4]For a multiple integral, the change of integration variables need to introduce the **Jacobian**, the determinant of the matrix formed by partial derivatives.

respectively. From a practical standpoint, this fact shows that we can make use of these integrals for defining new functions. Indeed, the above integrals (1.5.9) are used to define the **di-logarithm function** (or the **Spence function**), the **error function** and (by taking its inverse function) the **Jacobi elliptic function**.

1.5.3 *Definition of integration as the limit of a sum*

The second way of defining integration is not based on the concept of differentiation. It is, naively speaking, the limit of the sum over slices parallel to the y axis. There are various ways of defining this procedure mathematically, the most popular being the **Riemann integral**. Its precise definition is rather cumbersome. It is stated as follows.

Let $f(x)$ be a function of x defined in an interval including a closed interval $[a, b]$. We choose $n - 1$ points in such a way that $a = a_0 < a_1 < \cdots < a_{n-1} < a_n = b$ and decompose $[a, b]$ into n intervals $[a_0, a_1], [a_1, a_2], \ldots, [a_{n-1}, a_n]$. Furthermore, in each small interval $[a_{k-1}, a_k]$, we choose $x = x_k$ $(k = 1, 2, \ldots, n)$ in an arbitrary way. Now, we consider the $n \to \infty$ limit of the sum

$$\sum_{k=1}^{n} (a_k - a_{k-1}) f(x_k). \tag{1.5.10}$$

If it converges to a certain value independently of the choices of a_k and x_k, as long as $\max_k(a_k - a_{k-1}) \to 0$, then the $n \to \infty$ limit of (1.5.10) is called the **definite integral** of $f(x)$ between a and b. We write it as

$$\int_a^b dx\, f(x). \tag{1.5.11}$$

Although the above definition looks complicated, it is nothing but the limit of the area of the histogram approximating the graph $y = f(x)$, if $f(x)$ is a continuous function. Since the definite integral evidently obeys the simple addition law,

$$\int_a^b dx\, f(x) + \int_b^c dx\, f(x) = \int_a^c dx\, f(x). \tag{1.5.12}$$

It is all right to extend to the case that $f(x)$ is piecewise continuous. A problematic situation arises if $f(x)$ has an infinite number of discontinuous points in $[a, b]$. The origin of this trouble is the pointwise definition of the concept of the function discussed in §1.2.

To overcome this, the concept of the **Lebesgue integral** has been introduced. Its idea is essentially to neglect the contributions from the "countably infinite number" of exceptional points. The Lebesgue integral has the good property that it is commutative with the limiting procedure concerning the integrand. But it is delicate for the limiting procedure concerning the integration interval.

1.5.4 *Comments on the definite integral*

The definite integral is defined in a closed interval. If one wants to define the definite integral in an open interval, one needs to introduce the limiting procedure concerning the interval.

For example, we consider two integrals:

$$\lim_{\varepsilon \to +0} \int_{a-\varepsilon}^{b} dx\, f(x), \quad \lim_{b \to \infty} \int_{a}^{b} dx\, f(x). \tag{1.5.13}$$

If the limit exists, it is called the **improper integral**. Then the two examples presented in (1.5.13) are simply denoted by

$$\int_{a}^{b} dx\, f(x), \quad \int_{a}^{\infty} dx\, f(x). \tag{1.5.14}$$

But one must be careful for the fact that (1.5.14) doubly contains the limiting procedures. Therefore, when one considers another limiting procedure, the interchangeability of their order is often violated.

By definition, the definite integral gives us a constant value. But just as we have considered the derivative from the differential coefficient, we define a function by replacing the upper end point b by a variable x, Then many people write the definite integral as the form

$$\int_{a}^{x} dx\, f(x). \tag{1.5.15}$$

But this simplified notation is quite misleading. The variable x here is a newly-introduced quantity, but not the integration variable which was denoted by x. The letter should be replaced by x', say. Thus the correct notation for the definite integral as a function of x is written as

$$\int_{a}^{x} dx'\, f(x'). \tag{1.5.16}$$

Without using (1.5.16), we encounter confusion if the integrand contains x as a parameter.

1.5.5 *Fundamental theorem of the theory of analysis*

Now, from the simple additivity (1.5.12), we have

$$\int_a^{x+\Delta x} dx'\, f(x') - \int_a^x dx'\, f(x') = \int_x^{x+\Delta x} dx'\, f(x'). \qquad (1.5.17)$$

We divide both sides by Δx and then take the limit of $\Delta x \to 0$. The definition of the Riemann integral and that of the differentiation (1.3.1) imply

$$\frac{d}{dx}\int_a^x dx'\, f(x') = f(x). \qquad (1.5.18)$$

This is the **fundamental theorem of the theory of analysis**. Writing the primitive function of $f(x)$ as $F(x)$, we have

$$\int_a^x dx'\, f(x') = F(x) - F(a), \qquad (1.5.19)$$

where the integration constant C has been determined to be equal to $-F(a)$ by the requirement that the right-hand side of (1.5.19) should vanish for $x = a$.

Thus, the calculation of the definite integral is reduced to that of the indefinite integral. But the charming point of the definite integral is the fact that there are some special cases in which it is exactly calculable without knowing the corresponding indefinite integrals (see §1.9).

1.5.6 *Integral representation as an all-round weapon*

In the case of the derivative, we often encounter the higher-order derivative, and for it the generalized Leibniz rule (1.4.9) holds. On the contrary, there scarcely appears a higher-order integral. Why? The reason is that the higher-order integral is readily reduced to the expression written in terms of the first-order integrals only. Indeed, using formula (1.5.8) for the integration by parts, we can reduce the second-order integral in the following way:

$$\int_a^x dx' \int_a^{x'} dx''\, f(x'') = \int_a^x dx' \left[\frac{d(x'-a)}{dx'}\int_a^{x'} dx''\, f(x'')\right]$$

$$= (x-a)\int_a^x dx'\, f(x') - \int_a^x dx'\, (x'-a)f(x').$$

$$(1.5.20)$$

By repeating this procedure, it is obvious that the n-th-order integral is reduced to a sum of the first-order integrals. This is a mysterious property of the integral.

As we will see later, various non-elementary functions are expressible in terms of the integrals of elementary functions. Furthermore, the functions which are characterized by certain properties can be expressed by the integral of a particular form in a unified way. Owing to the above facts, it is very convenient to express a particular function or a set of particular functions in terms of the certain integral. Such an expression is generally called the **integral representation**. The integral representation is not only convenient for practical calculations but also useful for making the difficult proof of theorems simpler and/or transparent. As is seen later, even the concept of differentiation can be expressed in terms of the integral representation. This fact is used to extend the n-th-order of the derivative to the non-integer order one.

1.6 Taylor Expansion

1.6.1 *Derivation of the Taylor expansion*

Given a function continuous in a closed interval, we can approximate it by a polynomial in any high accuracy, provided that the degree n of the polynomial is not restricted. Hence, it is natural to expect that if we take the limit $n \to \infty$, then any function $f(x)$ continuous in a closed interval could be expanded into the power series $f(x) = \sum_{n=0}^{\infty} c_n x^n$, where c_n is a constant. Of course, such an expectation does not hold as it is, because there is a problem of the convergence of series. But, as is seen below, this expectation is realized for the "heavenly-given function", i.e., the analytic function (see §1.8).

If a function $f(x)$ is continuous in a certain interval I, it is said that $f(x)$ belongs to C^0-**class** in I. If $f(x)$ is n times differentiable in I and if $f^{(n)}(x)$ is continuous in I, then $f(x)$ is said to belong to C^n-**class**. Furthermore, if $f(x)$ is any times differentiable in I, it is said to belong to C^∞-**class** in I. The analytic function belongs to C^∞-class, but the converse is not necessarily true as long as x is restricted to be real.

Let $f(x)$ be a function belonging to C^∞-class in a closed interval $N(a)$, which includes a neighborhood of the point $x = a$. We rewrite (1.5.19) as

$$f(x) = f(a) + \int_a^x dx_1 f'(x_1), \qquad (1.6.1)$$

where $F(x)$ in (1.5.19) has been replaced by $f(x)$. Next, in (1.6.1), by replacing x by x_1, x_1 by x_2 and $f(x)$ by $f'(x)$, we have

$$f'(x_1) = f'(a) + \int_a^{x_1} dx_2 f''(x_2). \tag{1.6.2}$$

Substituting (1.6.2) into (1.6.1), we obtain

$$f(x) = f(a) + f'(a)(x - a) + \int_a^x dx_1 \int_a^{x_1} dx_2 f''(x_2). \tag{1.6.3}$$

Again in (1.6.1), by replacing x by x_2, x_1 by x_3 and $f(x)$ by $f''(x)$, and substituting into (1.6.3), we have

$$f(x) = f(a) + f'(a)(x - a) + f''(a)\frac{(x - a)^2}{2}$$
$$+ \int_a^x dx_1 \int_a^{x_1} dx_2 \int_a^{x_2} dx_3 f'''(x_3). \tag{1.6.4}$$

Repeating the above procedure, we obtain

$$f(x) = f(a) + f'(a)\frac{x - a}{1!} + f''(a)\frac{(x - a)^2}{2!} + \cdots$$
$$+ f^{(n-1)}(a)\frac{(x - a)^{n-1}}{(n - 1)!} + R_n, \tag{1.6.5}$$

where

$$R_n \equiv \int_a^x dx_1 \int_a^{x_1} dx_2 \cdots \int_a^{x_{n-1}} dx_n f^{(n)}(x_n). \tag{1.6.6}$$

Since $f^{(n)}(x)$ is a continuous function, it takes a maximum $M_n \equiv \max|f^{(n)}(x)|$, in the closed interval $N(a)$. Therefore, we obtain the inequality

$$|R_n| \leq M_n \frac{|x - a|^n}{n!}. \tag{1.6.7}$$

It is easily shown that $M^n/n! \to 0$ as $n \to \infty$ for any fixed M. Accordingly, if $(M_n)^{1/n}$ is bounded, that is, if there is a constant M such that $M_n < M^n$, then we find that $R_n \to 0$ as $n \to \infty$. Thus, the power series of (1.6.5) converges to $f(x)$; therefore, we have

$$f(x) = \sum_{n=0}^{\infty} \frac{f^{(n)}(a)}{n!}(x - a)^n. \tag{1.6.8}$$

The series on the right-hand side of (1.6.8) is called the **Taylor expansion** of $f(x)$ at $x = a$. In particular, its Taylor expansion at $x = 0$

$$f(x) = \sum_{n=0}^{\infty} \frac{f^{(n)}(0)}{n!} x^n, \qquad (1.6.9)$$

is called the **Maclaurin expansion**, which is a power series of x.

The series of the Taylor expansion is shown to be absolutely and uniformly convergent in $N(a)$. This fact implies that the Taylor series is differentiable and integrable term by term. Hence the Maclaurin expansion may be regarded as the "polynomial of infinite degree".

In general, given a power series $\sum_{n=0}^{\infty} c_n(x - a)^n$, where $\lim_{n \to \infty} |c_n|^{1/n} \equiv 1/R$ exists, R is called the **radius of convergence** of the above series. Then it is absolutely and uniformly convergent in $|x - a| \leq R - \varepsilon$, $(\varepsilon > 0)$.

1.6.2 *Taylor expansions of elementary functions*

In order to obtain the Taylor expansion of a given function, we have only to calculate its n-th-order derivative.

The Taylor expansion of the binomial function of an arbitrary power is given by[5]

$$(1 + x)^\alpha = \sum_{n=0}^{\infty} \frac{\alpha(\alpha - 1)\cdots(\alpha - n + 1)}{n!} x^n. \qquad (1.6.10)$$

Its radius of convergence is $R = 1$, but if α is a non-negative integer, the summation stops at $n = \alpha$, and (1.6.10) reduces to the ordinary binomial expansion.

For $\alpha = -1$, (1.6.10) becomes

$$\frac{1}{1 + x} = \sum_{n=0}^{\infty} (-1)^n x^n. \qquad (1.6.11)$$

Integrating (1.6.11) term by term, we obtain the Taylor expansion of the logarithmic function:

$$\log(1 + x) = \sum_{n=0}^{\infty} (-1)^n \frac{x^{n+1}}{n + 1} = \sum_{n=1}^{\infty} \frac{(-1)^{n-1}}{n} x^n. \qquad (1.6.12)$$

---

[5]The $n = 0$ term of the right-hand side of (1.6.10) should be understood to be equal to 1.

We can easily calculate the n-th-order derivatives of the exponential function and the trigonometric functions (sine and cosine); therefore, we obtain

$$e^x = \sum_{n=0}^{\infty} \frac{1}{n!} x^n = 1 + x + \frac{x^2}{2!} + \frac{x^3}{3!} + \cdots, \qquad (1.6.13)$$

$$\sin x = \sum_{n=0}^{\infty} \frac{(-1)^n}{(2n+1)!} x^{2n+1} = x - \frac{x^3}{3!} + \frac{x^5}{5!} - \cdots,$$

$$\cos x = \sum_{n=0}^{\infty} \frac{(-1)^n}{(2n)!} x^{2n} = 1 - \frac{x^2}{2!} + \frac{x^4}{4!} - \cdots. \qquad (1.6.14)$$

Their radii of convergence are $R = \infty$.

As for the inverse trigonometric functions (arcsine and arccosine), the Taylor expansions are obtained by integrating the Taylor expansion of $(1 - x^2)^{-1/2}$ term by term. (We omit writing their explicit expressions.)

Unfortunately, however, it is only for rather exceptionally simple functions that the explicit expression of the n-th-order derivative can be obtained.[6]

Changing our viewpoint, we make use of the Taylor expansion of an elementary function $f(x)$ in order to express a given sequence $\{\alpha_n\}$ as that of the expansion coefficients. In this situation, the function $f(x)$ is called the **generating function** of the sequence $\{\alpha_n\}$.

For example, consider the famous Fibonacci sequence $\{1, 1, 2, 3, 5, 8, 13, 21, \ldots\}$, which is defined by the **recurrence formula** $\alpha_{n+1} = \alpha_n + \alpha_{n-1}$, $(n \geq 1)$ with $\alpha_0 = \alpha_1 = 1$. Its generating function is shown to be $(1 - x - x^2)^{-1}$. Various interesting sequences are often expressed by generating functions. For example, the Bernoulli numbers are described by $\tan x$.

1.7 Euler Formula

1.7.1 *Complex plane*

Initially, the theory of analysis was the theory of the real function. The solutions to the 3-degree algebraic equation was given by the Tartaglia–Cardano method, in which the imaginary number $\sqrt{-1} \equiv i$ was used inevitably, even for writing real solutions. But $\sqrt{-1}$ was not recognized as a number in the

---

[6]If one makes use of the free software, *Maxima*, one can easily find the Taylor expansion of a somewhat complicated function up to a certain finite order, but the n-th-order derivative cannot be given.

middle ages. The complex number $a + ib$ had no "civil rights" in mathematics until the end of the 18th century.

Today, a "number" simply means a complex number. The concept of the complex number is not only very natural in pure mathematics but also surprisingly useful in practical calculations. This is the feeling of people who have learned mathematics and physics. Furthermore, quantum theory, which is the most fundamental theory of natural science, cannot be reasonably formulated without using complex numbers.

A complex number z is uniquely expressed as $z = x + iy$, where x and y are real numbers. The x is called the **real part** of z, and is denoted by $x = \Re z$. Similarly, y is called the **imaginary part** of z, and is denoted by $y = \Im z$. The $x - iy$ is called the **complex conjugate** of z, and is denoted by $\bar{z}$ in mathematics and z^* in most cases of physics. The "bar" notation is not only inconvenient for a lengthy expression but also somewhat ugly when it is used in the denominator of a long fractional formula. The "star" notation is free from such inconveniences. Some people may complain about the use of the "star" notation, because it is used in a different sense in mathematics. But even the "bar" notation is used as the symbol for closure. Therefore, both are equally unsuitable.

Given a complex number $z = x + iy$, (x, y) can be regarded as the Cartesian coordinates on a plane. This plane is called the **complex plane** or the **Gauss plane**. In the polar coordinate system, the radial length $r \equiv \sqrt{x^2 + y^2}$ and azimuthal angle $\theta \equiv \arctan y/x$, we have $z = r(\cos\theta + i\sin\theta)$ because $x = r\cos\theta$ and $y = r\sin\theta$. The r is called the **absolute value** or the **modulus** of z, and denoted by $|z|$. The θ is called the **argument** of z, and denoted by $\arg z$.

1.7.2 *Relationship between trigonometric functions and exponential functions*

It is rather difficult to remember the multiple-angle formulas for the trigonometric functions, sine and cosine, but one can easily reproduce them if one uses de Moivre's formula

$$(\cos\theta + i\sin\theta)^n = \cos(n\theta) + i\sin(n\theta), \qquad (1.7.1)$$

by expanding the left-hand side of (1.7.1) and by taking each of the real parts and the imaginary parts on both sides.

In (1.7.1), n is originally a non-negative integer. The proof of (1.7.1) then is easily made by means of mathematical induction and the addition theorem of the trigonometric functions. In the following, we show that n can be extended to any real number.

First, we consider the case of n being a negative integer. By taking the complex conjugate and the inverses on both sides, the formula to be proved is reduced to (1.7.1) for n being a positive integer. Second, we consider the case where $n' = 1/n$ is an integer. By setting $n\theta = \theta'$ and taking the n'-th power on both sides, the formula to be proved is reduced to the formula (1.7.1) with θ' and n' instead of θ and n, respectively. From the above result, we see that (1.7.1) holds for n being an arbitrary rational number. Then the continuity with respect to n implies that (1.7.1) should hold for n real.

Now, the background of de Moivre's formula is the Euler formula

$$e^{i\theta} = \cos\theta + i\sin\theta. \tag{1.7.2}$$

If we use (1.7.2), (1.7.1) is nothing more than the exponential rule $(e^{i\theta})^n = e^{in\theta}$. The proof of (1.7.2) can be made by comparing (1.6.13) and (1.6.14), where they are valid for the case of x being a complex number (see §1.8).

Taking the complex conjugate of (1.7.2), we have

$$e^{-i\theta} = \cos\theta - i\sin\theta. \tag{1.7.3}$$

From (1.7.2) and (1.7.3), we obtain

$$\cos\theta = \frac{e^{i\theta} + e^{-i\theta}}{2}, \qquad \sin\theta = \frac{e^{i\theta} - e^{-i\theta}}{2i}. \tag{1.7.4}$$

Thus, we have seen that the trigonometric functions are essentially nothing but the exponential function. This fact is very important.

From (1.7.2), we find that an arbitrary complex number is expressed as $z = re^{i\theta}$. In particular, we see $e^{2\pi i} = 1$ and $e^{\pm\pi i} = -1$. The addition theorem (1.1.3) for the trigonometric functions is reduced to that for the exponential function.

1.7.3 *Many-valuedness of the logarithmic function*

Since the trigonometric functions (sine and cosine) are periodic functions of a period 2π, (1.7.2) implies that the exponential function is also a periodic function of a period $2\pi i$. Therefore, the logarithmic function, which is the inverse function of the exponential function, is an infinitely many-valued function. Explicitly writing, we have

$$\log z = \log r + i(\theta + 2\pi n), \tag{1.7.5}$$

where n is an arbitrary integer.

In the case of the real integral, there has arisen a curious formula (1.5.7) containing $|x|$. This unnatural formula can be understood when we write the real part of (1.7.5). Therefore, in the discussion of solving differential equations (see next chapter), we need not take care of adopting $\log |x|$ instead of $\log x$.

1.8 Analytic Functions

1.8.1 *Holomorphic functions*

The analytic function is, so to speak, what we imagine as the natural extension of the elementary function. Given a complex function $w = f(z)$ of a complex independent variable z, we define its derivative by

$$\lim_{\Delta z \to 0} \frac{\Delta w}{\Delta z} = \lim_{\Delta z \to 0} \frac{f(z + \Delta z) - f(z)}{\Delta z}. \tag{1.8.1}$$

If (1.8.1) exists, we say $w = f(z)$ is differentiable at z, as is in the case of the real function. But it is also called **holomorphic** (or analytic) at z. The reason why such a special name is given is that the holomorphic function is always not only any times differentiable but also expandable into the Taylor series, as is seen below.

This wonderful property is the consequence of the deep implication of (1.8.1). Since $\Delta z \to 0$ is the synonym of $|\Delta z| \to 0$, $\arg \Delta z$ is arbitrary (of course, $0 \le \arg \Delta z < 2\pi$). That is, (1.8.1) requires that infinitely different limits must coincide with each other. The **Cauchy–Riemann differential equations** express this requirement more explicitly: $w = f(z)$ is differentiable at z, if and only if

$$\frac{\partial u}{\partial x} = \frac{\partial v}{\partial y}, \quad \frac{\partial u}{\partial y} = -\frac{\partial v}{\partial x}, \tag{1.8.2}$$

hold, where $z = x + iy$ and $w = u + iv$. We may rewrite (1.8.2) as

$$\left(\frac{\partial}{\partial x} + i \frac{\partial}{\partial y} \right)(u + iv) = 0. \tag{1.8.3}$$

If we regard $w = f(z)$ as a function of two variables z and z^* instead of x and y, (1.8.3) is nothing but the function $f(z)$ being independent of z^*, namely,

$$\frac{\partial w}{\partial z^*} = 0. \tag{1.8.4}$$

Thus we see that

$$dw = \frac{\partial w}{\partial z}dz + \frac{\partial w}{\partial z^*}dz^* = \frac{\partial w}{\partial z}dz. \tag{1.8.5}$$

In short, holomorphy means the condition that the two-dimensional total differential dz coincides with the one-dimensional differential.

1.8.2 *Cauchy theorem*

Various formulas for the differentiation of the holomorphic function are the same as those in the case of the real (C^∞) function. For the integration, however, a new problem arises: in order to define the definite integral between two points $z = \alpha$ and $z = \beta$, we must also indicate along what path P is connecting α and β. We define the complex version of the Riemann integral. The path P is called the **contour** or **integration path**. The definite integral along P is expressed as (see Fig. 1.1)

$$\int_P dz f(z) = \lim_{\delta \to 0} \sum_{j=1}^{n} f(\zeta_j)(z_j - z_{j-1}),$$

$$\zeta_j \in [z_{j-1}, z_j] \text{ on path } P, \quad \delta = \max|z_j - z_{j-1}|. \tag{1.8.6}$$

The important question is whether or not the definite integral between α and β changes when P varies. The following theorem, known as the **Cauchy theorem**, answers it:

Theorem 1.8.1. *Let $f(z)$ be a function holomorphic in a domain D, and C be a simple closed curve (Jordan curve) lying in D. If $f(z)$ is holomorphic in the inside of C, then*

$$\oint_C dz f(z) = 0, \tag{1.8.7}$$

holds.

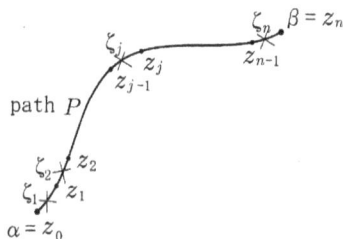

Fig. 1.1 Integration path in (1.8.6)

Fixing two points α and β on C, we divide C into two paths P (α to β) and P' (β to α); then (1.8.7) is rewritten as

$$\int_P dz f(z) = - \int_{P'} dz f(z) = \int_{-P'} dz f(z), \qquad (1.8.8)$$

where $-P'$ denotes the same curve as that of P', but in the reversed direction, i.e., α to β. Since C is arbitrary as long as the conditions for the Cauchy theorem are satisfied, (1.8.8) implies that the definite integral from α to β is independent of the choice of the path.[7]

Hence we may write

$$\int_\alpha^\beta dz f(z) = F(\beta) - F(\alpha). \qquad (1.8.9)$$

1.8.3 *Cauchy integral representation*

The point where the function $f(z)$ is not holomorphic is called the **singularity** or **singular point** of $f(z)$. If there are singular points in the domain encircled by C, then the integral on the left-hand side of (1.8.7) no longer vanishes.

Given a point $z = a$, the (small) domain which includes a disc $|z - a| < \varepsilon$ ($\varepsilon > 0$, arbitrarily small), is called the **neighborhood** of $z = a$. If $f(z)$ is holomorphic in a neighborhood of its singularity $z = a$ except for $z = a$, then it is called the **isolated singularity**. If $z = a$ is an isolated singularity of $f(z)$ such that $(z - a)^n f(z)$ for some positive integer n is holomorphic at $z = a$, then it is called the **pole**. The pole is the simplest example of the isolated singularity. The minimum value m of the above integer n is called the degree of the pole $z = a$. The $m = 1$ pole is called the **simple pole**, while the $m \geq 2$ pole is called the **multiple pole**.

Theorem 1.8.2. *Let $f(z)$ be the function satisfying the conditions of the Cauchy theorem. If $z = a$ lies in the inside of C, then the formula*

$$\frac{1}{2\pi i} \oint_C dz \frac{f(z)}{z - a} = f(a) \qquad (1.8.10)$$

holds, where the contour C is taken in the positive direction (i.e., anticlockwise).

---

[7]Note the similarity to Newtonian mechanics: if potential exists, the total change of kinetic energy is independent of the intermediate states.

This theorem can be easily proved. Because of the Cauchy theorem, C can be replaced by a small circle, $C_\varepsilon(a)$, defined by $|z - a| = \varepsilon$. Then, by taking the limit $\varepsilon \to 0$, the left-hand side of (1.8.10) becomes $f(a)\lim_{\varepsilon \to 0} I_\varepsilon$, where

$$I_\varepsilon \equiv \frac{1}{2\pi i} \oint_{C_\varepsilon(a)} \frac{dz}{z - a}. \tag{1.8.11}$$

Setting $z = a + \varepsilon e^{i\theta}$, we have

$$I_\varepsilon = \frac{1}{2\pi i} i \int_0^{2\pi} d\theta = 1. \tag{1.8.12}$$

Thus the left-hand side of (1.8.10) equals $f(a)$.

In (1.8.10), rewriting z and a into ζ and z, respectively, and interchanging both sides, we obtain

$$f(z) = \frac{1}{2\pi i} \oint_C d\zeta \frac{f(\zeta)}{\zeta - z}. \tag{1.8.13}$$

This formula is nothing but the integral representation of the function holomorphic in D. Formula (1.8.13) is called the **Cauchy integral representation**. Differentiating it n times, we obtain the **Goursat formula**

$$f^{(n)}(z) = \frac{n!}{2\pi i} \oint_C d\zeta \frac{f(\zeta)}{(\zeta - z)^{n+1}}. \tag{1.8.14}$$

We thus see that the holomorphic function is always arbitrary times differentiable.

1.8.4 *Taylor expansion*

Let $f(z)$ be a function holomorphic in a domain including $z = a$. Since the Taylor expansion (1.6.11) for $1/(1 + x)$ for x real can be extended to the case of complex values, hence we have

$$\frac{1}{\zeta - z} = \frac{1}{\zeta - a} \cdot \frac{1}{1 - \dfrac{z - a}{\zeta - a}} = \sum_{n=0}^{\infty} \frac{(z - a)^n}{(\zeta - a)^{n+1}}. \tag{1.8.15}$$

Multiplying both sides of (1.8.15) by $f(\zeta)$, and integrating them along the contour $C_\varepsilon(a)$ term by term, we obtain

$$\oint_{C_\varepsilon(a)} d\zeta \frac{f(\zeta)}{\zeta - z} = \sum_{n=0}^{\infty} (z - a)^n \oint_{C_\varepsilon(a)} d\zeta \frac{f(\zeta)}{(\zeta - a)^{n+1}}. \tag{1.8.16}$$

Then (1.8.13) and (1.8.14) imply

$$f(z) = \sum_{n=0}^{\infty} \frac{f^{(n)}(a)}{n!}(z-a)^n. \qquad (1.8.17)$$

Thus $f(z)$ can be expanded into the Taylor series. The radius of convergence R is the distance between the point $z = a$ and its nearest singularity.

1.8.5 *Theorem of identity*

From the Taylor expansion, we see that the zero points of a holomorphic function $f(z)$ ($\not\equiv 0$) are isolated. This is because (1.8.17) implies that, assuming $f(a) = 0$, in a small neighborhood of the zero point $z = a$, $f(z)$ behaves like $f(z) \simeq c(z-a)^n$, ($n \geq 1$), where c is a non-zero constant unless $f(z) \equiv 0$. Therefore, there is no other zero points near $z = a$. From this fact, a surprising theorem, called the **theorem of identity**, follows.

Theorem 1.8.3. *Let two functions $f(z)$ and $g(z)$ be holomorphic in a domain D. If both functions coincide on a curve[8] lying in D, then $f(z) = g(z)$ holds throughout D, that is, both functions are the same. This is because the function $f(z) - g(z)$ is holomorphic in D and vanishes at a set of non-isolated points. Therefore it must be identically zero everywhere in D, that is, $f(z) \equiv g(z)$ in D.*

Let $\tilde{f}(x)$ be a function defined in an interval I on the real axis. If a complex function, $f(z)$, holomorphic in a domain D including I is equal to $\tilde{f}(x)$ in I, then $f(z)$ is unique throughout D, owing to the above theorem. Accordingly, any equality proved for the real function $\tilde{f}(x)$ is automatically extended to the case of the complex function $f(z)$.

Let D_1 and D_2 be two domains such that their intersection $D_1 \cap D_2$ is non-empty and connected, and $f_1(z)$ and $f_2(z)$ be two functions such that $f_k(z)$ ($k = 1, 2$) is holomorphic in D_k, and $f_1(z) = f_2(z)$ in $D_1 \cap D_2$. Then the theorem of identity tells us that there is a holomorphic function $f(z)$ in the union $D \equiv D_1 \cup D_2$ such that $f(z) \equiv f_k(z)$ in D_k. The function $f(z)$ is called the **analytic continuation** of $f_k(z)$ into D. Evidently, the procedure of the analytic continuation can be extended to the case of n domains if they are topologically chain-like. Any equality property proved in one domain remains valid in all domains. Of course, the property which is kept

---

[8]More precisely, it is sufficient that they coincide on a set of non-isolated points.

by analytic continuation is *equality* only; it is *not* true for approximation, inequality and asymptotic behavior.[9]

It is important to note that the ways of the analytic continuation are *not* necessarily unique in the non-trivial topological situation of domains. For example, suppose that $D_1 \cap D_2$ is disconnected and consists of two domains $D_{12}^{(1)}$ and $D_{12}^{(2)}$, so that the union $D \equiv D_1 \cup D_2$ is not simply connected, that is, it has a hole. The analytic continuation from D_1 to D_2 can be made in two different ways, through $D_{12}^{(1)}$ and through $D_{12}^{(2)}$. In general, D may have many holes. But if we assume that the singularities which we encounter are isolated singularities only, D has a finite number of small holes, each of which has only one singularity. Hence the problem reduces to the case in which D has only one hole which contains only one singularity.

1.8.6 *Laurent expansion*

We first consider the case in which the analytic continuations from both sides around a ring-like domain coincide. In this case, in the ring-like domain around the isolated singularity $z = a$, the function $f(z)$ can be expanded in the following form:

$$f(z) = \sum_{n=-\infty}^{+\infty} c_n (z-a)^n, \qquad (1.8.18)$$

where c_n's are some constants. This expansion is called the **Laurent expansion**. Since $z = a$ is a singularity of $f(z)$, at least one c_n $(n < 0)$ is non-zero. If $m = \max(-n)$ such that $c_n \neq 0$ exists, the singularity is a pole of order m, because $(z-a)^m f(z)$ is holomorphic at $z = a$. If m does not exist, that is, if there are infinite number of terms of $c_n \neq 0$ $(n < 0)$, then the singularity $z = a$ is called the **essential singularity**. For example, $e^{1/z}$ has an essential singularity at $z = 0$. It is known that, in any neighborhood of an essential singularity, $f(z)$ can approach to any value. (However, there may exist a value which cannot take in a neighborhood of an essential singularity; for example, $e^{1/z}$ is never equal to zero.)

The function which is holomorphic in the whole complex plane is called the **entire function**. Polynomials and exponential functions are examples of entire functions. The behavior of the function $f(z)$ at infinity can be investigated by considering the behavior near $\zeta = 0$ of the function

---

[9]Especially, one must be careful for asymptotic expansion.

$g(\zeta) \equiv f(1/\zeta)$. If $f(z)$ is an entire function, the Laurent expansion of $g(\zeta)$ at $\zeta = 0$ has at least one $n < 0$ term, unless $f(z)$ is identically a constant. Hence $g(\zeta)$ has a pole or an essential singularity at $\zeta = 0$, and therefore it is not bounded. Accordingly, a bounded entire function is identically a constant. This result is called the **Liouville theorem**.

Now, setting $f(z) \equiv 1$ in (1.8.14) and rewriting ζ into z and z into a together with replacing $n + 1$ by m, we have

$$\frac{1}{2\pi i} \oint_C \frac{dz}{(z-a)^m} = 1, \quad (m = 1)$$
$$= 0, \quad (m \geq 2), \tag{1.8.19}$$

where the contour C lies in a neighborhood of $z = a$. Integrating (1.8.18) term by term, we find

$$c_{-1} = \frac{1}{2\pi i} \oint_C dz f(z), \tag{1.8.20}$$

provided that $z = a$ is not an essential singularity. The coefficient c_{-1} is called the **residue** of $f(z)$ at $z = a$. The function which has no singularity other than poles is called the **meromorphic function**.

Theorem 1.8.4. *The integral of a meromorphic function $f(z)$ along a contour C (simple closed curve) is equal to $2\pi i$ times the sum over the residues of all poles of $f(z)$ encircled by C.*

This theorem is called the **residue theorem**. The residue theorem is extremely useful for calculating the definite integral of a real function whose indefinite integral is unknown. (See §1.9 for a concrete example.)

1.8.7 *Riemann sheet*

In a ring-like domain, if the analytic continuations from both sides do not coincide, the function $f(z)$ is not single-valued. If there is only one isolated singularity in the hole, it is called the **branch point**. For example, $z^{1/n}$ ($n \geq 2$, integer) and $\log z$ have a branch point at $z = 0$. The former is n-valued, because the value of $z^{1/n}$ returns to its original when rotating around $z = 0$ n times. On the other hand, the latter has no such finite number n, whence the logarithmic function has infinitely many values (see (1.7.5)).

In order to restore the one-valuedness, we must introduce cuts. Here the **cut** is a line connecting two branch points or between one branch point and the infinity. If one forbids the analytic continuation going over any cut, $f(z)$ becomes one-valued. Such a cut plane is called the **Riemann sheet**. In order to see the whole structure of $f(z)$, we can join Riemann sheets along the cuts so as to recover the original analytic continuation. Then the Riemann sheets make a topologically non-trivial surface, which has no artificial boundary like a cut. This surface is called the **Riemann surface** of $f(z)$. The name of the **analytic function** is apt for indicating the whole structure of $f(z)$ on its Riemann surface. If non-isolated singularities are present, it is possible that even the Riemann surface may have a boundary. The boundary beyond which no analytic continuation is possible is called the **natural boundary**.

1.9 Gamma Function and Beta Function

1.9.1 *Euler's gamma function*

As seen in the previous section, the analytic function has wonderful properties. Therefore, it is often convenient to extend integer parameters to complex quantities. For example, the factorial, $n! = 1 \cdot 2 \cdots n$, is originally defined only for the positive integer n (and $0! = 1$ for $n = 0$). Following Euler, we extend the factorial to that for complex value, that is, we define the **gamma function** by the definite integral

$$\Gamma(\nu) \equiv \int_0^\infty dx\, x^{\nu-1} e^{-x}. \tag{1.9.1}$$

Evidently, (1.9.1) is convergent for $\Re\nu > 0$, because $|x^{i\Im\nu}| = 1$. Especially, for $\nu > 0$, $\Gamma(\nu)$ is positive.

In (1.9.1), replacing ν by $\nu + 1$ and then integrating by parts, we have

$$\Gamma(\nu+1) = \int_0^\infty dx\, x^\nu e^{-x} = \nu \int_0^\infty dx\, x^{\nu-1} e^{-x} = \nu\Gamma(\nu), \tag{1.9.2}$$

that is, we obtain the recurrence formula

$$\Gamma(\nu+1) = \nu\Gamma(\nu). \tag{1.9.3}$$

Since $\Gamma(1) = \int_0^\infty dx\, e^{-x} = 1$, (1.9.3) implies

$$\Gamma(n+1) = n(n-1)\cdots 2 \cdot 1 \cdot \Gamma(1) = n!, \tag{1.9.4}$$

when ν equals a positive integer n. Then we see that the gamma function is indeed an analytic extension of the factorial.[10]

---

[10]Of course, the analytic extension of the factorial is not unique. But the gamma function is the unique extension satisfying the recurrence formula.

Since, as is obvious from (1.9.1), the gamma function $\Gamma(\nu)$ is differentiable with respect to ν, it is a function of ν holomorphic in $\Re\nu > 0$. Hence we consider its analytic continuation to $\Re\nu \leq 0$. Since (1.9.3) is a functional equality, it remains valid in the analytic continuation. Using (1.9.3) repeatedly, we have

$$\Gamma(\nu) = \frac{\Gamma(\nu+n)}{\nu(\nu+1)(\nu+2)\cdots(\nu+n-1)}. \tag{1.9.5}$$

From (1.9.5), we see that $\Gamma(\nu)$ has simple poles at $\nu = 0, -1, -2, \ldots$; but it is holomorphic at any other points. Thus $\Gamma(\nu)$ is a meromorphic function of ν.

1.9.2 *Euler's beta function*

By the transformation $x = t^2$ (whence $dx = 2tdt$), (1.9.1) becomes

$$\Gamma(\nu) = 2\int_0^\infty dt\, t^{2\nu-1}e^{-t^2}. \tag{1.9.6}$$

By using different letters, (1.9.6) is rewritten as

$$\Gamma(\mu) = 2\int_0^\infty ds\, s^{2\mu-1}e^{-s^2}. \tag{1.9.7}$$

The simple product of (1.9.6) and (1.9.7) is

$$\Gamma(\mu)\Gamma(\nu) = 4\int_0^\infty ds\int_0^\infty dt\, s^{2\mu-1}t^{2\nu-1}e^{-(s^2+t^2)}. \tag{1.9.8}$$

The right-hand side of (1.9.8) is the integration over the first quadrant of the (s,t) plane. We transform the variables s and t into the polar coordinates (r,θ) by $s = r\cos\theta$, $t = r\sin\theta$. Since[11]

$$\int_0^\infty ds\int_0^\infty dt\cdots = \int_0^{\pi/2} d\theta\int_0^\infty rdr\cdots, \tag{1.9.9}$$

(1.9.8) becomes

$$\Gamma(\mu)\Gamma(\nu) = 4\int_0^{\pi/2} d\theta(\cos\theta)^{2\mu-1}(\sin\theta)^{2\nu-1}\int_0^\infty dr\, r^{2\mu+2\nu-1}e^{-r^2}. \tag{1.9.10}$$

The right-hand side of (1.9.10) is a simple product of the θ integral and the r integral. Owing to (1.9.6), the latter is equal to $\frac{1}{2}\Gamma(\mu+\nu)$. As for the

---

[11]The Jacobian of this transformation is $J = \frac{\partial s}{\partial r}\frac{\partial t}{\partial\theta} - \frac{\partial s}{\partial\theta}\frac{\partial t}{\partial r} = r$.

former, we set $\cos^2 \theta = x$; then $\sin^2 \theta = 1 - x$ and $-2\cos\theta \sin\theta d\theta = dx$. Then we find

$$\int_0^{\pi/2} d\theta (\cos\theta)^{2\mu-1} (\sin\theta)^{2\nu-1} = \frac{1}{2} \int_0^1 dx \, x^{\mu-1} (1-x)^{\nu-1}. \qquad (1.9.11)$$

Now, following Euler, we define the **beta function** by the definite integral,

$$B(\mu, \nu) \equiv \int_0^1 dx \, x^{\mu-1} (1-x)^{\nu-1}. \qquad (1.9.12)$$

Comparing (1.9.12) with (1.9.10) and (1.9.11), we obtain the important formula,

$$B(\mu, \nu) = \frac{\Gamma(\mu)\Gamma(\nu)}{\Gamma(\mu+\nu)}. \qquad (1.9.13)$$

In particular, setting $\mu = \nu = 1/2$ in (1.9.12), (1.9.11) and (1.9.13), we find

$$B(\tfrac{1}{2}, \tfrac{1}{2}) = \pi = \frac{\left(\Gamma(\tfrac{1}{2})\right)^2}{\Gamma(1)}, \qquad (1.9.14)$$

and therefore

$$\Gamma(\tfrac{1}{2}) = \sqrt{\pi}. \qquad (1.9.15)$$

Combining (1.9.15) and (1.9.3), we can calculate the values of $\Gamma(n+1/2), (n = \pm 1, \pm 2, \ldots)$.

1.9.3 *Important formulas for the gamma function*

The $\mu = \nu$ case of (1.9.11) is rewritten as

$$\frac{1}{2} \int_0^1 dx \, [x(1-x)]^{\nu-1} = \int_0^{\pi/2} d\theta \left(\frac{\sin(2\theta)}{2}\right)^{2\nu-1}. \qquad (1.9.16)$$

With $2\theta = \theta'$, the right-hand side of (1.9.16) becomes

$$= 2^{-2\nu+1} \int_0^{\pi/2} d\theta' \, (\sin\theta')^{2\nu-1} = 2^{-2\nu+1} \cdot \frac{1}{2} \int_0^1 dx \, x^{-1/2} (1-x)^{\nu-1}, \qquad (1.9.17)$$

where we have used $\sin\theta' = \sin(\pi - \theta')$ for $\theta' \geq \pi/2$ and the $\mu = 1/2$ case of (1.9.11). Since both the left-hand side of (1.9.16) and the right-hand side of (1.9.17) are special cases of the beta function, we can substitute (1.9.13) in them. Then we have

$$\frac{\left(\Gamma(\nu)\right)^2}{2\Gamma(2\nu)} = 2^{-2\nu}\frac{\Gamma(\frac{1}{2})\Gamma(\nu)}{\Gamma(\nu + \frac{1}{2})}. \tag{1.9.18}$$

Moreover, substituting (1.9.15) into (1.9.18), we obtain a functional relation for the gamma function:

$$\Gamma(\nu)\Gamma(\nu + \tfrac{1}{2}) = 2^{-2\nu+1}\sqrt{\pi}\Gamma(2\nu). \tag{1.9.19}$$

There is another important functional relation for the gamma function:

$$\Gamma(\nu)\Gamma(1 - \nu) = \frac{\pi}{\sin(\pi\nu)}. \tag{1.9.20}$$

In what follows, we prove (1.9.20).

Since we can later make analytic continuation to all values of ν, we may assume $0 < \Re\nu < 1$. Then from (1.9.13) we have

$$\Gamma(\nu)\Gamma(1 - \nu) = \int_0^1 dx\, x^{\nu-1}(1 - x)^{-\nu}. \tag{1.9.21}$$

By setting $x = t/(1 + t)$ (then $dx = (1 + t)^{-2}dt$), (1.9.21) becomes

$$\Gamma(\nu)\Gamma(1 - \nu) = \int_0^\infty dt\frac{t^{\nu-1}}{1 + t} \equiv I(\nu). \tag{1.9.22}$$

For calculating such a definite integral as (1.9.22), it is convenient to employ the residue theorem given in §1.8. We regard the integrand, $t^{\nu-1}/(1 + t)$, of $I(\nu)$ as an analytic function of the complex variable t. Since it has a branch point at $t = 0$, we consider the Riemann sheet which has a cut of the positive real axis. On this Riemann sheet the integrand function is holomorphic except for $t = -1$, where it has a simple pole. We introduce a contour integral, where the contour C is a closed curve constructed in the following way: start from the origin $t = 0$, proceed along the infinitesimally upper path of the positive real axis until $t = R + i0$ ($R > 1$), where $+i0$ means an infinitesimally positive imaginary part, go around along the circle $|t| = R$ anticlockwise to $t = R - i0$, and finally return to the origin $t = 0$ along the infinitesimally lower path of the positive real axis (see Fig. 1.2).

Since there is only one pole at $t = -1$ in the inside of C, the residue theorem implies

$$\int_C dt\frac{t^{\nu-1}}{1 + t} = 2\pi i e^{\pi i(\nu-1)} = \pi(-2i)e^{i(\pi\nu)}. \tag{1.9.23}$$

We take the $R \to \infty$ limit of the left-hand side of (1.9.23). Then, because of the assumption $\Re\nu < 1$, the contribution from the circle $|t| = R$ tends

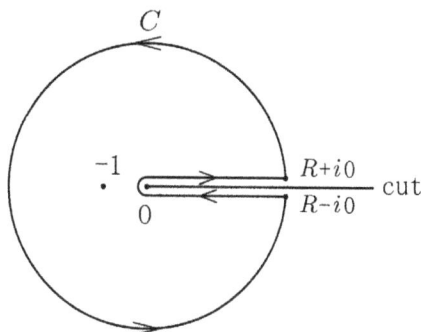

Fig. 1.2 Integration contour C

to zero as $R \to \infty$. The remaining contribution is the difference of the integral along the infinitesimally upper and lower paths of the whole positive real axis. The former is nothing but $I(\nu)$, while the latter equals $e^{2\pi i\nu} I(\nu)$ because of the increase of $\arg t$ by 2π. Thus the left-hand side of (1.9.23) equals $(1 - e^{2\pi i\nu}) I(\nu)$. Substituting this result in (1.9.23) and using the Euler formula (1.7.4), we obtain (1.9.20), the formula to be proved. □

1.10 Generalized Functions

1.10.1 *Dirac's delta function*

We consider the system of the orthonormalized n-dimensional vectors e_k ($k = 1, 2, \ldots, n$). Their inner product is defined by $(e_j, e_k) = \delta_{jk}$, where δ_{jk} is the **Kronecker delta**, defined by $\delta_{jk} = 1$ for $j = k$ and $\delta_{jk} = 0$ for $j \neq k$. Of course this notation can be used when n is countably infinite.

Paul Dirac, one of the founders of quantum theory, wished to extend the Kronecker delta to the case of continuous infinity. First, note that $\delta_{jk} = \delta_{j-k,0}$, namely δ_{jk} depends only on $j - k$. The continuously infinite version of δ_{jk} would be a "function" of a continuous variable $x - y$. It must be zero for $x - y \neq 0$. But correspondingly to the property $\sum_k \delta_{jk} a_k = a_j$, its "value" at $x - y = 0$ must be infinite. Dirac called it the "delta function", but mathematicians cannot accept it as a function.

The definition of the **delta function** can be stated more concretely as follows: for an arbitrarily "good" function $\varphi(x)$, the delta function satisfies

$$\int_{-\infty}^{+\infty} dy\, \delta(x - y)\varphi(y) = \varphi(x). \tag{1.10.1}$$

Mathematicians, however, wished to regard Dirac's delta function as a measure, where the **measure** is essentially the weight, that is, it is a derivative of a non-decreasing function. We introduce a simple measure, the Heaviside **step function**. Mathematicians denote it by $Y(x)$, but physicists denote it by $\theta(x)$ and therefore call it the **theta function**. It is defined by

$$\theta(x) = 1 \quad (x > 0),$$
$$= 0 \quad (x < 0) \tag{1.10.2}$$

where the value of $\theta(x)$ at $x = 0$ is undefined; but we require that the identities $\theta(x) + \theta(-x) = 1$ and $\theta(x)\theta(-x) = 0$ hold. Then we have $d\theta(x) = \delta(x)dx$, or more concisely $\theta'(x) = \delta(x)$.

1.10.2 *Schwartz's distribution*

Physicists were not satisfied with regarding the delta function as a measure, because they successfully used $\delta(x)$ as if it were a C^∞-class function in analytic calculations.

The person who succeeded in satisfactorily defining the delta function is Laurent Schwartz. He introduced a new concept, which he called the distribution. This rather curious name came from the fact that the delta function can be regarded as the density distribution of a unit massive point.

The **distribution** $f[\varphi]$ is a linear functional, where the **functional** is the function whose independent variable is a function. Explicitly, it is given by

$$f[\varphi] \equiv \int_{-\infty}^{+\infty} dx \, f(x)\varphi(x), \tag{1.10.3}$$

where $\varphi(x)$ is a C^∞-class function such that it and all its derivatives tend to zero very rapidly [almost exponentially (denoted by $\mathcal{S}$) or identically (denoted by $\mathcal{D}$)] as $|x| \to \infty$. The differentiation acting on $f(x)$ can be transferred to that of $\varphi(x)$ by integrating by parts because

$$\int_{-\infty}^{+\infty} dx \left[\left(\frac{d}{dx} \right)^n f(x) \right] \varphi(x) = (-1)^n \int_{-\infty}^{+\infty} dx \, f(x) \left(\frac{d}{dx} \right)^n \varphi(x), \tag{1.10.4}$$

where the integrated term always vanishes due to the assumed asymptotic behavior of $\varphi(x)$.

Thus the distribution is essentially nothing but some-order derivative of a (partly non-differentiable) continuous function. But it is very important to note that the pointwise value of a distribution is meaningless.

Hence it seems to be more natural to call the distribution the **generalized function**.

The function $\varphi(x)$ is called the **test function**. This is because $\varphi(x)$ is necessary to test how $f(x)$ behaves as a functional.

If $f[\varphi] \geq 0$ for any positive definite test function $\varphi(x)$, it is called the **positive distribution**, which is shown to be a measure.

The delta function $\delta(x)$ is a positive distribution, but its derivatives are no longer so. For example, $\delta'(x)$ is physically corresponding to a dipole of zero length.

It is easy to see that $\delta^{(n)}(-x) = (-1)^n \delta^{(n)}(x)$ and $\delta(cx) = |c|^{-1}\delta(x)$ for $c \neq 0$. As for $\delta(g(x))$, one should apply the above formula to each zero point of $g(x)$. For example, $\delta(x^2 - a^2)$ $(a \neq 0)$ equals $[\delta(x-a) + \delta(x+a)]/(2|a|)$. Furthermore, differentiating the self-evident formula $x\delta(x) = 0$, we find $x\delta'(x) = -\delta(x)$.

1.10.3 *Interchangeability of the order of limit and integration*

In the ordinary formulation of the analysis, it is important to introduce the concept of uniform convergence in order to assume the interchangeability of the order of limit and integration.

For example, consider a sequence of functions,

$$f_n(x) \equiv 2n^2 x \, e^{-n^2 x^2}. \tag{1.10.5}$$

It is evident that

$$\lim_{n \to \infty} f_n(x) = 0, \tag{1.10.6}$$

where one should note that the reasons why (1.10.6) holds for $x = 0$ and why it does for $x \neq 0$ are different. With some $a > 0$, therefore, we, of course, have

$$\int_0^a dx \lim_{n \to \infty} f_n(x) = 0. \tag{1.10.7}$$

However, since

$$\int_0^a dx \, f_n(x) = 1 - e^{-n^2 a^2}, \tag{1.10.8}$$

we have

$$\lim_{n \to \infty} \int_0^a dx \, f_n(x) = 1, \tag{1.10.9}$$

which is *not* equal to (1.10.7).

The origin of this disagreement arises from the range near $x \approx 1/n$, where $f_n(x) \approx 2e^{-1}n$ is big. Such a contribution is neglected when the range collapses into zero width in the ordinary pointwise approach.

If we reconsider the above problem on the basis of the generalized function, the "true" limit of the sequence (1.10.5) is

$$\lim_{n\to\infty} f_n(x) = 2x\delta(x^2) \qquad (1.10.10)$$

instead of (1.10.6). If seen pointwisely, (1.10.10) would be zero both for $x = 0$ and for $x \neq 0$. But the integral is given by

$$\int_0^a dx \, 2x\delta(x^2) = \int_0^{a^2} du \, \delta(u) = 1, \qquad (1.10.11)$$

which coincides with (1.10.9). Thus we see that the pathological behavior arising from the infinitesimal range can be taken into account by using the generalized function. The pointwise approach is, so to speak, to always give the answer "0" for any indefinite form "0×∞" without calculating the limiting expression.

1.10.4 *Delta function and the Cauchy integral representation*

The definition (1.10.1) of the delta function is quite similar to the Cauchy integral representation (1.8.10). That is, for real a, the operator

$$\frac{1}{2\pi i} \oint_C \frac{dz}{z - a} \cdots \qquad (1.10.12)$$

is precisely the delta function $\delta(x - a)$. The contour C can be deformed into infinitesimally near the real axis in the neighborhood of $z = a$. Then (1.10.12) is rewritten as

$$\frac{1}{2\pi i} \int_{-\infty}^{+\infty} dx \left(\frac{1}{x - (a + i0)} - \frac{1}{x - (a - i0)} \right) \cdots . \qquad (1.10.13)$$

Therefore, we have

$$\frac{1}{x - a - i0} - \frac{1}{x - a + i0} = 2\pi i \delta(x - a), \qquad (1.10.14)$$

or equivalently,

$$\Im \frac{1}{x - a - i0} = \pi \delta(x - a). \qquad (1.10.15)$$

1.10.5 *Cauchy principal value*

The formula $1/(x-a)$ is not well-defined at $x = a$, and hence it cannot be integrated over an interval including $x = a$. But, if $\varphi(x)$ is continuous in the interval $[b, c]$, where $b < a < c$, then the integral

$$\lim_{\varepsilon \to +0} \left(\int_b^{a-\varepsilon} dx \, \frac{\varphi(x)}{x-a} + \int_{a+\varepsilon}^c dx \, \frac{\varphi(x)}{x-a} \right) \tag{1.10.16}$$

is well-defined. This is called the **Cauchy principal value**, and is denoted by

$$\mathrm{P} \int_b^c dx \, \frac{\varphi(x)}{x-a}. \tag{1.10.17}$$

Indeed, for $\varphi(x) \equiv 1$, from (1.5.7) we have

$$\mathrm{P} \int_b^c \frac{dx}{x-a} = \lim_{\varepsilon \to +0} \left[\left(\log|-\varepsilon| - \log|b-a| \right) + \left(\log(c-a) - \log \varepsilon \right) \right]$$

$$= \log(c-a) - \log(a-b), \tag{1.10.18}$$

that is, it is finite. In the general case, in the neighborhood of $x = a$, $\varphi(x)$ can be approximated by $\varphi(a)$, and therefore the Cauchy principal value is finite.

The Cauchy principal value can be understood from the point of view of the analytic function. It is given by

$$\Re \frac{1}{x-a-i0} = \mathrm{P} \frac{1}{x-a}. \tag{1.10.19}$$

Indeed, we have

$$\Re \int_b^c \frac{dx}{x-a-i0} = \Re \left(\log(c-a) - \log(a-b) + \pi i \right)$$

$$= \log(c-a) - \log(a-b), \tag{1.10.20}$$

which coincides with (1.10.18).

In general, it is called the **finite part**, i.e. the finite value taken from a divergent integral. The finite part can be defined as a generalized function. From this point of view, the **Hadamard finite part** is naturally defined by

$$\mathrm{Pf} \frac{1}{x^n} \equiv \frac{(-1)^{n-1}}{(n-1)!} \left(\frac{d}{dx} \right)^{n-1} \mathrm{P} \frac{1}{x} = \Re \frac{1}{(x-i0)^n}, \tag{1.10.21}$$

where, for simplicity, we have written the formula in the $a = 0$ case. The symbol "Pf" is the abbreviation of *partie finie* (= finite part). It should be noted that it is *not* a positive distribution even if n is an even integer.

1.10.6 *Y distribution*

Schwartz introduced a distribution containing a complex parameter ν as a useful generalization of the Heaviside step function and the delta function. Since its standard name is non-existent, we call it the **Y distribution**. It is defined by

$$Y_\nu(x) \equiv \frac{x^{\nu-1}}{\Gamma(\nu)}\theta(x), \tag{1.10.22}$$

where, of course, $\Gamma(\nu)$ denotes the gamma function defined by (1.9.1). Since $\Gamma(-n)$ $(n = 0, 1, 2, \ldots)$ is infinite, $Y_{-n}(x)$ looks to be zero, but this is *not* so as a generalized function. Since $x^{-n-1}\theta(x)$ is singular at $x = 0$, we have to take account of this fact. As a generalized function, we have a remarkable relation

$$Y_{-n}(x) = \delta^{(n)}(x). \tag{1.10.23}$$

We present two proofs of (1.10.23). From (1.10.3), we have

$$Y_\nu[\varphi] = \int_{-\infty}^{+\infty} dx\, Y_\nu(x)\varphi(x). \tag{1.10.24}$$

We assume that the test function can be expressed in terms of the Laplace transform.[12] Then we may assume that the test function is

$$\varphi(x) = e^{-sx} \quad (s > 0). \tag{1.10.25}$$

Substituting (1.10.22) and (1.10.25) into (1.10.24), and making the transformation $sx = y$, we obtain

$$\int_{-\infty}^{+\infty} dx\, Y_\nu(x)\, e^{-sx} = \frac{1}{\Gamma(\nu)} \int_0^\infty dx\, x^{\nu-1}\, e^{-sx}$$

$$= \frac{s^{-\nu}}{\Gamma(\nu)} \int_0^\infty dy\, y^{\nu-1}\, e^{-y} = s^{-\nu}. \tag{1.10.26}$$

On the other hand, as a distribution (1.10.23) implies

$$\int_{-\infty}^{+\infty} dx\, Y_{-n}(x)\, e^{-sx} = \int_{-\infty}^{+\infty} dx\, \delta^{(n)}(x)\, e^{-sx} = s^n. \tag{1.10.27}$$

This result is precisely the analytic continuation of (1.10.26) to $\nu = -n$.

$$\square$$

---

[12] See §3.1 of Chapter 3 for more on the Laplace transform.

The second proof is more direct. From the definition (1.10.22) of the Y distribution, we can easily derive the fundamental formula

$$\frac{d}{dx}Y_\nu(x) = Y_{\nu-1}(x) \tag{1.10.28}$$

for $\Re\nu > 0$. By analytic continuation with respect to ν, (1.10.28) holds for any value of ν. Setting $\nu = -n$, we see that $Y_{-n}(x)$ is the $(n+1)$th-order derivative of $Y_1(x) \equiv \theta(x)$, that is, we have shown (1.10.23). □

In the beta function formula (1.9.12) with (1.9.13), transforming x by y/a and replacing a by x, we have

$$x^{-\mu-\nu+1} \int_0^x dy\, y^{\mu-1}(x-y)^{\nu-1} = \frac{\Gamma(\mu)\Gamma(\nu)}{\Gamma(\mu+\nu)}. \tag{1.10.29}$$

Hence, we obtain the following convolution-type formula for the Y distribution:

$$\int_{-\infty}^{+\infty} dy\, Y_\mu(x-y)Y_\nu(y) = Y_{\mu+\nu}(x). \tag{1.10.30}$$

This formula is very interesting because the simple addition rule holds for the parameters. Formula (1.10.28) is nothing but the $\mu = -1$ case of (1.10.30). The $\mu = +1$ case of (1.10.30) is the integration formula

$$\int_{-\infty}^x dy\, Y_\nu(y) = Y_{\nu+1}(x). \tag{1.10.31}$$

The Y distribution is often used in this book.

1.10.7 *Boundary value of the analytic function*

The Hadamard finite part (1.10.21) has been given as $\Re(x-i0)^{-n}$, where n is a positive integer. We want to extend n to a complex number $-\nu$ as in the case of the Y distribution. For this purpose, we consider an analytic function of z, namely

$$X_\nu(z) \equiv \int_{-\infty}^{+\infty} dy\, \frac{Y_\nu(y)}{y-z}. \tag{1.10.32}$$

If $\nu \neq -n+1$, $(n = 1, 2, \ldots)$, $X_\nu(z)$ has a branch point at $z = 0$, and is holomorphic in the Riemann sheet which has a cut on the positive real axis. At $\nu = -n+1$, (1.10.23) implies

$$X_{-n+1}(z) = \frac{(n-1)!}{(-z)^n}, \tag{1.10.33}$$

and therefore

$$\Re X_{-n+1}(x+i0) = (-1)^n(n-1)! \, \mathrm{Pf}\frac{1}{x^n}. \tag{1.10.34}$$

Thus, it is a constant multiple of the Hadamard finite part. On the other hand, from (1.10.15) and (1.10.32), the imaginary part of $X_\nu(z)$ for $z = x+i0$ becomes

$$\Im X_\nu(x+i0) = \pi Y_\nu(x). \tag{1.10.35}$$

It is possible to carry out the integration of (1.10.32) explicitly. From (1.9.22) with the transformation $t = x/a$, we have

$$a^{-\nu+1} \int_0^\infty dx \, \frac{x^{\nu-1}}{x+a} = \Gamma(\nu)\Gamma(1-\nu). \tag{1.10.36}$$

Setting $a = -z$ in (1.10.36) we obtain

$$X_\nu(z) = \frac{1}{\Gamma(\nu)} \int_0^\infty dx \, \frac{x^{\nu-1}}{x-z} = \Gamma(1-\nu)(-z)^{\nu-1}. \tag{1.10.37}$$

As is seen in (1.10.32), in general, given a generalized function $f(x)$, we can construct an analytic function by the formula

$$F(z) \equiv \frac{1}{\pi} \int_{-\infty}^{+\infty} dy \, \frac{f(y)}{y-z}, \tag{1.10.38}$$

provided that the improper integral is well defined. Then we obtain

$$\Im F(x \pm i0) = \pm f(x). \tag{1.10.39}$$

Thus one half of the difference of the boundary values (on the real axis) of the analytic function reproduces the original generalized function.

By reversing the direction of the above consideration, we can make use of an analytic function in order to define a generalized function as the difference of its boundary values on the real axis. The mathematical formation of this idea is known as Sato's **hyperfunction**. In the one-variable case, the hyperfunction is almost the same as Schwartz's distribution, but in the several-variable case, the former is a beautiful mathematical theory based on cohomology theory.

Extensions of the Distribution and my Honorable Friend, Mikio Sato

In 1957, I was investigating the mathematical treatment of the state of an unstable particle (e.g., neutron) in the framework of quantum field theory. The state of a stable particle (e.g., proton) corresponds to a simple pole, lying on the real axis, of the analytic function given by the scattering amplitude which describes particle reaction. On the other hand, the state of an unstable particle corresponds to a simple pole *not* lying on the real axis. Since the scattering amplitude is usually holomorphic off the real axis on the Riemann sheet with cuts lying on the real axis, the pole of the unstable particle lies on the next (unphysical) Riemann sheet, which is attainable by analytic continuation through a cut from the original (physical) Riemann sheet. In order to construct the state corresponding to the pole lying on the unphysical Riemann sheet, it is necessary to extend the delta function $\delta(x - a)$ to that for the complex value of a.

For this purpose, I introduced an extension of Schwartz's distribution, that is, I generalized (1.10.3) by replacing the real integration by the complex integration along a path from $-\infty$ to $+\infty$. The test function should be holomorphic in a sufficiently large domain including the real axis. The functional $f[\varphi]$ depends not only on $\varphi(x)$ but also on the choice of the integration path. Then (1.10.12) is realized as the "complex delta function" even for the complex value of a. I call this complexification of Schwartz's distribution the **complex distribution**.

In 1959, I was invited to a symposium of mathematical sciences, organized by Japan's leading mathematicians and held at a hot spring resort in Akakura. When I arrived there, I saw a young mathematician in intense discussion with a renown mathematician, Mikio Sato. Later I told Sato my theory of complex distribution. He said he had considered a similar concept a few years back, which he called analytic distribution. I requested for a reprint of his work. After the symposium, he sent me reprints of his papers, but they were not on analytic distribution but on the theory of hyperfunction. The theory of the former has never been published.

Sato's hyperfunction in the one-variable case is the difference of the boundary values of an analytic function on the real axis of a

Riemann sheet. The hyperfunction in the several-variable case is not formulated in such an elementary way but constructed with the use of the theory of cohomology. The theory of hyperfunction has been developed to the microlocal analysis, in which the detailed structure of singularity is investigated. Sato and his collaborators investigated the theory of the several-variable analytic function by means of the algebraic approach. They call their theory the algebraic analysis.

In my young age, I was involved in investigating the mathematical structure of the Feynman integral. In his famous 1949 paper, Richard Feynman proposed today's standard method of calculating scattering amplitude perturbatively. In higher-order approximation, one encounters Feynman integrals, which are complicated multiple improper integrals over a product of generalized functions determined by Feynman diagrams (Feynman graphs). I first succeeded in giving the mathematically well-defined expression for the general form of the Feynman integral. Then I extensively investigated the general theory of Feynman integral on the basis of graph theory. In 1959, Landau and I independently investigated the analytic structure of the Feynman integral. Later, on the basis of microlocal analysis, Sato and his collaborators made a mathematically rigorous formulation of the analytic structure of the Feynman integral.

Both Sato and I were at the Institute for Advanced Study, Princeton in 1961–1962, and were members of the Research Institute for Mathematical Sciences, Kyoto University for more than two decades.

Chapter 2 Differential Equations

Various methods for analytically solving ordinary differential equations are presented. Second-order linear differential equations are solved by means of power series expansion. Special functions such as Gauss' hypergeometric function are discussed. The boundary value problem of second-order linear differential equations is considered as an eigenvalue problem. Partial differential equations are briefly mentioned.

2.1 What is a Differential Equation?

2.1.1 *Galilei and Kepler, and then Newton*

It is said that modern physics begins with Newton. But its pioneers are Galilei and Kepler.

Galileo Galilei is famous for his discovery of the four satellites of Jupiter, which gave Copernicus's heliocentric theory affirmative evidence. But his most important contribution to modern physics is discovering the importance of performing quantitative experiments on earth. He discovered a fundamental law, the **law of inertia**. He found serious defects in Aristotle's time-honored laws, which were advocated on the basis of daily experiences. A moving body usually finally stops if no force continues to act on it. What is important is that he regarded resistance as a force.[1] In an ideal world in which no forces, including resistance, act, a body can keep moving eternally; that is, a body moves with a constant velocity (it may be equal to zero) along a straight line in the system of inertia. Once this

---

[1]Note that this is not self-evident. In electric-current theory, the Ohm law contains a resistance that is not a force. The Ohm law is a phenomenological law but not a fundamental one.

viewpoint is accepted, everything becomes clear; mathematical formulation of kinematics becomes possible.

Galilei also discussed the principle of equivalence in the sense that the movement of a body is independent of its mass if only gravitational force acts on it.

While Galilei was a positivist, Johannes Kepler was an idealist like Plato. He attempted to explain the five orbital intervals of the six known planets by using the composition of five regular polyhedrons. But after he obtained the accurate observational data of the movement of planets due to Brahe, he analyzed them in detail and discovered three phenomenological laws, called **Kepler's laws**. Kepler's first law states that the orbit of every planet is an ellipse at whose focus the sun is located. Therefore, one no longer needs to consider Ptolemy's complicated epicycle. Kepler's result gave Newton the confidence that gravitational force is inversely proportional to the square of distance. Kepler's second law is essentially nothing but the conservation law of angular momentum, which implies that the gravitational force is centrifugal towards the sun.

Isaac Newton found that the elliptic orbit of a planet and the parabolic one of a ball thrown on earth are essentially of the same mechanism. Thus, Newton unified the law that is in the heavens with that which is on earth. But the direct verification of this unification was almost three hundred years later, except for the uncontrollable falling of meteors.

In 1957, the Soviet Union's "Sputnik", the world's first artificial satellite, moved around the earth without consuming power. Upon hearing this news, my professor, Hideki Yukawa,[2] said, "It is a great miracle! There is no better evidence than this which confirms the validity of Newtonian mechanics."

2.1.2 *Newton's equation of motion*

The representative of the differential equation is **Newton's equation of motion**. Of course, it is an equation in a 3-dimensional space, but, for simplicity, we consider it here in the 1-dimensional case. Let the mass of a massive point P be m, the acceleration be α, and the force acting on P be F; then Newton's equation of motion is written as

$$m\alpha = F. \tag{2.1.1}$$

---

[2] Hideki Yukawa predicted the existence of the pion (Yukawa meson) as the quantum of nuclear force, and was awarded the Nobel Prize in Physics in 1949.

According to Newton's second law in mechanics, (2.1.1) holds in any system of inertia. Let the time be t, the coordinate of the position of P be x, and the velocity of the motion of P be v; then we have

$$v = \frac{dx}{dt}, \quad \alpha = \frac{dv}{dt}, \tag{2.1.2}$$

by definition. We assume that F is the conservative force (in a 1-dimensional case, this assumption means that F is dependent only on x but not on v.) Then (2.1.1) becomes

$$m\frac{d^2x}{dt^2} = F(x). \tag{2.1.3}$$

If we regard $F(x)$ as a given function and $x = x(t)$ as an unknown function of the independent variable t, (2.1.3) is an equation for $x(t)$ involving its derivative.

In general, an equation for unknown functions including their derivatives is called a **differential equation**. Given a differential equation, to find unknown functions means to solve the differential equation. It is generally very difficult to solve a differential equation analytically.

But (2.1.3) can be solved in the following way. Its left-hand side multiplied by dx can be rewritten as

$$m\frac{d^2x}{dt^2}dx = m\frac{dv}{dt}vdt = mvdv = mv\frac{dv}{dx}dx. \tag{2.1.4}$$

We can, therefore, easily integrate (2.1.3) with respect to x, and obtain

$$\frac{1}{2}mv^2 + U(x) = E, \tag{2.1.5}$$

where E is an integration constant, and we have set

$$U(x) \equiv -\int_0^x du F(u). \tag{2.1.6}$$

We call $\frac{1}{2}mv^2$ the **kinetic energy**, $U(x)$ the **potential energy**, and (2.1.5) the **conservation law of mechanical energy**. Since (2.1.5) is a quadratic algebraic equation, it can be solved with respect to v. Integrating v with respect to t, we obtain $x = x(t)$, though without carrying out the integration explicitly. In general, to find the solution concretely is not required. Indeed, even if $F(x)$ is a simple function, it is very difficult to carry out the integration explicitly.

2.1.3 *Ordinary differential equations*

A differential equation is called an **ordinary differential equation** if its independent variable is only one. Otherwise, it is called a **partial differential equation** (see §2.8). For the time being, since we deal with ordinary differential equations only, we omit writing the word "ordinary".

In Newtonian mechanics the independent variable is the time and is denoted by t, but in the mathematical theory of differential equations, the independent variable is usually denoted by x. If the number of the unknown functions is also one, it is usually denoted by y. Thus, in this case, the differential equation is generally written as

$$f(x, y, y', y'', \dots, y^{(n)}) = 0. \tag{2.1.7}$$

If $y^{(n)}$ is actually involved in f, (2.1.7) is called the n-th-order differential equation. If we can solve (2.1.7) explicitly with respect to $y^{(n)}$, that is, if we have

$$y^{(n)} = g(x, y, y', y'', \dots, y^{(n-1)}), \tag{2.1.8}$$

this form is called the **normal form** (one may transfer some parts of the right-hand side to the left-hand side). Otherwise, the form in (2.1.7) is called the **non-normal form**.[3]

The functions usually used in the expression for differential equations are almost limited to the elementary functions. In particular, the dependence on the derivatives of y is usually explicitly expressed algebraic functions.

If the number of differential equations is more than one, they are called **simultaneous differential equations**. They may not have the solution. As in the case of simultaneous algebraic equations, the number n of the equations is usually equal to the number of the unknown functions. The n-th-order differential equation (2.1.7) is equivalent to the following simultaneous first-order differential equations:

$$f(x, y_1, y_2, \dots, y_n, y_n') = 0,$$
$$y_1' = y_2, \quad y_2' = y_3, \quad \dots \quad, y_{n-1}' = y_n. \tag{2.1.9}$$

In a similar way, we see that simultaneous high-order differential equations can be reduced to simultaneous first-order ones. If we can rewrite the simultaneous first-order differential equations in the normal form, we can regard them as one differential equation for an unknown *vector* function.

---

[3] We give a non-trivial solvable example of a differential equation of the non-normal form at the end of §2.2.

2.1.4 *Arbitrary constants*

When one solves a first-order differential equation, one has to integrate once. Accordingly, the solution involves one integration constant C, which is an arbitrary constant. Hence, the solution to m simultaneous first-order differential equations generically contains m arbitrary constants. In particular, the solution to (2.1.7) generically contains n arbitrary constants. The solution is called the **general solution** if it contains n independent arbitrary constants. The solution which is obtained by setting all or some arbitrary constants to particular values is called the **particular solution**. The solution which is not expressed as a special case of the general solution, if any, is called the **singular solution**.

The general solution to (2.1.7), containing n arbitrary constants $C_1, C_2, \ldots, C_n$ is written as

$$y = \phi(x; C_1, C_2, \ldots, C_n). \tag{2.1.10}$$

Differentiating (2.1.10) k times, we have

$$y^{(k)} = \left(\frac{\partial}{\partial x}\right)^k \phi(x; C_1, C_2, \ldots, C_n), \quad (k = 1, 2, \ldots, n). \tag{2.1.11}$$

If we eliminate n arbitrary constants $C_1, C_2, \ldots, C_n$ from $n+1$ equations (2.1.10) and (2.1.11), then the resultant should reproduce the original differential equation (2.1.7). Thus (2.1.7) is nothing but the equation characterizing the set of "curves" expressed by (2.1.10).

For example, consider a differential equation

$$xy' - 2y = 0. \tag{2.1.12}$$

The general solution to (2.1.12) is the set of parabolas given by

$$y = Cx^2. \tag{2.1.13}$$

Differentiating (2.1.13), we have

$$y' = 2Cx. \tag{2.1.14}$$

From (2.1.13) and (2.1.14), we eliminate C; then we return to the original equation (2.1.12).

In most cases of applications to physics (and other sciences), the arbitrary constants are determined by imposing **initial conditions**, which are usually stated as $y = b_0$, $y' = b_1, \ldots, y^{(n-1)} = b_{n-1}$ at $x = a$, where $a, b_0, \ldots, b_{n-1}$ are definite constants given by physical situations.

Thus we have n simultaneous algebraic equations for n unknown quantities $C_1, C_2, \ldots, C_n$. Therefore, they are determined generically.

For example, in the case of Newton's equation of motion (2.1.3), it is usual that the initial position $x = x_0$ and the initial velocity $v = v_0$ at $t = 0$ are given. Then the solution is uniquely determined.

2.1.5 *Invariance under transformation*

If a given differential equation is invariant under some transformation of the variables x and/or y, the form of the solutions is restricted, and therefore we often have a better chance of solvability. One of the representative example is the scale invariance. Here, the **scale invariance** is the invariance under the **scale transformation** $\{x \mapsto c^j x,\ y \mapsto c^k y\}$, where j and k are certain fixed constants and c is an arbitrary constant (we may set $k = 1$, without loss of generality, unless $k = 0$). If (2.1.7) is invariant under this scale transformation, then what is obtained by the scale transformation of (2.1.10), namely,

$$c^k y = \phi(c^j x; \tilde{C}_1, \tilde{C}_2, \ldots, \tilde{C}_n) \tag{2.1.15}$$

is also a solution. Here, $\tilde{C}_1, \tilde{C}_2, \ldots, \tilde{C}_n$ are *new* arbitrary constants.

For example, (2.1.12) is invariant under two independent scale transformations $x \mapsto cx$ and $y \mapsto c'y$, so we can infer without solving it, that the solution must be of the form of $x^\alpha y = C$, where α is an undetermined constant.

The transformation like $x \mapsto x + c$ is called the **translation**. It can be reduced to a scale transformation by means of the change of the variable $x = \log x'$, namely, $x' = e^x$.

The fundamental laws of physics should be valid at any time, that is, they should be invariant under the time translation. Indeed, Newton's equation of motion is invariant under the time translation. The effect of this fact is absorbed into the change of the integration constant.

The equations which appear in physics are generally invariant under various kinds of transformations. The existence of the invariance under a transformation group yields the existence of a conservation law. This property is called **Noether's theorem**, whose proof is done on the basis of the Lagrangian formalism. The temporal translation invariance, the spatial translation invariance and the spatial rotation invariance imply the **energy conservation law**, the **momentum conservation law** and the **angular momentum conservation law**, respectively.

The energy conservation (2.1.5) is nothing but the consequence of the temporal translation invariance of (2.1.3) together with Noether's theorem.

2.1.6 *Remarks when solving a differential equation*

It is generally very difficult to solve a differential equation. It is normal that the differential equation cannot be solved analytically. In this case, we calculate the solution numerically and/or investigate its qualitative properties. In the former half of this chapter, we present various methods for solving the differential equation. In contrast with textbook exercises, the method for solving differential equations that we encounter is actually often unknown, and therefore we must devise a method. Essentially it is important to find a "good" transformation of the variables. If one happens to find a particular solution, then it is often helpful to set the general solution y equal to a sum or a product of the particular solution and an unknown function.

If one has succeeded in finding the solution, one should always check the correctness of the obtained solution by substituting it in the original differential equation. While it is generally very hard to solve a differential equation, checking is easy and, therefore, it is advisable not to forget to check. The solution which one has found may not be the same expression as the one written in a textbook. In this situation, one might immediately judge the former incorrect. But to do so is not right, because the ways of expressing the solution are not unique; the same solutions often look different. In particular, if the solution involves multi-valued functions, special care must be taken. Hence it is advisable to express the solution in a form which does not involve multi-valued functions as far as possible.

Furthermore, when the solution involves an arbitrary constant, one must take care in transforming that arbitrary constant. The arbitrary constant involved in the obtained solution may *not* be the same as that in a textbook solution.

Symmetry and Invariance

Symmetry is, as is seen from right-left symmetry, spherical symmetry, etc., the property that a geometrical object is invariant under the mapping of all its points to itself based on a certain rule. This procedure is called the transformation because it can be expressed in terms of the transformation of coordinate variables. Generally, the symmetry is the invariance under transformations. Thus "symmetry" and "invariance" are regarded as almost synonymous. According to Klein's "Erlangen Catalogue", geometry is the mathematics for investigating the invariance properties under transformation.

One can define a product T_2T_1 of two transformations T_1 and T_2 by their successive operations. The identity transformation I is defined by doing nothing, and the inverse transformation of T, denoted by T^{-1}, is to cancel the effect of T, namely, $T^{-1}T = I$. Under these definitions, the totality of transformations constitutes a group in the group-theoretical sense. Thus the invariance under transformations implies the existence of the group of symmetry.

Conversely, suppose that a group G of symmetry is given abstractly (namely, forget about geometry). If a system of equations is invariant under G, then the totality of its solutions is invariant under a group $\tilde{G}$, where $\tilde{G}$ is homomorphic to G, that is, the mapping $G \to \tilde{G}$ keeps the product rule (if $T_j \mapsto \tilde{T}_j$, then $T_2T_1 \mapsto \tilde{T}_2\tilde{T}_1$). The group $\tilde{G}$ is called a **representation** of G. In particular, suppose that a system of differential equations is invariant under a group G; then its particular solutions are mutually transformed under $\tilde{G}$.

The physical laws are invariant under some transformations. The fundamental laws should not depend on the time, nor on the spatial position, and therefore, they should be independent of the choice of the coordinate origin. Accordingly, the fundamental laws are invariant under the temporal and spatial translations. Furthermore, since the space where we live is isotropic in the directions, they are also invariant under the spatial rotations. In Newtonian mechanics, the time is dealt with as a special variable. In the theory of relativity, however, the time and the space are treated on an equal footing. Accordingly, the rotation is extended to the Lorentz transformation, which is the "rotation" in the (3+1)-dimensional space called the Minkowski space. The fourth axis of the Minkowski space is the temporal axis, which takes imaginary values.

The theory of elementary particles is formulated in such a way that it is manifestly consistent with Einstein's theory of special relativity. It is, therefore, invariant under the 4-dimensional translations and under the Lorentz transformations. The combined invariance under them is called the Poincaré symmetry. Each of the elementary particles is described as a representation of the Poincaré group.

As everybody knows well, nature is not invariant under the scale transformations. This fact is a consequence of the existence of the (rest) mass. All elementary particles, except for the photon, are known to be massive, where the mass is an index of the representation of the Poincaré group.

Nowadays, fundamental physics has a very successful theory of elementary particles, called the Standard Theory. It is based on a symmetry, called the chiral invariance, which is very similar to the scale invariance. If the theory is invariant under the chiral transformation, then it implies that all particles are massless. In the Standard Theory, this dilemma is resolved by using the concept of the "spontaneous breakdown" of the symmetry. By means of a special mechanism, called the Higgs mechanism, all elementary particles, except for the photon, acquire a non-zero mass. Another consequence of the Higgs mechanism is the prediction that there must exist a special heavy particle, called the Higgs boson. In 2013, the existence of the Higgs boson was experimentally confirmed by the Large Hadron Collider (LHC) built by CERN in Geneva. Thus the validity of the Standard Theory has been established.

2.2 First-Order Differential Equations

2.2.1 *Phenomenological models*

The equation of motion in Newtonian mechanics is a *second-order* differential equation. This fact originates from the existence of the law of inertia. If one discusses, not such a fundamental problem, but a more phenomenological model, one often encounters first-order differential equations.

For example, a radioactive element decays at its own decay rate with no relevance to the circumstances surrounding it. The (approximate) number N of the atoms of the radioactive element obeys the first-order differential equation $dN/dt = -\alpha N$, where α is a positive constant. Its solution is, of course, $N = Ce^{-\alpha t}$. The time interval in which a half of the atoms decay is called the half-life, which is given by $(\log 2)/\alpha$. If several radioactive elements successively decay, one obtains simultaneous first-order differential equations.

In biology, one encounters first-order differential equations when discussing the ecology of animals. The number of each kind of animal increases if their food sources increase and decreases if they are more likely to be eaten by other animals. In a system of such animals, their numbers satisfy simultaneous first-order differential equations. From the analysis of these numbers, one can predict which animal is in danger of extinction and which animal will flourish.

2.2.2 *Separation of the variables*

If the first-order differential equation is of the form called the **variable-separation type**

$$y' = f(x)g(y), \tag{2.2.1}$$

one can easily find the general solution. Indeed, (2.2.1) can be rewritten as

$$\frac{1}{g(y)}\frac{dy}{dx} = f(x). \tag{2.2.2}$$

Hence by integrating both sides of (2.2.2) with respect to x, we find the solution

$$\int \frac{dy}{g(y)} = \int dx\, f(x). \tag{2.2.3}$$

Although integration constants appear on both sides, they are not independent.

It should be noted that in writing (2.2.2), we have divided both sides by $g(y)$. If there is a zero point of $g(y)$ at $y = b$, namely, if $g(b) = 0$, the division is unjustified at $y = b$. Indeed, $y \equiv b$ is a solution to (2.2.1). This solution is not regarded as a singular solution, because it is obtained as the infinite limit of the integration constant of the general solution.

2.2.3 *Homogeneous scale-invariant equations*

The scale transformation for $j = k = 1$ (see §2.1) keeps both y/x and dy/dx invariant. Hence

$$y' = f\left(\frac{y}{x}\right) \tag{2.2.4}$$

is invariant under it. With

$$y = xv, \tag{2.2.5}$$

(2.2.4) becomes

$$xv' + v = f(v), \qquad (2.2.6)$$

which is rewritten as

$$v' = \frac{1}{x}(f(v) - v). \qquad (2.2.7)$$

Thus (2.2.4) is reduced to the variable-separation type.

The differential equation

$$y' = f\left(\frac{ax + by + c}{\alpha x + \beta y + \gamma}\right) \qquad (2.2.8)$$

can be reduced to (2.2.4) by making appropriate translations of x and y.

2.2.4 *Linear differential equations*

We consider the differential equation linear with respect to both y and y', that is,

$$y' + p(x)y = f(x). \qquad (2.2.9)$$

If $f(x) \equiv 0$, then (2.2.9) becomes the variable-separation type, and the solution in this case is given by

$$y_{f \equiv 0} = P(x) \equiv \exp\left(-\int dx\, p(x)\right). \qquad (2.2.10)$$

For $f(x) \neq 0$, we make the variable transformation

$$y = P(x)v. \qquad (2.2.11)$$

Since $P'(x) = -p(x)P(x)$, (2.2.9) is transformed into

$$P(x)v' = f(x). \qquad (2.2.12)$$

Hence we have

$$v = \int dx\, \frac{f(x)}{P(x)}. \qquad (2.2.13)$$

From (2.2.11) and (2.2.13), we find that the general solution to (2.2.9) is given by

$$\begin{aligned}
y &= P(x)\int dx\, \frac{f(x)}{P(x)} \\
&= \exp\left(-\int dx\, p(x)\right) \cdot \int dx\left[f(x)\exp\left(\int dx\, p(x)\right)\right]. \quad (2.2.14)
\end{aligned}$$

The integration constant involved in $P(x)$ is canceled in (2.2.14). Thus the integration constant involved in (2.2.14) is that of (2.2.13) only, as it should be.

2.2.5 *Complete differential equations*

Remember the total derivative stated in (1.3.5) of Chapter 1, that is, the total derivative of $z = F(x, y)$ is

$$dz = \frac{\partial F}{\partial x} dx + \frac{\partial F}{\partial y} dy. \tag{2.2.15}$$

Hence, if the given differential equation can be rewritten as

$$\frac{\partial F}{\partial x} dx + \frac{\partial F}{\partial y} dy = 0, \tag{2.2.16}$$

then its solution is given by

$$F(x, y) = C. \tag{2.2.17}$$

The equation of the form (2.2.16) is called the **complete differential equation**.

Given a differential equation in the form

$$P(x, y)dx + Q(x, y)dy = 0, \tag{2.2.18}$$

the necessary and sufficient condition for its completeness is

$$\frac{\partial P(x, y)}{\partial y} = \frac{\partial Q(x, y)}{\partial x}. \tag{2.2.19}$$

The necessity of (2.2.19) is evident owing to the commutativity of $\partial/\partial x$ and $\partial/\partial y$. The proof of sufficiency is a little cumbersome; we omit it. The variable-separation type is nothing but a special case in which both sides of (2.2.19) is zero.

If $F(x, y)$ is decomposable into a sum of $F_j(x, y)$, then the total derivative dz becomes a sum of dz_j. By using this property, the complete differential equation is decomposable into a formal sum of simpler complete differential equations. If this is the case, it becomes easier to find $F(x, y)$. The term consisting of x alone involved in $P(x, y)$ and the term consisting of y alone involved in $Q(x, y)$ can easily be separated. For example, the differential equation

$$(2xy + 3x^2)dx + (x^2 + 2y)dy = 0 \tag{2.2.20}$$

is decomposable into three complete differential equations. Hence we find the solution

$$x^2y + x^3 + y^2 = C. \tag{2.2.21}$$

2.2.6 *Integrating factors*

It is a rather rare case that the given differential equation is already complete as it is. It happens more frequently that if we multiply it by a certain function $\varphi(x,y)$, then it becomes a complete differential equation. In this case, $\varphi(x,y)$ is called the **integrating factor**. Unfortunately, however, there is no general algorithm for finding the integrating factor. We must find it by intuition. If $\varphi(x,y)$ is a function of either x alone or y alone, then the above stated separation is kept, and therefore the integrating factor can easily be found. For example, the linear differential equation (2.2.9) is rewritten as

$$\big(p(x)y - f(x)\big)dx + dy = 0. \tag{2.2.22}$$

Hence the term of $f(x)$ can be separated. The integrating factor is $[P(x)]^{-1}$, defined in (2.2.10). Multiplying (2.2.22) by it, we have

$$\Big(\big([P(x)]^{-1}\big)'y - f(x)[P(x)]^{-1}\Big)dx + [P(x)]^{-1}dy = 0. \tag{2.2.23}$$

The solution to (2.2.22) is seen to be

$$[P(x)]^{-1}y - \int dx\, f(x)[P(x)]^{-1} = 0, \tag{2.2.24}$$

which, of course, coincides with (2.2.14).

2.2.7 *Differential equations of the non-normal form*

If the differential equation

$$f(x, y, y') = 0 \tag{2.2.25}$$

is not solved with respect to y', it is of the non-normal form.

We consider the simultaneous equations

$$f(x, y, p) = 0, \quad \frac{\partial}{\partial p}f(x, y, p) = 0. \tag{2.2.26}$$

If it is possible to eliminate p from (2.2.26), the resultant $\phi(x, y) = 0$ is the singular solution to (2.2.25). Geometrically, it is nothing but the envelope of the set of the curves, $f(x, y, p) = 0$, parametrized by p.

The simplest example of the non-normal form is Clairaut's differential equation:

$$y = y'x + f(y'). \tag{2.2.27}$$

Its general solution is evidently given by

$$y = Cx + f(C), \tag{2.2.28}$$

because then $y' = C$. Geometrically, (2.2.28) is the set of straight lines. The singular solution to (2.2.27) is obtained by eliminating C from (2.2.28) and $x + f'(C) = 0$. For example, the singular solution to the differential equation $y = y'x + \frac{1}{2}y'^2$ is obtained by substituting $y' = -x$ in it, that is, $y = -\frac{1}{2}x^2$. Geometrically, it is a parabola standing upright on the origin.

2.3 Higher-Order Differential Equations

It is almost impossible to solve higher-order nonlinear differential equations analytically. What we present here is merely the methods for lowering the order of differential equations in some special cases.

2.3.1 *Translationally invariant case*

We consider the case in which the given differential equation is invariant under the translation either $x \mapsto x + c$ or $y \mapsto y + c$. This implies that x or y is not *explicitly* involved in the equation. For example, Newton's equation of motion (2.1.3) does not explicitly involve the variable t.

We first consider the differential equation which does not explicitly depend on y:

$$f(x, y', \ldots, y^{(n)}) = 0. \tag{2.3.1}$$

If we set $y' = z$, (2.3.1) becomes an $(n-1)$th-order differential equation of z.

The differential equation which does not explicitly depend on x can be reduced to the above case by adopting y as the independent variable instead of x. For example, we consider the $n = 2$ case:

$$f(y, y', y'') = 0. \tag{2.3.2}$$

Since we have

$$y' = \frac{dy}{dx} = \left(\frac{dx}{dy}\right)^{-1} = \frac{1}{x'}$$

$$y'' = \frac{dy'}{dx} = \frac{dy}{dx}\frac{d}{dy}\left(\frac{1}{x'}\right) = \frac{1}{x'}(-1)\frac{x''}{x'^2} = -\frac{x''}{x'^3}, \tag{2.3.3}$$

(2.3.2) becomes

$$f\left(y, \frac{1}{x'}, -\frac{x''}{x'^3}\right),\tag{2.3.4}$$

which does not explicitly involve x, that is, (2.3.4) is written in terms of $\{y, x', x''\}$.

2.3.2 Scale transformation invariant case

If the given differential equation is invariant under either $x \mapsto cx$ or $y \mapsto cy$, it reduces to the translation invariance case by setting either $x = e^u$ or $y = e^v$, correspondingly. For example, the differential equation

$$f(y, xy', x^2 y'') = 0\tag{2.3.5}$$

becomes

$$f\left(y, \frac{dy}{du}, \frac{d^2 y}{du^2} - \frac{dy}{du}\right) = 0\tag{2.3.6}$$

by setting $x = e^u$, (2.3.6) does not involve u explicitly.

Next, we consider the case in which the given differential equation is invariant under the homogeneous transformation $\{x \mapsto cx, \ y \mapsto cy\}$:

$$f(x^{-1}y, y', xy'', \dots, x^{n-1}y^{(n)}) = 0.\tag{2.3.7}$$

By setting $y = xv$, (2.3.7) becomes the equation which is invariant for the scale transformation with respect to variable x.

2.3.3 Integrating factors

We rewrite Newton's equation of motion (2.1.3) as

$$my'' - F(y) = 0,\tag{2.3.8}$$

by replacing t by x and x by y. By multiplying both sides of (2.3.8) by y', the left-hand side becomes

$$y'(my'' - F(y)) = \frac{d}{dx}\left(\frac{my'^2}{2} - \int dy\, F(y)\right).\tag{2.3.9}$$

Therefore, as seen in §2.1, it is possible to integrate (2.3.8). Similarly, if one can find a certain integrating factor, the given differential equation may be integrated. Such an interesting example is the **Liouville equation**:

$$y'' + f(x)y' - g(y)y'^2 = 0.\tag{2.3.10}$$

By multiplying (2.3.10) by $1/y'$, we see that its left-hand side becomes

$$\frac{y''}{y'} + f(x) - g(y)y' = \frac{d}{dx}\left(\log y' + \int dx\, f(x) - \int dy\, g(y)\right). \qquad (2.3.11)$$

Hence it can be integrated.

2.4 Linear Differential Equations

The **linear differential equation** is the differential equation in which its expression is linear with respect to y and all its derivatives. In the following general discussions, we assume that it is written in the normal form, that is, the coefficient of the highest-order term is normalized to 1.

2.4.1 *Homogeneous linear differential equations*

The n-th-order linear differential equation is generally written as

$$y^{(n)} + p_1(x)y^{(n-1)} + \cdots + p_{n-1}(x)y' + p_n(x)y = 0. \qquad (2.4.1)$$

It is convenient to use the vector notation. Introducing the n-dimensional column vector $\boldsymbol{y} \equiv {}^{\mathrm{t}}(y, y', \ldots, y^{(n-1)})$, where the symbol t means transposition, and the n-dimensional row vector $\boldsymbol{p} \equiv (p_n, p_{n-1}, \ldots, p_1)$, we rewrite (2.4.1) as

$$y^{(n)} + (\boldsymbol{p} \cdot \boldsymbol{y}) = 0, \qquad (2.4.2)$$

where $(\boldsymbol{p} \cdot \boldsymbol{y})$ is the inner product of the two vectors $\boldsymbol{p}$ and $\boldsymbol{y}$.

The principle of superposition applies to (2.4.1), that is, if we find n linearly-independent solutions $y_1, \ldots, y_n$, the general solution is given by

$$y = C_1 y_1 + C_2 y_2 + \cdots + C_n y_n, \qquad (2.4.3)$$

where $C_1, \ldots, C_n$ are integration constants. The set $\{y_1, \ldots, y_n\}$ is called the **fundamental system of solutions**. The choice of the fundamental system of solutions is not unique, of course.

2.4.2 *Wronskian*

Owing to (2.4.1), $y^{(n)}$ is completely expressed in terms of the lower-order derivatives $y, y', \ldots, y^{(n-1)}$. Hence the formula of the Taylor expansion, (1.6.5) of Chapter 1, implies that the linear independence of the solution *as functions* is equivalent to that of the solutions *as vectors*.

The determinant composed by n n-dimensional vectors, $\boldsymbol{y}_1, \boldsymbol{y}_2, \ldots, \boldsymbol{y}_n$ in the component representation,

$$W(y_1, \ldots, y_n) \equiv \det(\boldsymbol{y}_1, \boldsymbol{y}_2, \ldots, \boldsymbol{y}_n) \equiv \begin{vmatrix} y_1 & y_2 & \cdots & y_n \\ y_1' & y_2' & \cdots & y_n' \\ \vdots & \vdots & \ddots & \vdots \\ y_1^{(n-1)} & y_2^{(n-1)} & \cdots & y_n^{(n-1)} \end{vmatrix}$$

$$(2.4.4)$$

is called the **Wronskian**, or **Wronski's determinant**. For example, in the $n = 2$ case, the Wronskian is given by $W = y_1 y_2' - y_1' y_2$. From the property of the determinant, the necessary and sufficient condition for the linear independence of n solutions $y_1, \ldots, y_n$ is $W \neq 0$.

We calculate the derivative of the Wronskian W. According to the Leibniz rule, we have $W' = \sum_{j=1}^{n} V_j$, where V_j is the determinant which is obtained from W by differentiating its j-th row only. Evidently, for $j = 1, 2, \ldots, n-1$, the j-th row of V_j is the same as the $(j+1)$-th row of V_j, and therefore $V_j = 0$, namely, $W' = V_n$. The n-th row of V_n is $\{y_1^{(n)}, \ldots, y_n^{(n)}\}$, where (2.4.1) implies

$$y_k^{(n)} = -p_1 y_k^{(n-1)} - p_2 y_k^{(n-2)} - \cdots - p_{n-1} y_k' - p_n y_k. \qquad (2.4.5)$$

According to the property of the determinant, the contributions from all terms, except for the first term, vanish. Taking out the common factor of the first term, $-p_1$, the remainder is just equal to W, that is, we have $V_n = -p_1 W$. We thus find that the Wronskian W satisfies the differential equation

$$\frac{d}{dx} W = -p_1 W. \qquad (2.4.6)$$

For $W \neq 0$, the solution to (2.4.6) is

$$W = \exp\left(-\int dx\, p_1(x)\right). \qquad (2.4.7)$$

In particular, if $p_1(x) \equiv 0$, then W is a constant.

2.4.3 *Inhomogeneous linear differential equations*

We consider the case in which the right-hand side of (2.4.1) is not zero but a given function $f(x)$, that is, we consider

$$y^{(n)} + p_1(x) y^{(n-1)} + \cdots + p_{n-1}(x) y' + p_n(x) y = f(x), \qquad (2.4.8)$$

which is called the **inhomogeneou linear differential equation**. As is evident from the principle of superposition, if the fundamental system $\{y_1, \ldots, y_n\}$ of the solutions to the corresponding homogeneous equation (2.4.1) and a particular solution y_0 to (2.4.8) are found, then the general solution to (2.4.8) is given by

$$y = y_0 + C_1 y_1 + C_2 y_2 + \cdots + C_n y_n. \tag{2.4.9}$$

2.4.4 General solution to the inhomogeneous linear differential equation

If a fundamental system, $\{y_1, \ldots, y_n\}$, of the solutions to (2.4.1) is known, we can always find the general solution to (2.4.8). The method in this case is to replace the constant coefficients $C_1, \ldots, C_n$ by unknown functions $v_1(x), \ldots, v_n(x)$ and then determine them. This method is called the **variation of constants**. We set

$$y = y_1 v_1 + y_2 v_2 + \cdots + y_n v_n, \tag{2.4.10}$$

where the derivative $\{v_1', \ldots, v_n'\}$ of $\{v_1, \ldots, v_n\}$ is assumed to satisfy the simultaneous linear algebraic equations

$$\begin{aligned} y_1^{(j)} v_1' + y_2^{(j)} v_2' + \cdots + y_n^{(j)} v_n' &= 0, \quad (j = 0, 1, \ldots, n-2) \\ y_1^{(n-1)} v_1' + y_2^{(n-1)} v_2' + \cdots + y_n^{(n-1)} v_n' &= f. \end{aligned} \tag{2.4.11}$$

With the help of (2.4.11), we differentiate (2.4.10) successively. Then we find

$$\begin{aligned} y^{(j)} &= y_1^{(j)} v_1 + y_2^{(j)} v_2 + \cdots + y_n^{(j)} v_n, \quad (j = 0, 1, \ldots, n-1) \\ y^{(n)} &= y_1^{(n)} v_1 + y_2^{(n)} v_2 + \cdots + y_n^{(n)} v_n + f. \end{aligned} \tag{2.4.12}$$

By taking the inner product of the first equation of (2.4.12) and the vector $\boldsymbol{p}$ defined in (2.4.2), we have

$$(\boldsymbol{p} \cdot \boldsymbol{y}) = (\boldsymbol{p} \cdot \boldsymbol{y_1}) v_1 + (\boldsymbol{p} \cdot \boldsymbol{y_2}) v_2 + \cdots + (\boldsymbol{p} \cdot \boldsymbol{y_n}) v_n, \tag{2.4.13}$$

in terms of the vector notation. Adding (2.4.13) to the second equation of (2.4.12), and using (2.4.5), namely, $y_k^{(n)} + (\boldsymbol{p} \cdot \boldsymbol{y_k}) = 0$, we obtain

$$y^{(n)} + (\boldsymbol{p} \cdot \boldsymbol{y}) = f, \tag{2.4.14}$$

which is nothing but (2.4.8). Thus we have shown that the quantity y given by (2.4.10) is the solution to (2.4.8). Since we have to integrate $\{v_1', \ldots, v_n'\}$

once in order to obtain $\{v_1, \ldots, v_n\}$, we encounter n integration constants. Thus (2.4.10) is the general solution to (2.4.8).

According to Cramer's method, we can give the explicit expressions for $\{v'_1, \ldots, v'_n\}$. Introducing the n-dimensional column vector $\boldsymbol{f} \equiv {}^t(0, \ldots, 0, f)$, we set

$$W_k \equiv \det(\boldsymbol{y}_1, \ldots, \boldsymbol{y}_{k-1}, \boldsymbol{f}, \boldsymbol{y}_{k+1}, \ldots, \boldsymbol{y}_n). \tag{2.4.15}$$

Then the solution to (2.4.11) is given by

$$v'_k = \frac{W_k}{W}, \tag{2.4.16}$$

where W is the Wronskian defined by (2.4.4). From (2.4.10), therefore, the general solution to (2.4.8) is given by

$$y = \sum_{k=1}^{n} y_k \int dx \, \frac{W_k}{W}. \tag{2.4.17}$$

For example, in the $n = 2$ case, we have

$$y = -y_1 \int dx \, \frac{f \, y_2}{y_1 y'_2 - y_2 y'_1} + y_2 \int dx \, \frac{f \, y_1}{y_1 y'_2 - y_2 y'_1}. \tag{2.4.18}$$

2.4.5 *Constant coefficient homogeneous linear differential equations*

According to the above discussion, the problem of solving the inhomogeneous equation reduces to the problem of finding the fundamental system of the solution to the corresponding homogeneous equation. Unfortunately, there is no general way of finding the latter. But, if all coefficients p_k are constants, that is, if the given equation is

$$y^{(n)} + p_1 y^{(n-1)} + \cdots + p_{n-1} y' + p_n y = 0, \tag{2.4.19}$$

we can find explicit solutions.

Setting $y = e^{\alpha x}$, substituting it into (2.4.19), and then multiplying the resultant by $e^{-\alpha x}$, we obtain

$$\alpha^n + p_1 \alpha^{n-1} + \cdots + p_{n-1} \alpha + p_n = 0, \tag{2.4.20}$$

which is an algebraic equation for α of degree n. It is called the **characteristic equation** of (2.4.19). If $\alpha = \alpha_k$ is a solution to (2.4.20), $y = e^{\alpha_k x}$ is a particular solution to the differential equation (2.4.19). Hence, if (2.4.20) has no multiple solutions, then the general solution to (2.4.19) is given by

$$y = C_1 e^{\alpha_1 x} + \cdots + C_n e^{\alpha_n x}. \tag{2.4.21}$$

Of course, α_k is not necessarily real. If one needs to have the real representation, one should rewrite (2.4.21) in terms of the trigonometric functions by using the Euler formula (1.7.2) of Chapter 1. For example, the solution to the characteristic equation of $y'' + a^2 y = 0$ (a is real) is $\alpha = \pm ia$. Hence the general solution in the real representation is $y = C_1 \cos(ax) + C_2 \sin(ax)$.

For the case in which (2.4.20) has multiple solutions, we have only to take the limit of coincidence. For example, in the $\alpha_2 \to \alpha_1$ limit of the two solutions $e^{\alpha_1 x}$ and $e^{\alpha_2 x}$, two independent solutions are given by $e^{\alpha_1 x}$ and

$$\lim_{\alpha_2 \to \alpha_1} \frac{e^{\alpha_2 x} - e^{\alpha_1 x}}{\alpha_2 - \alpha_1} = \frac{\partial e^{\alpha_1 x}}{\partial \alpha_1} = x e^{\alpha_1 x}. \tag{2.4.22}$$

Likewise, for the m_1-ple solutions to (2.4.20), the independent solutions to (2.4.19) are $e^{\alpha_1 x}, x e^{\alpha_1 x}, \ldots, x^{m_1 - 1} e^{\alpha_1 x}$. Therefore, in general, if α_k is the m_k-ple solution to (2.4.20), then the general solution to (2.4.19) is given by

$$y = P_1(x) e^{\alpha_1 x} + \cdots + P_\ell(x) e^{\alpha_\ell x}, \tag{2.4.23}$$

where $P_k(x)$ is an arbitrary polynomial of degree $m_k - 1$ and $\sum_{k=1}^{\ell} m_k = n$.

For solving the constant coefficient linear differential equation, it is more convenient to employ the operator calculus. This is discussed in §3.1 of Chapter 3.

2.5 Second-Order Linear Differential Equations

The second-order linear differential equation is extremely important for practical applications. Expanding both coefficient functions and solution into a power series, we determine the coefficients of the power series expansion of the solution from lower degree to higher degree in a successive way. Then we obtain the solution in the form of a power series. In general, the convergence radius of the Taylor series is equal to the distance between the expansion point and its nearest singularity. Hence, we choose a branch point of the solution as the expansion point; by taking out the branch point factor from the solution, the remainder factor becomes holomorphic at the expansion point.

2.5.1 *Representation of coefficient functions in terms of the solutions*

Suppose that the second-order homogeneous linear differential equation

$$y'' + p_1(x)y' + p_2(x)y = 0 \qquad (2.5.1)$$

has the fundamental system of solutions $\{y_1, y_2\}$. If we want y_1 and y_2 to be the solutions expressible as expected in the above, then we have to know what conditions the coefficient functions, $p_1(x)$ and $p_2(x)$, should satisfy. For this purpose, we express them in terms of y_1 and y_2.

Since y_1 and y_2 are solutions to (2.5.1), we have

$$y_1'' + p_1 y_1' + p_2 y_1 = 0, \quad y_2'' + p_1 y_2' + p_2 y_2 = 0. \qquad (2.5.2)$$

Multiplying the first and second formulas of (2.5.2) by y_2 and by y_1, respectively, and taking their difference, we obtain

$$p_1 = -\frac{y_1 y_2'' - y_2 y_1''}{y_1 y_2' - y_2 y_1'} = -\frac{W'}{W} = -\frac{d}{dx} \log W. \qquad (2.5.3)$$

Here W is the $n = 2$ Wronskian, which can be rewritten[4] as

$$W = y_1 y_2' - y_2 y_1' = y_1^2 \frac{d}{dx}\left(\frac{y_2}{y_1}\right). \qquad (2.5.4)$$

Substituting (2.5.4) into (2.5.3), we have

$$p_1 = -\frac{d}{dx} \log\left[y_1^2 \frac{d}{dx}\left(\frac{y_2}{y_1}\right)\right]$$

$$= -2\frac{d\log y_1}{dx} - \frac{d}{dx}\log\left[\frac{d}{dx}\left(\frac{y_2}{y_1}\right)\right]. \qquad (2.5.5)$$

Thus, we have succeeded in expressing p_1 in the form of $(d/dx)\log(*)$, where $(*)$ is expressed in terms of y_1 or the derivative of y_2/y_1. Similarly, knowing p_1 already, from the first formula of (2.5.2), we obtain

$$p_2 = -\frac{y_1''}{y_1} - p_1 \frac{y_1'}{y_1}$$

$$= -\frac{d}{dx}\left(\frac{d\log y_1}{dx}\right) - \left(\frac{d\log y_1}{dx}\right)^2 - p_1 \frac{d\log y_1}{dx}. \qquad (2.5.6)$$

Later, (2.5.5) and (2.5.6) are used in order to determine the conditions which p_1 and p_2 should satisfy.

---

[4] See (1.4.5) of Chapter 1.

2.5.2 *Regular singular points*

We regard x as a complex variable and $y(x)$ as an analytic function of x. The coefficient functions $p_1(x)$ and $p_2(x)$ are assumed to be holomorphic in a neighborhood D of $x = a$ *except for* $x = a$. In this case, (2.5.1) has a solution of the form

$$y_1(x) = (x - a)^\rho \phi(x), \tag{2.5.7}$$

where $\phi(x)$ is a holomorphic in D and $\phi(a) \neq 0$. Since ρ is a complex number, the point $x = a$ is generically a branch point of $y_1(x)$. This can be seen as follows.

Let $\{\tilde{y}_1, \tilde{y}_2\}$ be a fundamental system of solutions to (2.5.1). We analytically go around the point $x = a$ in D by the angle 2π. If $x = a$ is not a branch point, the resultant is invariant. But, if it is a branch point, their expression will change. Of course, since they are still the solutions to (2.5.1), we can write them as

$$\{\alpha_{11}\tilde{y}_1 + \alpha_{12}\tilde{y}_2, \ \alpha_{21}\tilde{y}_1 + \alpha_{22}\tilde{y}_2\} \tag{2.5.8}$$

in terms of $\{\tilde{y}_1, \tilde{y}_2\}$. We diagonalize the matrix formed by the coefficients of $\tilde{y}_1$ and $\tilde{y}_2$ in (2.5.8), that is, with some constants c_1 and c_2 such that $|c_1| + |c_2| \neq 0$, we set up the eigenvalue problem

$$c_1\alpha_{11} + c_2\alpha_{21} = c_1\lambda, \quad c_1\alpha_{12} + c_2\alpha_{22} = c_2\lambda. \tag{2.5.9}$$

The eigenvalue of λ is determined by

$$\begin{vmatrix} \alpha_{11} - \lambda & \alpha_{21} \\ \alpha_{12} & \alpha_{22} - \lambda \end{vmatrix} = (\alpha_{11} - \lambda)(\alpha_{22} - \lambda) - \alpha_{21}\alpha_{12} = 0. \tag{2.5.10}$$

We set

$$y_1 \equiv c_1\tilde{y}_1 + c_2\tilde{y}_2. \tag{2.5.11}$$

Then, by the above analytic continuation, y_1 is transformed into

$$c_1(\alpha_{11}\tilde{y}_1 + \alpha_{12}\tilde{y}_2) + c_2(\alpha_{21}\tilde{y}_1 + \alpha_{22}\tilde{y}_2) = \lambda(c_1\tilde{y}_1 + c_2\tilde{y}_2). \tag{2.5.12}$$

That is,

$$y_1 \mapsto \lambda y_1. \tag{2.5.13}$$

We, therefore, see that y_1 generically has a branch point at $x = a$, as shown in (2.5.7). Hence, as is evident from $(x - a)^\rho = \exp\big(\rho\log(x - a)\big)$,

$$\lambda = \exp(2\pi i \rho) \tag{2.5.14}$$

is obtained. $\qquad\qquad\qquad\qquad\qquad\qquad\qquad\qquad\qquad\qquad\qquad\square$

Since (2.5.10) is a quadratic equation, there is one more solution λ'. If $\lambda' \neq \lambda$,

$$y_2 = (x-a)^{\rho'} \varphi(x) \tag{2.5.15}$$

is a solution independent of y_1, where

$$\lambda' = \exp(2\pi i \rho') \tag{2.5.16}$$

and $\varphi(x)$ is holomorphic in D and $\varphi(a) \neq 0$. Then $\{y_1, y_2\}$ is a fundamental system of solutions to (2.5.1). If $\lambda' = \lambda$, as done in (2.4.22), we have only to take the partial derivative with respect to λ. Thus the solution independent of y_1 is given by

$$y_2(x) = y_1(x)\big(h \log(x-a) + \psi(x)\big), \tag{2.5.17}$$

where h is a complex number and $\psi(x)$ is a holomorphic function.

Now, we determine the conditions for $p_1(x)$ and $p_2(x)$ by using the results obtained in the beginning of this section. Substituting (2.5.7) and (2.5.15) [or (2.5.17)] into (2.5.5), we find that, owing to the operation $(d/dx) \log(*)$, $p_1(x)$ can have a pole of at most order 1 at $x = a$. Similarly, substituting them into (2.5.6), we find that $p_2(x)$ can have a pole of at most order 2 at $x = a$.

In general, if the point $x = a$ is an isolated singularity of an analytic function $f(x)$ such that it becomes holomorphic by multiplying a certain power of $(x-a)$, it is called the **regular singular point** of $f(x)$.

From the above consideration, we see that if the general solution to (2.5.1) has a regular singular point at $x = a$, then $q(x) \equiv (x-a)p_1(x)$ and $r(x) \equiv (x-a)^2 p_2(x)$ are holomorphic at $x = a$. Thus (2.5.1) is rewritten as

$$(x-a)^2 y'' + (x-a)q(x)y' + r(x)y = 0, \tag{2.5.18}$$

in terms of the coefficient functions holomorphic at $x = a$.

2.5.3 *Series expansion method*

The differential equation (2.5.18) can be solved in terms of the series expansion at the regular singular point $x = a$. Since $q(x)$ and $r(x)$ are holomorphic at $x = a$, they can be expressed into the Taylor series:

$$q(x) = \sum_{n=0}^{\infty} b_n(x-a)^n, \quad r(x) = \sum_{n=0}^{\infty} c_n(x-a)^n. \tag{2.5.19}$$

From (2.5.7), we know that the solution of the form

$$y(x) = \sum_{n=0}^{\infty} \alpha_n (x-a)^{\rho+n}, \quad (\alpha_0 \neq 0) \tag{2.5.20}$$

exists. Substituting (2.5.19) and (2.5.20) into (2.5.18), we obtain

$$\sum_{n=0}^{\infty} A_n (x-a)^{\rho+n} = 0, \tag{2.5.21}$$

where A_n denotes a sum over the terms such that the total sum of the lower scripts equals n. The necessary and sufficient condition for (2.5.21) is

$$A_n = 0, \quad (n = 0, 1, 2, \ldots). \tag{2.5.22}$$

In what follows, we calculate the concrete expression for A_n successively.

To begin with, for $n = 0$ we have

$$A_0 \equiv \rho(\rho-1)\alpha_0 + b_0\rho\alpha_0 + c_0\alpha_0 = \left(\rho^2 + (b_0-1)\rho + c_0\right)\alpha_0 = 0. \tag{2.5.23}$$

This is nothing but a quadratic algebraic equation for ρ, that is, (2.5.23) determines the value of ρ. Hence it is called the **determining equation**. The solutions to (2.5.23) are given by

$$\rho = \frac{-b_0 + 1 \pm \sqrt{D}}{2}, \quad D \equiv (b_0-1)^2 - 4c_0. \tag{2.5.24}$$

The two solutions to the determining equation correspond to the two solutions to the eigenvalue equation (2.5.10). Since the correspondence relation is given by (2.5.14), if the difference between both solutions to (2.5.23) is an integer, namely, if D is a square number, then both solutions correspond to the same value of λ. Therefore, in this case, special care must be taken. When $D > 0$, we denote the solutions to (2.5.23) by ρ_1 and ρ_2, so as to be $\rho_1 - \rho_2 > 0$. But in the situation in which such distinction is unnecessary, we simply write ρ.

Choosing one of the solutions as ρ, we proceed to the next step. For $n = 1$, we have

$$A_1 \equiv (\rho+1)\rho\alpha_1 + \left(b_1\rho\alpha_0 + b_0(\rho+1)\alpha_1\right) + (c_1\alpha_0 + c_0\alpha_1)$$
$$= \left(\rho^2 + (b_0+1)\rho + b_0 + c_0\right)\alpha_1 + (b_1\rho + c_1)\alpha_0 = 0. \tag{2.5.25}$$

In (2.5.25), the coefficient of α_1 becomes $2\rho + b_0$ owing to (2.5.23). Furthermore, (2.5.24) implies that it is equal to $1 \pm \sqrt{D}$. Hence it does not vanish except for ρ_2 with $D = 1$. In all other cases, we can divide (2.5.25) by it. We thus find

$$\alpha_1 = -\frac{b_1\rho + c_1}{2\rho + b_0}\alpha_0. \tag{2.5.26}$$

Likewise, in the $n = 2$ case of (2.5.22), namely,

$$A_2 \equiv \left(\rho^2 + (b_0 + 3)\rho + 2b_0 + c_0 + 2\right)\alpha_2$$
$$+ \left(b_1(\rho + 1) + c_1\right)\alpha_1 + (b_2\rho + c_2)\alpha_0 = 0, \tag{2.5.27}$$

the coefficient of α_2 becomes $2(2\rho + b_0 + 1)$ owing to (2.5.23). Furthermore, (2.5.24) implies that it does not vanish except for ρ_2 with $D = 4$. Hence, in all other cases, α_2 is uniquely determined.

For the general value of n, the coefficient of α_n in A_n is given by

$$\frac{\partial A_n}{\partial \alpha_n} = (\rho + n)(\rho + n - 1) + b_0(\rho + n) + c_0$$

$$= \rho^2 + (b_0 + 2n - 1)\rho + nb_0 + c_0 + n(n - 1)$$

$$= n(2\rho + b_0 + n - 1) = n(n \pm \sqrt{D}), \tag{2.5.28}$$

where we have used (2.5.23) and (2.5.24). Thus, unless $\rho_1 - \rho_2$ is an integer, the two independent solutions to (2.5.18) are obtained by the series expansion method. Even for the case in which $\rho_1 - \rho_2$ is an integer, we have one of the solutions, y_1.

2.5.4 *Exceptional case*

If $\rho_1 - \rho_2$ is an integer, it is harder to obtain y_2 by the expansion method (**Frobenius' method**). But, since, as stated above, this case is nothing but the case in which (2.5.10) has a double solution, y_2 should be written in the form (2.5.17). We consider a slightly modified differential equation by changing b_0 or c_0 by a very small amount ε. In this case, of course, $\sqrt{D}$ is not an integer. Therefore, by means of the above method, we can find the fundamental system of solutions $\{y_1(\varepsilon), y_2(\varepsilon)\}$. As $\varepsilon \to 0$, both solutions tend to y_1. Hence by choosing some appropriate constants $C_1(\varepsilon)$ and $C_2(\varepsilon)$ such that they go to infinity as $\varepsilon \to 0$, we construct

$$y_2 \equiv \lim_{\varepsilon \to 0} \left(C_1(\varepsilon)y_1(\varepsilon) - C_2(\varepsilon)y_2(\varepsilon)\right), \tag{2.5.29}$$

in such a way that (2.5.29) is well defined. Then y_2 is the solution independent of y_1.

2.6 Special Functions

Euler's gamma function and beta function, stated in §1.9 of Chapter 1, are, so to speak, functions of parameters, that is, although they are expressed in terms of the integral representations, the independent variables are those unrelated with the integration variable. The most well-known function of this kind is Riemann's **zeta function**, $\zeta(s) \equiv \sum_{n=1}^{\infty} n^{-s}$, which is the object of the central problem in number theory. But most special functions are defined as the solution to the differential equation.

2.6.1 *Hypergeometric differential equations*

A linear differential equation is said to be that of **Fuchs type**, if all its singularities, including the infinite point, are regular singular points. As stated in §2.5, since all its solutions can be expressed as a generalized power series such as (2.5.20), we may understand that special functions are defined by the generalized power series. Among such special functions, the most important one is Gauss' hypergeometric function.

The **hypergeometric differential equation** is

$$x(1-x)y'' + \big(c - (a+b+1)x\big)y' - ab\,y = 0, \qquad (2.6.1)$$

which contains three independent parameters, denoted by a, b and c. The special way of introducing a and b is for later convenience. As is evident from (2.6.1), the two points $x = 0$ and $x = 1$ are regular singular points. Moreover, the infinite point $x = \infty$ is also a regular singular point. This can be seen by introducing the independent variable $u = 1/x$ and then by checking the analyticity of the coefficient functions of the transformed differential equation.

Thus (2.6.1) is a differential equation of Fuchs type, characterized by the three regular singular points $\{0, 1, \infty\}$. This fact corresponds to the fact that the number of independent parameters is three. **Riemann's differential equation**, which is characterized by the existence of three regular singular points, has eight parameters, but it can be reduced to the hypergeometric differential equation by simple transformations.

2.6.2 *Hypergeometric functions*

According to the general theory of the series expansion developed in §2.5, we set up

$$y(x) = \sum_{n=0}^{\infty} \alpha_n x^{\rho+n}, \qquad (\alpha_0 \neq 0), \qquad (2.6.2)$$

and solve (2.6.1). Substituting (2.6.2) into (2.6.1), we have

$$\sum_{n=0}^{\infty} \left[(\rho+n)(\rho+n-1)\alpha_n(x^{\rho+n-1} - x^{\rho+n}) \right.$$

$$\left. + (\rho+n)\alpha_n(c\,x^{\rho+n-1} - (a+b+1)x^{\rho+n}) - ab\,\alpha_n x^{\rho+n} \right] = 0. \quad (2.6.3)$$

From the coefficient of $x^{\rho-1}$, we obtain the determining equation

$$\rho(\rho - 1 + c) = 0, \quad (2.6.4)$$

and therefore $\rho = 0$ or $\rho = 1 - c$. From the coefficient of $x^{\rho+n}$, $(n \geq 0)$, we obtain the recurrence formula

$$(\rho+n+1)(\rho+n+c)\alpha_{n+1} - \left((\rho+n)(\rho+n+a+b) + ab \right)\alpha_n = 0. \quad (2.6.5)$$

In the $\rho = 0$ case, assuming $c \neq 0, -1, -2, \ldots$, from (2.6.5) we obtain

$$\alpha_{n+1} = \frac{n(n+a+b) + ab}{(n+1)(n+c)}\alpha_n = \frac{(a+n)(b+n)}{(c+n)(n+1)}\alpha_n. \quad (2.6.6)$$

With the normalization $\alpha_0 = 1$, the solution to (2.6.6) is given by

$$\alpha_n = \frac{(a)_n(b)_n}{(c)_n n!}, \quad (2.6.7)$$

where the notation

$$(\lambda)_n \equiv \lambda(\lambda+1)\cdots(\lambda+n-1) = \frac{\Gamma(\lambda+n)}{\Gamma(\lambda)}, \quad (2.6.8)$$

has been used. Therefore, the solution to (2.6.1) corresponding to $\rho = 0$ is given by

$$y_1 = F(a, b;\, c;\, x) \equiv \sum_{n=0}^{\infty} \frac{(a)_n(b)_n}{(c)_n n!} x^n$$

$$= \sum_{n=0}^{\infty} \frac{\Gamma(a+n)\Gamma(b+n)\Gamma(c)}{\Gamma(a)\Gamma(b)\Gamma(c+n)n!} x^n. \quad (2.6.9)$$

The series on the right-hand side of (2.6.9) is called the **hypergeometric series**, and the analytic function defined by it is called the **hypergeometric function**. As is evident from the definition, $F(a, b;\, c;\, x)$ is symmetric in a and b, and its derivative is given by $(ab/c)F(a+1, b+1;\, c+1;\, x)$. Furthermore, from (2.6.8), for an integer m, we have $(-m)_n = 0$, $(n > m)$.

Hence if a or b is equal to $-m$, the hypergeometric series becomes a finite series, and the hypergeometric function reduces to a polynomial of degree m.

The hypergeometric function $F(a, b; c; x)$ is sometimes written as $_2F_1(a, b; c; x)$, where the subscripts 2 and 1 indicate the numbers of the factors of the type $(\lambda)_n$ appearing in the numerator and in the denominator, respectively, of the coefficient of the series expansion. This is because it is regarded as a special case of the **generalized hypergeometric function** $_pF_q(a_1, \ldots, a_p; c_1, \ldots, c_q; x)$. In particular, $_0F_0(x) \equiv e^x$, and $_1F_0(a; x) \equiv (1-x)^{-a}$.

Now, we proceed to investigate another solution to (2.6.1). Substituting $\rho = 1 - c$ into (2.6.5), we have

$$\alpha_{n+1} = \frac{(a - c + n + 1)(b - c + n + 1)}{(-c + n + 2)(n + 1)} \alpha_n. \tag{2.6.10}$$

With $\alpha_0 = 1$, therefore, we obtain

$$\alpha_n = \frac{(a - c + 1)_n (b - c + 1)_n}{(-c + 2)_n \, n!}, \tag{2.6.11}$$

where $c \neq 2, 3, \ldots$ from the condition that the denominator does not vanish. Thus the second solution is given by

$$y_2 = x^{1-c} F(a - c + 1, b - c + 1; -c + 2; x), \tag{2.6.12}$$

where $c \neq 1$ because for $c = 1$ we have $y_2 \equiv y_1$.

After all, unless c is an integer, the fundamental system of solutions $\{y_1, y_2\}$ is given by (2.6.9) and (2.6.12). When c is an integer, it is cumbersome to find the fundamental system of solutions. Hence we omit its presentation, but see (2.6.21) below.

2.6.3 *Integral representation of the hypergeometric function*

Since the nearest singularity of $x = 0$ is the point $x = 1$, the radius of convergence of the hypergeometric series (2.6.9) is $R = 1$. In order to see the behavior of the hypergeometric function in $|x| \geq 1$, we have to continue it analytically. For this purpose, as emphasized in Chapter 1, it is very powerful to construct the integral representation.

From the formula for the beta function given in (1.9.12) and (1.9.13) of Chapter 1, we have

$$\int_0^1 dt \, t^{b+n-1}(1-t)^{c-b-1} = \frac{\Gamma(b+n)\Gamma(c-b)}{\Gamma(c+n)}, \qquad (2.6.13)$$

where $\Re c > \Re b > 0$. Moreover, from the binomial expansion formula, we have

$$(1-tx)^{-a} = \sum_{n=0}^{\infty} \frac{\Gamma(a+n)}{\Gamma(a)\,n!} t^n x^n. \qquad (2.6.14)$$

We multiply both sides of (2.6.14) by $t^{b-1}(1-t)^{c-b-1}$ and then integrate the resultant from $t=0$ to $t=1$. Owing to (2.6.13), we obtain

$$\int_0^1 dt \, t^{b-1}(1-t)^{c-b-1}(1-tx)^{-a} = \sum_{n=0}^{\infty} \frac{\Gamma(a+n)\Gamma(b+n)\Gamma(c-b)}{\Gamma(a)\Gamma(c+n)\,n!} x^n. \qquad (2.6.15)$$

Comparing (2.6.15) with the definition (2.6.9) of the hypergeometric function, we find the integral representation

$$F(a,b;\,c;\,x) = \frac{\Gamma(c)}{\Gamma(b)\Gamma(c-b)} \int_0^1 dt \, t^{b-1}(1-t)^{c-b-1}(1-tx)^{-a}. \qquad (2.6.16)$$

This formula has been derived for $|x| < 1$, but, now, it can be analytically continued to the Riemann sheet with a cut from $x=1$ to $x=\infty$ along the positive real axis. The point $x=1$ is a branch point, and the value at $x=1$ can be calculated. Setting $x=1$ in (2.6.16) and using the formula for the beta function, we find

$$F(a,b;\,c;\,1) = \frac{\Gamma(c)\Gamma(c-a-b)}{\Gamma(c-a)\Gamma(c-b)}. \qquad (2.6.17)$$

In the usual sense, (2.6.16) is convergent only for $\Re c > \Re b > 0$, but we can remove this restriction if we employ the Y distribution introduced in (1.10.22) and (1.10.23) of Chapter 1. That is, the analytic continuation of (2.6.16) with respect to b and c is given by

$$F(a,b;\,c;\,x) = \Gamma(c) \int_{-\infty}^{\infty} dt \, Y_b(t) Y_{c-b}(1-t)(1-tx)^{-a}. \qquad (2.6.18)$$

Furthermore, dividing both sides of (2.6.18) by $\Gamma(c)$, we can use it even for the previously excluded cases $c = 0, -1, -2, \ldots$. For example, in (2.6.18), setting $b = 1$ and $c \to 0$, we have

$$\lim_{c \to 0} \frac{F(a, 1; c; x)}{\Gamma(c)} = \int_0^1 dt\, \delta'(1 - t)(1 - tx)^{-a} = ax(1 - x)^{-a-1}, \quad (2.6.19)$$

with the aid of (1.10.23) of Chapter 1. This result, apart from the multiplicative constant, coincides with the solution obtained from (2.6.12), namely

$$xF(a + 1, 2; 2; x) = x(1 - x)^{-(a+1)}. \qquad (2.6.20)$$

When c is an integer, y_1 and y_2 coincide, and therefore another independent solution is obtained by partially differentiating y_1 with respect to c and then setting $c = n$:

$$\int_{-\infty}^{+\infty} dt\, Y_b(t) \left[\frac{\partial}{\partial c} Y_{c-b}(1 - t) \right]_{c=n} (1 - tx)^{-a}. \qquad (2.6.21)$$

Of course, all the above considerations remain valid by interchanging b and a.

2.6.4 *Legendre's differential equation*

When one considers a 3-dimensional partial differential equation in the polar coordinate system, what becomes important is **Legendre's differential equation**

$$(1 - z^2)\frac{d^2 y}{dz^2} - 2z\frac{dy}{dz} + \nu(\nu + 1)y = 0, \qquad (2.6.22)$$

where ν is a parameter. The variable z is understood to be related to the azimuthal angle θ through $z = \cos\theta$. Therefore, for practical applications, it is usually considered in the interval $-1 \leq z \leq 1$.

By means of the transformation $\frac{1}{2}(1 - z) = x$ in (2.6.22), we have

$$x(1 - x)y'' + (1 - 2x)y' + \nu(\nu + 1)y = 0, \qquad (2.6.23)$$

which is nothing but the hypergeometric differential equation with $a = -\nu$, $b = \nu + 1$ and $c = 1$. From (2.6.9), therefore, we obtain one of the solutions to Legendre's differential equation. We write it as

$$P_\nu(z) = F\left(-\nu, \nu + 1; 1; \tfrac{1}{2}(1 - z)\right), \qquad (2.6.24)$$

which is called the **Legendre function of the first kind**. Because of the symmetric property of the hypergeometric function, we have

$$P_\nu(z) = P_{-\nu-1}(z). \tag{2.6.25}$$

Since (2.6.12) with $c = 1$ is the same as (2.6.9) with $c = 1$, we need to find another independent solution to (2.6.22). Since (2.6.22) is invariant under the sign change of z, $P_\nu(-z)$ is seen to be the desired one. While $P_\nu(z)$ is holomorphic at $z = 1$, $P_\nu(-z)$ is infinite at $z = 1$, as is seen from (2.6.17). Thus they are independent, provided that ν is *not* an integer. For $\nu = n = 0, 1, 2, \ldots$, we have $P_n(z) = (-1)^n P_n(-z)$ as is shown later. For $\nu = -1, -2, \ldots$, the situation reduces to the above one, because of (2.6.25).

We now define the **Legendre function of the second kind** in the following way:

$$Q_\nu(z) \equiv \frac{\pi}{2} \cdot \frac{\cos(\nu\pi)P_\nu(z) - P_\nu(-z)}{\sin(\nu\pi)}, \tag{2.6.26}$$

which is always the solution to Legendre's differential equation independent of $P_\nu(z)$. Of course, for $\nu = n$, we should take the limit $\nu \to n$.

2.6.5 *Legendre polynomials*

When $\nu = n = 0, 1, 2, \ldots$, $P_n(z)$ becomes a polynomial of degree n, which is called the **Legendre polynomial**. From its definition, its expansion is given by

$$P_n(z) = \sum_{k=0}^{n} \frac{(-n)_k(n+1)_k}{(1)_k \, k!} x^k = \sum_{k=0}^{n} \frac{(-1)^k(n+k)!}{(n-k)! \, (k!)^2} x^k, \tag{2.6.27}$$

where $z = 1 - 2x$. Rather than the above expansion formula, it is often more convenient to use the formula

$$P_n(z) = \frac{1}{2^n n!} \left(\frac{d}{dz} \right)^n (z^2 - 1)^n, \tag{2.6.28}$$

which is called **Rodrigues' formula for the Legendre polynomial**.[5] In what follows, we prove (2.6.28) on the basis of the binomial expansion formula.

---

[5] For special polynomials like the Legendre polynomial, Rodrigues' formulas usually exist. They are of compact expression containing n times differentiations.

The right-hand side of (2.6.27) becomes

$$\sum_{k=0}^{n} \frac{(-1)^k (n+k)!}{(n-k)! \, (k!)^2} x^k = \frac{1}{n!} \sum_{k=0}^{n} (-1)^k \frac{n!}{k!(n-k)!} \cdot \frac{(n+k)!}{k!} x^k$$

$$= \frac{1}{n!} \sum_{k=0}^{n} (-1)^k \, {}_nC_k \left(\frac{d}{dx}\right)^n x^{n+k} = \frac{1}{n!} \left(\frac{d}{dx}\right)^n \left(x^n \sum_{k=0}^{n} {}_nC_k(-x)^k\right)$$

$$= \frac{1}{n!} \left(\frac{d}{dx}\right)^n (x(1-x))^n. \tag{2.6.29}$$

Since $dx = -\frac{1}{2}dz$ and $x(1-x) = \frac{1}{4}(1-z^2)$, the last line of (2.6.29) equals the right-hand side of (2.6.28).   □

As is evident from (2.6.28), $P_n(z)$ is an even function for n even and an odd function for n odd, that is

$$P_n(-z) = (-1)^n P_n(z). \tag{2.6.30}$$

We have used this fact in (2.6.26).

It is instructive to reconfirm that $P_n(z)$ satisfies Legendre's differential equation for $\nu = n$, namely

$$\left((z^2 - 1)\frac{d^2}{dz^2} + 2z\frac{d}{dz} - n(n+1)\right) P_n(z) = 0, \tag{2.6.31}$$

by means of Rodrigues' formula. For this purpose, we calculate a quantity defined by

$$A \equiv \left(\frac{d}{dz}\right)^{n+1} \left[(z^2-1)\frac{d}{dz}(z^2-1)^n\right], \tag{2.6.32}$$

in two ways. On the one hand, carrying out the differentiation inside of the square brackets and then doing the differentiations $(n+1)$ times by using (1.4.9) of Chapter 1, we obtain

$$A = \left(\frac{d}{dz}\right)^{n+1} \left(2nz(z^2-1)^n\right)$$

$$= 2nz \left(\frac{d}{dz}\right)^{n+1} (z^2-1)^n + 2n(n+1) \left(\frac{d}{dz}\right)^n (z^2-1)^n. \tag{2.6.33}$$

On the other hand, performing the differentiations directly $(n+1)$ times, we obtain

$$A = (z^2-1) \left(\frac{d}{dz}\right)^{n+2} (z^2-1)^n + 2(n+1)z \left(\frac{d}{dz}\right)^{n+1} (z^2-1)^n$$

$$+ n(n+1) \left(\frac{d}{dz}\right)^n (z^2-1)^n. \tag{2.6.34}$$

Subtracting (2.6.33) from (2.6.34), we find

$$(z^2 - 1)\left(\frac{d}{dz}\right)^{n+2}(z^2 - 1)^n + 2z\left(\frac{d}{dz}\right)^{n+1}(z^2 - 1)^n$$

$$- n(n+1)\left(\frac{d}{dz}\right)^n(z^2 - 1)^n = 0. \tag{2.6.35}$$

This formula is nothing but Legendre's differential equation (2.6.31) with $y = P_n(z)$, as is seen from Rodrigues' formula (2.6.28). □

2.6.6 *Bessel's differential equations*

Bessel functions are very important in various applications (see PART II). In the analysis of wave equations in the cylindrical coordinate system, we encounter the Bessel function as the radial wave function. Usually, as the behavior in remote distant regions, while trigonometric functions describe the non-decreasing waves, Bessel functions describe the decreasing waves.

The differential equation which the Bessel function satisfies is called **Bessel's differential equation**, which is given by

$$x^2 y'' + x y' + (x^2 - \nu^2)y = 0, \tag{2.6.36}$$

where ν is a parameter. The regular singular points of (2.6.36) are $x = 0$ and $x = \infty$ only.

Now, we solve (2.6.36) by expanding y into the generalized power series at $x = 0$. Substituting

$$y = \sum_{n=0}^{\infty} \alpha_n x^{\rho+n}, \tag{2.6.37}$$

in (2.6.36), we obtain

$$\sum_{n=0}^{\infty}\left((\rho+n)(\rho+n-1) + (\rho+n) - \nu^2\right)\alpha_n x^{\rho+n} + \sum_{n=0}^{\infty}\alpha_n x^{\rho+n+2} = 0,$$
$$\tag{2.6.38}$$

namely,

$$\sum_{n=0}^{\infty}\left((\rho+n)^2 - \nu^2\right)\alpha_n x^n + \sum_{n=2}^{\infty}\alpha_{n-2} x^n = 0. \tag{2.6.39}$$

Therefore, the equations

$$(\rho^2 - \nu^2)\alpha_0 = 0, \quad (\alpha_0 \neq 0)$$
$$((\rho+1)^2 - \nu^2)\alpha_1 = 0, \tag{2.6.40}$$
$$((\rho+n)^2 - \nu^2)\alpha_n + \alpha_{n-2} = 0, \quad (n = 2, 3, \ldots)$$

must hold. The first equation of (2.6.40) is the determining equation. Its solutions are $\rho = \nu$ and $\rho = -\nu$.

In the $\rho = \nu$ case, the second and third equations reduce to

$$(2\nu + 1)\alpha_1 = 0,$$

$$n(2\nu + n)\alpha_n = -\alpha_{n-2}, \quad (n = 2, 3, \ldots),$$

$$(2.6.41)$$

respectively. From the first equation of (2.6.41), we have $\alpha_1 = 0$ unless $\nu = -1/2$.[6] Substituting $\alpha_1 = 0$ into the second equation of (2.6.41), we find $\alpha_n = 0$ for n odd. Therefore, the power series consists of even powers of x only.

Accordingly, we set $n = 2m$ and solve the second equation of (2.6.41). Unless ν is a negative integer, we obtain

$$\alpha_{2m} = -\frac{\alpha_{2m-2}}{(2m)(2\nu + 2m)} = \frac{-\alpha_{2(m-1)}}{2^2 m(\nu + m)}, \qquad (2.6.42)$$

and therefore

$$\alpha_{2m} = \frac{(-1)^m \alpha_0}{2^{2m} m! \, (\nu + 1)_m}. \qquad (2.6.43)$$

Thus, the solution to (2.6.36) is

$$y_1 = \alpha_0 \sum_{m=0}^{\infty} \frac{(-1)^m}{2^{2m} m! \, (\nu + 1)_m} x^{\nu + 2m}. \qquad (2.6.44)$$

2.6.7 *Bessel functions*

In (2.6.44), setting $\alpha_0 = 2^{-\nu}/\Gamma(\nu + 1)$, we define the **Bessel function of the first kind** by

$$J_\nu(x) \equiv \sum_{m=0}^{\infty} \frac{(-1)^m}{m! \, \Gamma(\nu + m + 1)} \left(\frac{x}{2}\right)^{\nu + 2m}. \qquad (2.6.45)$$

For m large, the absolute value of the coefficient in the series behaves like $1/(m!)^2$; hence the radius of convergence is $R = \infty$. This is due to the fact that there is no other finite singular point.

The solution corresponding to $\rho = -\nu$ is, of course, obtained from (2.6.45) by replacing ν by $-\nu$. Evidently, both solutions are independent, as long as ν is not an integer. The case of ν being an integer is discussed later.

---

[6] In the $\nu = -1/2$ case, by the transformation $y = v/\sqrt{x}$, (2.6.36) reduces to $v'' + v = 0$. Thus the solutions are trigonometric functions.

From (2.6.45), the following relations are derived:

$$\frac{d}{dx}\left(x^{-\nu}J_\nu(x)\right) = -x^{-\nu}J_{\nu+1}(x),$$

$$\frac{d}{dx}\left(x^\nu J_\nu(x)\right) = x^\nu J_{\nu-1}(x). \tag{2.6.46}$$

The proof of the first formula of (2.6.46) is as follows. First, differentiating (2.6.45), we find

$$x^\nu \frac{d}{dx}\left(x^{-\nu}J_\nu(x)\right) = \sum_{m=1}^{\infty} \frac{(-1)^m}{(m-1)!\,\Gamma(\nu+m+1)}\left(\frac{x}{2}\right)^{\nu+2m-1}. \tag{2.6.47}$$

Setting $m = n+1$, we have

$$x^\nu \frac{d}{dx}\left(x^{-\nu}J_\nu(x)\right) = \sum_{n=0}^{\infty} \frac{(-1)^{n+1}}{n!\,\Gamma(\nu+1+n+1)}\left(\frac{x}{2}\right)^{\nu+1+2n}. \tag{2.6.48}$$

Comparing (2.6.48) with (2.6.45), we see that the right-hand side of (2.6.48) is equal to $-J_{\nu+1}(x)$. The proof of the second formula of (2.6.46) is easier; we have only to use $\Gamma(\nu+m+1) = (\nu+m)\Gamma(\nu+m)$. □

We rewrite (2.6.46) as

$$x\frac{dJ_\nu(x)}{dx} - \nu J_\nu(x) = -xJ_{\nu+1}(x),$$

$$x\frac{dJ_\nu(x)}{dx} + \nu J_\nu(x) = xJ_{\nu-1}(x). \tag{2.6.49}$$

By taking the sum and the difference of both formulas of (2.6.49), we obtain the useful recurrence formulas of the Bessel function:

$$2\frac{dJ_\nu(x)}{dx} = J_{\nu-1}(x) - J_{\nu+1}(x),$$

$$\frac{2\nu}{x}J_\nu(x) = J_{\nu-1}(x) + J_{\nu+1}(x). \tag{2.6.50}$$

Now, we return to the case in which ν is an integer. For $\nu = n = 0, 1, 2, \ldots$, from (2.6.45) we have

$$J_n(x) = \sum_{m=0}^{\infty} \frac{(-1)^m}{m!\,(n+m)!}\left(\frac{x}{2}\right)^{n+2m}. \tag{2.6.51}$$

For $\nu = -n = -1, -2, \ldots$, since $1/\Gamma(-n+m+1)$ becomes zero for $m < n$, the summation begins with $m = n$, that is,

$$J_{-n}(x) = \sum_{m=n}^{\infty} \frac{(-1)^m}{m!\,(m-n)!} \left(\frac{x}{2}\right)^{-n+2m}. \tag{2.6.52}$$

With $m - n = k$, (2.6.52) is rewritten as

$$J_{-n}(x) = \sum_{k=0}^{\infty} \frac{(-1)^{n+k}}{(n+k)!\,k!} \left(\frac{x}{2}\right)^{n+2k}. \tag{2.6.53}$$

That is, we find the relation

$$J_{-n}(x) = (-1)^n J_n(x). \tag{2.6.54}$$

Since $J_1(x) = -J_0'(x)$ is obtained from the first formula of (2.6.50) and (2.6.54), by combining it with the second formula of (2.6.50), we can reconstruct all of $J_n(x)$'s from $J_0(x)$.

Even for the case in which ν is an integer, the solution to (2.6.36) independent of $J_\nu(x)$ is given by

$$N_\nu(x) \equiv \frac{\cos(\nu\pi)J_\nu(x) - J_{-\nu}(x)}{\sin(\nu\pi)}. \tag{2.6.55}$$

It is called the **Bessel function of the second kind** or **Neumann's Bessel function.**[7]

2.6.8 *Laguerre's differential equation*

We consider **Laguerre's differential equation**

$$xy'' + (1-x)y' + \nu y = 0, \tag{2.6.56}$$

which is similar to, but simpler than, Bessel's differential equation.

Setting

$$y = \sum_{n=0}^{\infty} a_n x^{\rho+n} \tag{2.6.57}$$

---

[7] This "Neumann" is not the famous J. von Neumann (20th century) but K. G. Neumann (19th century).

and substituting it into (2.6.56), we obtain

$$x^\rho \sum_{n=0}^{\infty} \left((\rho+n)^2 \alpha_n x^{n-1} + (-\rho - n + \nu)\alpha_n x^n \right) = 0. \tag{2.6.58}$$

Therefore, the determining equation is

$$\rho^2 = 0, \tag{2.6.59}$$

that is, $\rho = 0$. For $n \geq 1$, we have

$$n^2 \alpha_n + \left(-(n-1) + \nu \right)\alpha_{n-1} = 0. \tag{2.6.60}$$

With the normalization $\alpha_0 = 1$, we obtain

$$\alpha_n = \frac{(-\nu)_n}{(n!)^2}. \tag{2.6.61}$$

Hence the solution to Laguerre's differential equation is given by[8]

$$y = \sum_{n=0}^{\infty} \frac{(-\nu)_n}{(n!)^2} x^n. \tag{2.6.62}$$

We omit considering another independent solution to (2.6.56).

2.6.9 *Laguerre polynomials*

If $\nu = n = 0, 1, 2, \ldots$, (2.6.62) becomes a finite series, that is,

$$L_n(x) \equiv \sum_{k=0}^{n} \frac{(-1)^k n!}{(n-k)!\,(k!)^2} x^k, \tag{2.6.63}$$

is a polynomial of degree n. It is called the **Laguerre polynomial**. It is known to appear as the radial wave function satisfying the Schrödinger equation for a hydrogen atom.

As is evident from the above consideration, the Laguerre polynomial satisfies Laguerre's differential equation, that is,

$$xL_n''(x) + (1-x)L_n'(x) + nL_n(x) = 0. \tag{2.6.64}$$

---

[8]Note that if the denominator is not $(n!)^2$ but $n!$, then the series is that of $(1-x)^\nu$.

From (2.6.63), we obviously have

$$xL_n'(x) = \sum_{k=0}^{n} k \cdot \frac{(-1)^k n!}{(n-k)! \, (k!)^2} x^k. \tag{2.6.65}$$

Rewriting k as $n - (n - k)$, we see that the right-hand side of (2.6.65) is equal to $nL_n - nL_{n-1}$, that is,

$$xL_n'(x) = n\big(L_n(x) - L_{n-1}(x)\big). \tag{2.6.66}$$

Substituting (2.6.64) into the derivative of (2.6.66), we obtain

$$-xL_n'(x) + n\big(L_n'(x) + L_n(x) - L_{n-1}'(x)\big) = 0. \tag{2.6.67}$$

Substituting (2.6.66) into the first term of (2.6.67) yields

$$L_n'(x) + L_{n-1}(x) - L_{n-1}'(x) = 0. \tag{2.6.68}$$

Multiplying (2.6.68) by x and substituting (2.6.66) again, we obtain the recurrence formula of the Laguerre polynomial;

$$nL_n(x) = (2n - 1 - x)L_{n-1}(x) - (n-1)L_{n-2}(x). \tag{2.6.69}$$

Rodrigues' formula for the Laguerre polynomial is given by

$$L_n(x) = \frac{e^x}{n!} \left(\frac{d}{dx}\right)^n (e^{-x}x^n). \tag{2.6.70}$$

The proof of (2.6.70) is easy. Using the higher-order Leibniz rule (1.4.9) of Chapter 1, we carry out the differentiations; then we find

$$\left(\frac{d}{dx}\right)^n (e^{-x}x^n) = \sum_{k=0}^{n} {}_nC_k(-1)^k e^{-x} \frac{n!}{k!} x^k. \tag{2.6.71}$$

Hence, the right-hand side of (2.6.70) is equal to that of (2.6.63). $\qquad\square$

Finally, we prove that a special limit of the Laguerre polynomial becomes the Bessel function J_0, that is,

$$\lim_{n \to \infty} L_n\left(\frac{x}{n}\right) = J_0(2\sqrt{x}). \tag{2.6.72}$$

From (2.6.63), the left-hand side of (2.6.72) becomes

$$\lim_{n \to \infty} L_n\left(\frac{x}{n}\right) = \lim_{n \to \infty} \sum_{k=0}^{\infty} \frac{n!}{(n-k)! \, n^k} \frac{(-1)^k}{(k!)^2} x^k. \tag{2.6.73}$$

On the other hand, from (2.6.51), the right-hand side of (2.6.72) becomes

$$J_0(2\sqrt{x}) = \sum_{k=0}^{\infty} \frac{(-1)^k}{(k!)^2} x^k. \tag{2.6.74}$$

Since we have

$$\lim_{n\to\infty} \frac{n!}{(n-k)!\,n^k} = \lim_{n\to\infty} \frac{n(n-1)\cdots(n-k+1)}{n^k} = 1, \tag{2.6.75}$$

both sides of (2.6.72) coincide. □

Risky Bessel Function Formula

The well-known formula for the Bessel function (2.6.54),

$$J_{-n}(x) = (-1)^n J_n(x),$$

n being a positive integer, is correct but very *risky*.

Let a be an arbitrary positive constant and $\varphi(x)$ be an arbitrary C^∞-class function. Then, generically, we have

$$\int_0^a dx\,\varphi(x)J_{-n}(x) \neq (-1)^n \int_0^a dx\,\varphi(x)J_n(x).$$

This curious phenomenon does not seem to be generally recognized; at least I have never seen a mathematical book in which it is clearly pointed out.

The reason why this pathological phenomenon happens is that (2.6.45) is not meaningful at $x = 0$, because, $J_\nu(x)$ has a pole at $\nu = -1, -2, \ldots$. More precisely speaking, the two limits of $J_\nu(x)$, $\nu \to -n$ (analytic continuation) and $x \to 0$, are not commutative.

In order to discuss this more quantitatively, it is convenient to use the Y distribution introduced in §1.10 of Chapter 1. From (2.6.45), we have

$$J_\nu(x)\theta(x) = \sum_{m=0}^{\infty} \frac{(-1)^m 2^{-\nu-2m}\Gamma(\nu+2m+1)}{m!\,\Gamma(\nu+m+1)} Y_{\nu+2m+1}(x).$$

We calculate the difference

$$\Phi_n(x) \equiv J_{-n}(x)\theta(x) - (-1)^n J_n(x)\theta(x).$$

Since

$$\lim_{\nu\to-n} \frac{\Gamma(\nu+2m+1)}{\Gamma(\nu+m+1)} = (-1)^m \frac{(n-m-1)!}{(n-2m-1)!}, \quad \text{for } m \leq \frac{n-1}{2},$$

we find

$$\Phi_n(x) = \sum_{m=0}^{[(n-1)/2]} \frac{2^{n-2m}(n-m-1)!}{m!\,(n-2m-1)!} \delta^{(n-2m-1)}(x),$$

where $[\,]$ is the Gauss notation.

For example, we have

$$J_{-1}(x)\theta(x) = 2\delta(x) - J_1(x)\theta(x),$$

$$J_{-2}(x)\theta(x) = 4\delta'(x) + J_2(x)\theta(x),$$

and therefore

$$\int_0^a dx\,\varphi(x)J_{-1}(x) = 2\varphi(0) - \int_0^a dx\,\varphi(x)J_1(x),$$

$$\int_0^a dx\,\varphi(x)J_{-2}(x) = -4\varphi'(0) + \int_0^a dx\,\varphi(x)J_2(x).$$

$$****************$$

For $x < 0$, we see $J_n(-x) = (-1)^n J_n(x)$, $\;J_{-n}(-x) = (-1)^n J_{-n}(x)$, and $\Phi_n(-x) = (-1)^{n-1}\Phi_n(x)$, because $\delta^{(n-2m-1)}(-x) = (-1)^{n-1}\delta^{(n-2m-1)}(x)$. Hence from the defining formula of $\Phi_n(x)$, we have

$$-\Phi_n(x) = J_{-n}(x)\theta(-x) - (-1)^n J_n(x)\theta(-x).$$

By adding this formula and the definition of $\Phi_n(x)$, and by using $\theta(x) + \theta(-x) = 1$, the original formula

$$J_{-n}(x) - (-1)^n J_n(x) = 0$$

is reproduced.

2.7 Eigenvalue Problems

2.7.1 *Boundary value problems*

We consider the **boundary value problem** of the second-order homogeneous linear differential equation. While the initial value problem is the problem of finding the solution to the differential equation under the given values for y and y' at *one* point of x, the boundary value problem is the problem of finding the solution under the given condition between y and y' at each of *two* different points of x.

In the boundary value problem, the question is whether or not there is a non-trivial solution $y(x) \not\equiv 0$. In the generic situation, there exists no non-trivial solution. Therefore, we introduce a new parameter, denoted by λ, into the differential equation in the form of $\lambda w(x)y$. Then the boundary value problem has the non-trivial solutions for some particular values $\lambda = \lambda_k$. The value λ_k is called the **eigenvalue**, and the corresponding solution y_k is called the **eigenfunction** belonging to λ_k. (As for the uniqueness of y_k, see below.) The central problem is to find the eigenvalues and the eigenfunctions. It is called the **eigenvalue problem**, which arises, in addition to the boundary value problem, in various mathematical situations, such as matrix algebra, integral equation, more abstract operator calculus, etc. Furthermore, it plays a fundamentally important role in quantum mechanics.

2.7.2 *Self-adjoint differential equations*

The second-order homogeneous linear differential equation is generally written as

$$\left(p_0(x)\frac{d^2}{dx^2} + p_1(x)\frac{d}{dx} + p_2(x) \right) y(x) = 0. \tag{2.7.1}$$

As done in the definition of the distribution, (1.10.3) of Chapter 1, we multiply an arbitrary function $z(x)$ and integrate the resultant in the closed interval $[a, b]$. Then (2.7.1) becomes

$$\int_a^b dx\, z(x) \left(p_0(x)\frac{d^2}{dx^2} + p_1(x)\frac{d}{dx} + p_2(x) \right) y(x) = 0. \tag{2.7.2}$$

We integrate its left-hand side by parts. If the contribution from the boundary term vanishes, then we have

$$\int_a^b dx\, y(x) \left(\frac{d^2}{dx^2}p_0(x) - \frac{d}{dx}p_1(x) + p_2(x) \right) z(x) = 0, \tag{2.7.3}$$

where the differential operator acts also on $z(x)$. If we regard $y(x)$ as an arbitrary function, we obtain the differential equation for $z(x)$,

$$\left(\frac{d^2}{dx^2}p_0(x) - \frac{d}{dx}p_1(x) + p_2(x) \right) z(x) = 0, \tag{2.7.4}$$

which is called the **adjoint differential equation** of (2.7.1).

If a differential equation coincides with its adjoint one, it is called the **self-adjoint differential equation**. The self-adjoint differential equation is usually written as

$$\left(\frac{d}{dx} p(x) \frac{d}{dx} + r(x) \right) y(x) = 0, \tag{2.7.5}$$

so as to be able to see its self-adjointness manifestly. Here, of course, the leftmost differential operator acts not only on $p(x)$ but also on dy/dx. Any second-order homogeneous linear differential equation can be rewritten in the self-adjoint form by multiplying an appropriate factor. Indeed, as is easily confirmed, the necessary factor is given by

$$\frac{1}{p_0(x)} \exp \left(\int dx \, \frac{p_1(x)}{p_0(x)} \right). \tag{2.7.6}$$

We present some concrete examples. Legendre's differential equation (2.6.22) is rewritten in the form of (2.7.5) without multiplying any factor, that is,

$$\left(\frac{d}{dz} (1 - z^2) \frac{d}{dz} + \nu(\nu + 1) \right) y = 0. \tag{2.7.7}$$

Next, Bessel's differential equation (2.6.36) becomes

$$\left(\frac{d}{dx} x \frac{d}{dx} + x - \frac{\nu^2}{x} \right) y = 0, \tag{2.7.8}$$

by multiplying (2.6.36) by $1/x$. As for Laguerre's differential equation (2.6.56), we multiply it by e^{-x}; then we have

$$\left(\frac{d}{dx} x e^{-x} \frac{d}{dx} + \nu e^{-x} \right) y = 0. \tag{2.7.9}$$

2.7.3 *Theory of Strum–Liouville*

The introduction of a parameter λ is made by setting $r(x) = q(x) + \lambda w(x)$ in (2.7.5), that is, we have

$$\left(\frac{d}{dx} p(x) \frac{d}{dx} + q(x) + \lambda w(x) \right) y(x) = 0. \tag{2.7.10}$$

The function $w(x)$ is called the **weight** (in a mathematical sense). The **boundary value problem of Strum–Liouville** in the closed interval

$[a, b]$ is set up in the following way. The functions $p(x)$, $q(x)$ and $w(x)$ are continuous real functions, and

$$p(x) > 0, \quad w(x) > 0. \tag{2.7.11}$$

The boundary conditions are set up as

$$p(a)y'(a) \sin \alpha - y(a) \cos \alpha = 0,$$
$$p(b)y'(b) \sin \beta - y(b) \cos \beta = 0, \tag{2.7.12}$$

where α and β are some angular constants. In particular, if $\alpha = \beta = 0$, the boundary conditions are $y(a) = y(b) = 0$.

In the theory of Strum–Liouville, the following results are known:

(1) The number of eigenvalues is countably infinite. The eigenvalues are real, and can be expressed as a monotonically increasing sequence, having the infinite limit, that is,

$$\lambda_1 < \lambda_2 < \cdots < \lambda_n < \cdots; \quad \lim_{n \to \infty} \lambda_n = \infty. \tag{2.7.13}$$

(2) The eigenfunction $\varphi_n(x)$ belonging to the eigenvalue λ_n is unique (i.e. there is no degeneracy) apart from the constant coefficient. The eigenfunction $\varphi_n(x)$ is real, and has $n - 1$ zero points in the open interval (a, b).
(3) Any two different eigenfunctions are orthogonal, that is, with the weight $w(x)$, the **orthogonality condition**

$$\int_a^b dx\, w(x)\varphi_m(x)\varphi_n(x) = 0, \quad (m \neq n) \tag{2.7.14}$$

is satisfied.
(4) The totality of the eigenfunctions is complete, that is, for any real function $y(x)$ satisfying (2.7.12), if

$$\int_a^b dx\, w(x)y(x)\varphi_n(x) = 0 \tag{2.7.15}$$

holds for any n, then we have $y(x) \equiv 0$. This property is called the **completeness condition**.

2.7.4 *Proof of orthogonality*

To prove all the above propositions is too lengthy to present here. We present only the proof of the orthogonality condition (3), which makes clear the meaning of the boundary conditions.

From (2.7.10), we have

$$
\left(\frac{d}{dx} p(x) \frac{d}{dx} + q(x) + \lambda_n w(x) \right) \varphi_n(x) = 0,
$$

$$
\left(\frac{d}{dx} p(x) \frac{d}{dx} + q(x) + \lambda_m w(x) \right) \varphi_m(x) = 0.
$$

$$(2.7.16)$$

Multiplying the first formula by $\varphi_m(x)$ and the second one by $\varphi_n(x)$, and taking the difference of both, we obtain

$$
K_{mn}(x) + (\lambda_n - \lambda_m) w(x) \varphi_m(x) \varphi_n(x) = 0, \tag{2.7.17}
$$

where

$$
K_{mn}(x) \equiv \varphi_m(x) \frac{d}{dx} \Big(p(x) \varphi_n'(x) \Big) - \varphi_n(x) \frac{d}{dx} \Big(p(x) \varphi_m'(x) \Big). \tag{2.7.18}
$$

This quantity can be rewritten as

$$
K_{mn}(x) = \frac{d}{dx} \Big(\varphi_m(x) p(x) \varphi_n'(x) - \varphi_n(x) p(x) \varphi_m'(x) \Big). \tag{2.7.19}
$$

Integrating (2.7.19) from a to b, we have

$$
\int_a^b dx \, K_{mn}(x) = \Big(\varphi_m(b) p(b) \varphi_n'(b) - \varphi_n(b) p(b) \varphi_m'(b) \Big)
$$

$$
- \Big(\varphi_m(a) p(a) \varphi_n'(a) - \varphi_n(a) p(a) \varphi_m'(a) \Big). \tag{2.7.20}
$$

Eliminating $p(c) \varphi_n'(c)$ and $p(c) \varphi_m'(c), (c = a, b)$ by using the boundary conditions (2.7.12), we find that the right-hand side of (2.7.20) is equal to zero. Therefore, (2.7.17) implies

$$
(\lambda_n - \lambda_m) \int_a^b dx \, w(x) \varphi_m(x) \varphi_n(x) = 0. \tag{2.7.21}
$$

Since $\lambda_m - \lambda_n \neq 0$ for $n \neq m$, we obtain (2.7.14). $\square$

2.7.5 *Complete orthogonal systems*

For $n = m$, of course, the value of (2.7.14) is positive. Hence we set

$$\int_a^b dx \, w(x) |\varphi_n(x)|^2 \equiv N_n^2, \quad (N_n > 0), \tag{2.7.22}$$

where N_n is called the **normalization constant**: Then, the normalized eigenfunction is defined by

$$\hat{\varphi}_n(x) \equiv \frac{1}{N_n} \varphi_n(x). \tag{2.7.23}$$

Thus the **orthonormality condition** is

$$\int_a^b dx \, w(x) \hat{\varphi}_m(x) \hat{\varphi}_n(x) = \delta_{mn}. \tag{2.7.24}$$

Here δ_{mn} is the **Kronecker delta**, and as stated in §1.10 of Chapter 1, is defined by

$$\begin{aligned} \delta_{mn} &= 1, \quad \text{(for } m = n) \\ &= 0, \quad \text{(for } m \neq n). \end{aligned} \tag{2.7.25}$$

Next, the completeness condition (2.7.15) of the system of the eigenfunctions implies that any function $\psi(x)$, satisfying (2.7.12), can be expanded into

$$\psi(x) = \sum_n c_n \hat{\varphi}_n(x), \quad c_n \equiv \int_a^b du \, w(u) \psi(u) \hat{\varphi}_n(u), \tag{2.7.26}$$

at least formally. This is because the difference of both sides of (2.7.26) satisfies the condition (2.7.15). But since this series is not necessarily convergent pointwise, we understand that (2.7.26) holds in the sense of

$$\lim_{N \to \infty} \int_a^b dx \, w(x) \left| \psi(x) - \sum_{n=1}^N c_n \hat{\varphi}_n(x) \right|^2 = 0. \tag{2.7.27}$$

In (2.7.26) with c_n being substituted, we interchange the order of the integration and the summation. Then the resultant can be regarded as the identity operator acting on any arbitrary function $\psi(x)$. Thus we may write

$$\sum_{n=1}^\infty w(u) \hat{\varphi}_n(u) \hat{\varphi}_n(x) = \delta(u - x), \tag{2.7.28}$$

by using Dirac's delta function. Since the expression for (2.7.28) is similar to the one for the orthonormality condition (2.7.24), the completeness condition is usually expressed in the form of (2.7.28) in physics.

The system of the functions which satisfy both (2.7.24) and (2.7.28) is called the **complete orthonormal system**. In the boundary value problem of Strum–Liouville, the system of eigenfunctions is a complete orthonormal one.

2.7.6 *Fourier series*

The simplest case of the boundary value problem of Strum–Liouville is the case of $\{p(x) \equiv 1, \ q(x) \equiv 0, \ w(x) \equiv 1\}$. We set $a = 0$ and $b = \pi$,[9] and $\alpha = \beta = 0$, that is, we set up the boundary value problem

$$y''(x) + \lambda y(x) = 0, \quad y(0) = y(\pi) = 0. \tag{2.7.29}$$

The eigenvalues are given by

$$\lambda_n = n^2, \quad (n = 1, 2, \ldots). \tag{2.7.30}$$

The eigenfunction belonging to λ_n is

$$\varphi_n(x) = \sin(nx). \tag{2.7.31}$$

The orthogonality condition is easily calculated by using the properties of the trigonometric functions; we have

$$\int_0^\pi dx \, \sin(mx) \sin(nx)$$

$$= \frac{1}{2} \int_0^\pi dx \left[\cos\left((m-n)x\right) - \cos\left((m+n)x\right) \right] = \frac{\pi}{2} \delta_{mn}. \tag{2.7.32}$$

Hence, the normalized eigenfunction is given by

$$\hat{\varphi}_n(x) = \sqrt{\frac{2}{\pi}} \sin(nx). \tag{2.7.33}$$

The completeness condition implies that, for any function $\psi(x)$ satisfying $\psi(0) = \psi(\pi) = 0$, it can be expanded into

$$\psi(x) = \sum_{n=1}^{\infty} b_n \sin(nx), \quad b_n = \frac{2}{\pi} \int_0^\pi dx \, \psi(x) \sin(nx). \tag{2.7.34}$$

The right-hand side is called the **Fourier sine series**.

---

[9]If we want to keep $b > 0$ arbitrary, we have only to make a scale transformation $x \mapsto (\pi/b)x$.

If we set $\alpha = \beta = \pi/2$, namely, $y'(0) = y'(\pi) = 0$, then the eigenfunction is given by

$$\varphi_n(x) = \cos(nx), \quad (n = 0, 1, 2, \ldots). \tag{2.7.35}$$

The orthonormality is similar to the above. The completeness condition implies the **Fourier cosine series**

$$\psi(x) = a_0 + \sum_{n=1}^{\infty} a_n \cos(nx), \tag{2.7.36}$$

where

$$
\begin{aligned}
a_0 &= \frac{1}{\pi} \int_0^\pi dx\, \psi(x), \\
a_n &= \frac{2}{\pi} \int_0^\pi dx\, \psi(x) \cos(nx), \quad (n = 1, 2, \ldots).
\end{aligned}
\tag{2.7.37}
$$

2.7.7 *Orthogonal polynomials*

Orthogonal polynomials are the polynomials obtainable as the eigenfunctions of the boundary value problem of Strum–Liouville. As examples, we consider the Legendre polynomials and the Laguerre polynomials.

Legendre's differential equation written as in (2.7.7) implies $p(z) \equiv 1 - z^2$, $q(z) \equiv 0$ and $w(z) \equiv 1$. The boundary conditions are set up as $y(-1) = y(1) = 0$. If ν is not an integer, the eigenfunctions are singular at either of the boundary points $z = \pm 1$, and therefore they do not satisfy the boundary conditions. Thus, if and only if ν is an integer, the eigenfunction is a polynomial. The eigenfunction belonging to the eigenvalue $\lambda = n(n+1)$ is given by the Legendre polynomial $P_n(z), (n = 0, 1, 2, \ldots)$. The orthogonality condition is

$$\int_{-1}^1 dz\, P_m(z) P_n(z) = 0, \quad (m \neq n). \tag{2.7.38}$$

We present a direct proof of (2.7.38). Without loss of generality, we may assume $n > m$. Substituting Rodrigues' formula (2.6.28), we integrate by parts $m + 1$ times, where there are no contributions from the boundary values. Since the $(m+1)$th-order derivative of any polynomial of degree m is zero, we obtain (2.7.38). □

With a similar consideration, we find the normalization constant; it is given by $N_n^2 = 2/(2n + 1)$.

Next, we consider the Laguerre polynomial. Since Laguerre's differential equation can be written in the form of (2.7.9), we have $p(x) \equiv xe^{-x}$, $q(x) \equiv 0$ and $w(x) \equiv e^{-x}$. The boundary points $a = 0$ and $b = +\infty$, and the boundary conditions are (2.7.12) with $\alpha = \beta = \pi/2$, that is, $\lim_{x \to 0} xe^{-x}y'(x) = 0$ and $\lim_{x \to +\infty} xe^{-x}y'(x) = 0$. As is seen from (2.6.62), if ν is not an integer, then, as $x \to +\infty$, $y'(x)$ behaves like e^x; therefore the second boundary condition is not satisfied. Thus the eigenfunctions are the Laguerre polynomials only. The eigenfunction belonging to $\lambda = n$ is given by $L_n(x)$ ($n = 0, 1, 2, \ldots$). The orthogonality condition is

$$\int_0^\infty dx\, e^{-x} L_m(x) L_n(x) = 0, \quad (m \neq n). \tag{2.7.39}$$

Its proof is similar to the above. Substituting Rodrigues' formula (2.6.70), we integrate by parts $m + 1$ ($\leq n$) times. The normalization constant is also easily obtained; $N_n{}^2 = 1$.

The **associated Laguerre polynomial** is defined by $L_n{}^k(x) \equiv (d/dx)^k L_n(x)$, which arises in the energy eigenvalue problem of the hydrogen atom in quantum mechanics.

2.7.8 *Multi-weight pendulums*

The boundary value problem of Strum–Liouville is quite similar to the eigenvalue problem of the real symmetric matrix. As a concrete example of realizing the relation between them, we consider the problem of a multi-weight pendulum and that of a chain pendulum.

It is said that the isochronal character of a pendulum was discovered by Galilei. A pendulum consists of a string and a weight. One end of the string is fixed and the other end is free but attached with the weight. A **multi-weight pendulum** is a pendulum which has several weights not only at the end point but also at different points of the string.

In what follows, we consider a multi-weight pendulum with n equal-mass weights placed at equal spacing along the string. The following facts are known for this multi-weight pendulum.

Let the length of the string be $n\ell$, and the weights with a mass m are placed at the positions where the string length from the free end is $k\ell$ ($k = 0, 1, 2, \ldots, n-1$). We consider the small oscillation of the pendulum in the linear approximation. We denote the angular frequency by ω and the gravitational acceleration by g, and set $\lambda = \ell\omega^2/g$. Then the eigenvalues of ω are given by the n solutions λ_j ($= \ell\omega_j^2/g$) to the equation $L_n(\lambda) = 0$, where $L_n(x)$ denotes the Laguerre polynomial of degree n. The amplitude

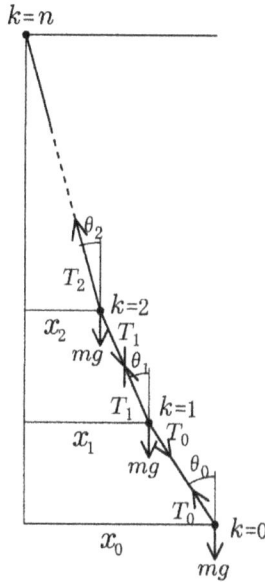

Fig. 2.1 Multi-weight pendulum

of the k-th weight corresponding to the j-th eigen frequency is proportional to $L_k(\lambda_j)$.

This is a simple problem of classical mechanics. It is not difficult to confirm the above results. Let x_k be the horizontal distance between the position of the k-th weight and the perpendicular line starting from the fixed end of the string (see Fig. 2.1). Since we always work in the linear approximation, x_k can be identified with the amplitude of the oscillation of the k th weight.

Case of n = 1

This case is nothing but the ordinary pendulum. The gravitational force is $-mg$ in the perpendicular direction. The tension T_0 of the string is decomposed into the perpendicular component and the horizontal one. Let θ_0 be the angle between the string and the perpendicular direction. Then, the former component of T_0 is $mg\cos\theta_0 \cong mg$ and the latter component of T_0 is $-mg\sin\theta_0 \cong -mgx_0/\ell$. In the linear approximation, the equation of horizontal motion is given by

$$m\frac{d^2x_0}{dt^2} = -mg\frac{x_0}{\ell}. \tag{2.7.40}$$

The general solution to (2.7.40) can be written as $x_0 = u_0 \sin(\omega t + \alpha)$, where $u_0 (\neq 0)$ and α are some constants. Substituting it into (2.7.40), we obtain $\omega^2 u_0 = (g/\ell)u_0$. The eigenvalue equation for $\lambda = \ell\omega^2/g$ is $D_1(\lambda) \equiv -\lambda + 1 = 0$. Since $L_1(\lambda) = D_1(\lambda)$, the eigenvalue equation can be written as $L_1(\lambda) = 0$.

Case of n = 2

For the $(k = 0)$th weight, the equation of motion is the same as in the above case except for $\sin\theta_0 \cong (x_0 - x_1)/\ell$. For the $(k = 1)$th weight, the forces acting in the perpendicular direction are the gravitational force $-mg$, the downward string tension $T_0 \cong -mg$ and the upward string tension $T_1 \cong 2mg$. Therefore, the equation of horizontal motion is given by

$$m\frac{d^2 x_0}{dt^2} = -mg\frac{x_0 - x_1}{\ell},$$
$$m\frac{d^2 x_1}{dt^2} = -2mg\frac{x_1}{\ell} + mg\frac{x_0 - x_1}{\ell}. \tag{2.7.41}$$

Setting $x_k = u_k \sin(\omega t + \alpha)$ $(k = 0, 1)$, we have

$$(1 - \lambda)u_0 - u_1 = 0,$$
$$-u_0 + (3 - \lambda)u_1 = 0. \tag{2.7.42}$$

Thus the eigenvalue equation is

$$D_2(\lambda) \equiv \begin{vmatrix} 1 - \lambda & -1 \\ -1 & 3 - \lambda \end{vmatrix} = \lambda^2 - 4\lambda + 2 = 2L_2(\lambda) = 0, \tag{2.7.43}$$

that is, we have obtained $L_2(\lambda) = 0$. Furthermore, from the first formula of (2.7.42), we see that $u_1 = L_1(\lambda)u_0$.

Case of n general

Likewise, the equation of the horizontal motion of the k $(=0, 1, 2, \ldots, n-1)$th weight is given by

$$m\frac{d^2 x_k}{dt^2} = -(k+1)mg\frac{x_k - x_{k+1}}{\ell} + kmg\frac{x_{k-1} - x_k}{\ell}, \tag{2.7.44}$$

where $x_n = 0$ (the fixed end point). Setting $x_k = u_k \sin(\omega t + \alpha)$, we have

$$-\lambda u_k + (k+1)(u_k - u_{k+1}) - k(u_{k-1} - u_k) = 0, \tag{2.7.45}$$

that is,

$$-ku_{k-1} + (2k + 1 - \lambda)u_k - (k+1)u_{k+1} = 0, \quad (k = 0, 1, \ldots, n-1),$$

$$u_n \equiv 0. \tag{2.7.46}$$

In order for the above n simultaneous equations to have a non-trivial solution, the determinant

$$D_n(\lambda) \equiv \begin{vmatrix} 1-\lambda & -1 & 0 & 0 & \cdots & 0 & 0 \\ -1 & 3-\lambda & -2 & 0 & \cdots & 0 & 0 \\ 0 & -2 & 5-\lambda & -3 & \cdots & 0 & 0 \\ \vdots & \vdots & \vdots & \vdots & \ddots & \vdots & \vdots \\ 0 & 0 & 0 & 0 & \cdots & 2n-3-\lambda & -n+1 \\ 0 & 0 & 0 & 0 & \cdots & -n+1 & 2n-1-\lambda \end{vmatrix} \tag{2.7.47}$$

must vanish. In what follows, we show that $D_n(\lambda) = n!L_n(\lambda)$ by mathematical induction. We have already shown its validity for $n = 1, 2$.

We expand the determinant $D_n(\lambda)$ with respect to the n-th row. It has only two non-vanishing elements, $2n - 1 - \lambda$ and $-n + 1$. The former element is multiplied by $D_{n-1}(\lambda)$ and the latter element is multiplied by $(-n+1)D_{n-2}(\lambda)$. We, therefore, obtain the recurrence formula

$$D_n(\lambda) = (2n - 1 - \lambda)D_{n-1}(\lambda) - (n-1)^2 D_{n-2}(\lambda). \tag{2.7.48}$$

Dividing (2.7.48) by $(n-1)!$ and setting $D_n(\lambda) = n!L_n(\lambda)$, we have

$$nL_n(\lambda) = (2n - 1 - \lambda)L_{n-1}(\lambda) - (n-1)L_{n-2}(\lambda), \tag{2.7.49}$$

which is nothing but the recurrence formula (2.6.69) of the Laguerre polynomial. Hence, by mathematical induction, $L_n(\lambda)$ is the Laguerre polynomial of degree n. □

Thus the eigen angular frequency is $\omega_j = \sqrt{g\lambda_j/\ell}$, where λ_j satisfies

$$L_n(\lambda_j) = 0. \tag{2.7.50}$$

Substituting $\lambda = \lambda_j$ into (2.7.46), we obtain

$$-ku_{k-1}^{(j)} + (2k + 1 - \lambda_j)u_k^{(j)} - (k+1)u_{k+1}^{(j)} = 0, \quad u_n^{(j)} = 0. \tag{2.7.51}$$

This shows that $u_{k+1}^{(j)}$ satisfies the same recurrence formula as (2.6.69) with $n = k + 1$. Therefore, we have

$$u_k^{(j)} = \frac{1}{N_j} L_k(\lambda_j), \tag{2.7.52}$$

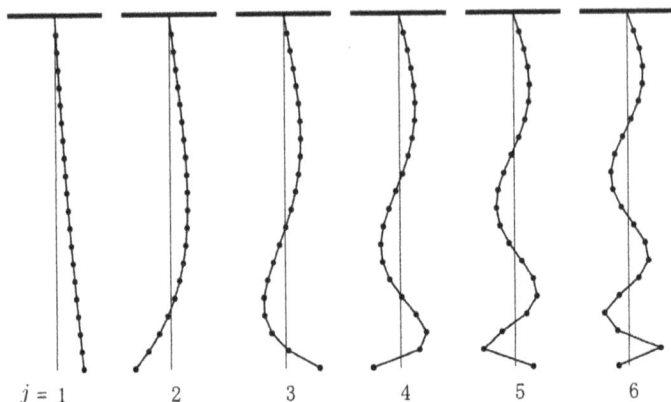

Fig. 2.2 Amplitudes of eigen modes of the multi-weight pendulum

$$n = 20, \ u_k^{(j)} (j = 1, 2, \ldots, 6)$$

where N_j is the normalization constant, which is given by $N_j^2 = \left[n L_{n-1}(\lambda_j) \right]^2 / \lambda_j$. (The proof is omitted.)

By introducing a matrix

$$\mathcal{H} \equiv \begin{pmatrix} 1 & -1 & 0 & 0 & \cdots & 0 & 0 \\ -1 & 3 & -2 & 0 & \cdots & 0 & 0 \\ 0 & -2 & 5 & -3 & \cdots & 0 & 0 \\ \vdots & \vdots & \vdots & \vdots & \ddots & \vdots & \vdots \\ 0 & 0 & 0 & 0 & \cdots & 2n-3 & -n+1 \\ 0 & 0 & 0 & 0 & \cdots & -n+1 & 2n-1 \end{pmatrix}, \tag{2.7.53}$$

we can rewrite the determinant (2.7.47), which determines the eigenvalues, as

$$D_n(\lambda) = \det \left(\mathcal{H} - \lambda \mathcal{E} \right), \tag{2.7.54}$$

where $\mathcal{E}$ denotes the unit matrix (δ_{jk}). Since $\mathcal{H}$ is a regular (i.e. $\det \mathcal{H} \neq 0$) real symmetric matrix, the equation

$$\mathcal{H} u^{(j)} = \lambda_j u^{(j)}, \quad u^{(j)} \equiv {}^t(u_0^{(j)}, u_1^{(j)}, \ldots, u_{n-1}^{(j)}) \tag{2.7.55}$$

is the standard eigenvalue problem for n-dimensional vectors. Hence, we know that all eigenvalues λ_j are real positive, and the eigenvectors $u^{(j)}$ are orthogonal and complete.

2.7.9 *Chain pendulums*

A **chain pendulum** is a pendulum which has no discrete weights but a continuous mass distribution. As is expected, a chain pendulum of uniform mass distribution is obtained as the $n \to \infty$ limit of the above multi-weight pendulum, keeping the total length of the string $\hat{L} = n\ell$ constant (whence $\ell = \hat{L}/n \to 0$). We set $k = z/\ell$; z is the length from the free end along the string. We set $u_k = f(z)$ and $\lambda = \ell\omega^2/g$. Then, (2.7.45) becomes

$$\frac{\ell\omega^2}{g}f(z) - \frac{z+\ell}{\ell}(f(z) - f(z+\ell)) + \frac{z}{\ell}(f(z-\ell) - f(z)) = 0. \quad (2.7.56)$$

After dividing (2.7.56) by ℓ, we take the limit of $\ell \to 0$. Then we have

$$\frac{\omega^2}{g}f(z) + z\frac{d^2 f(z)}{dz^2} + \frac{df(z)}{dz} = 0. \quad (2.7.57)$$

This is the differential equation for a chain pendulum.

The boundary conditions for the amplitude $f(z)$ of a chain pendulum becomes as follows. Since the point $z = \hat{L}$ is the fixed end, we have $f(\hat{L}) = 0$. The other end point $z = 0$ is completely free, but there is a restriction, that is, the chain is not torn off. Writing it in a mathematical expression, we have

$$0 = \lim_{\ell \to 0}(u_1 - u_0) = \lim_{\ell \to 0}(f(\ell) - f(0)) = \lim_{\ell \to 0}\ell f'(\theta\ell), \quad (0 < \theta < 1).$$
$$(2.7.58)$$

Here we have used the mean-value theorem. Thus, the boundary condition at $z = 0$ is $\lim_{z\to 0} zf'(z) = 0$.

From the above consideration, we obtain the boundary value problem of Strum–Liouville

$$\frac{d}{dz}\left(z\frac{d}{dz}f(z)\right) + \frac{\omega^2}{g}f(z) = 0,$$
$$\lim_{z\to 0} zf'(z) = 0, \quad f(\hat{L}) = 0. \quad (2.7.59)$$

This is a rare example of the case of $\alpha \neq \beta$ in (2.7.12).

By the transformation of the variable $z = (g/4\omega^2)x^2$, (2.7.59) becomes the $\nu = 0$ case of Bessel's differential equation (2.6.36) [or (2.7.8)]. Since the Bessel function of the second kind does not satisfy the boundary condition at $z = 0$, the solution to (2.7.59) is given by

$$f_j(z) = \frac{1}{\tilde{N}_j} J_0\left(2\omega_j \sqrt{\frac{z}{g}}\right), \tag{2.7.60}$$

where $\tilde{N}_j$ is the normalization constant. The eigen angular frequencies $\omega = \omega_j$ $(j = 1, 2, \ldots)$ are determined by the eigenvalue equation

$$J_0\left(2\omega\sqrt{\frac{\hat{L}}{g}}\right) = 0. \tag{2.7.61}$$

This result can also be reproduced as the limit of the solution to the multi-weight pendulum. For $k = z/\ell$, $L_k(\lambda)$ behaves as

$$L_k(\lambda) = L_k\left(\frac{\ell\omega^2}{g}\right) = L_k\left(\frac{\omega^2 z/g}{k}\right). \tag{2.7.62}$$

Formula (2.6.72), which shows that the limit of the Laguerre polynomial becomes the Bessel function, implies

$$\lim_{k\to\infty} L_k(\lambda) = J_0\left(2\omega\sqrt{\frac{z}{g}}\right). \tag{2.7.63}$$

Therefore, the limit of (2.7.50) is (2.7.61) and that of (2.7.52) is (2.7.60). The normalization constant is given by

$$\tilde{N}_j{}^2 = \hat{L} \cdot \left[J_1\left(2\omega_j\sqrt{\frac{\hat{L}}{g}}\right)\right]^2 = \lim_{n\to\infty} \frac{\hat{L}}{n} N_j{}^2. \tag{2.7.64}$$

Thus the boundary value problem of Strum–Liouville can be understood as the $n \to \infty$ limit of the matrix eigenvalue problem of n dimensions. Compare Fig. 2.3 with Fig. 2.2 where $n = 20$; they are almost the same.

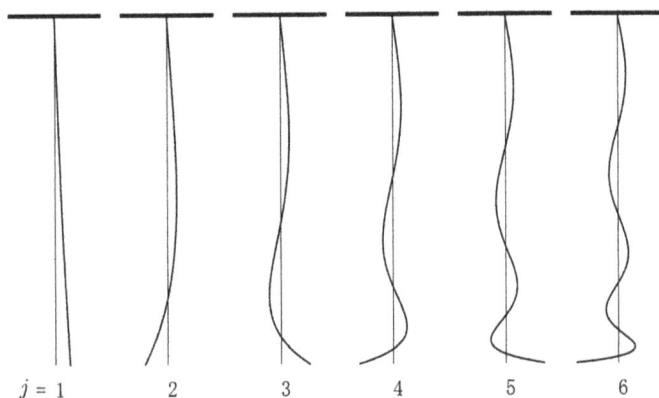

Fig. 2.3   Amplitudes of eigen modes of the chain pendulum $f_j(z)(j = 1, 2, \ldots, 6)$

Multi-Weight Pendulum of Unequal Masses

In §2.7, we have discussed the multi-weight pendulum consisting of n equal-mass weights attached to a string at equal spacing. We have confirmed the remarkable result, due to Nobuo Watabe (1943), that the eigenvalue problem is describable by means of the Laguerre polynomial L_k. We have, then, considered the $n \to \infty$ limit, keeping the string length $\hat{L}$. This limit is nothing but the chain pendulum, which was first discussed by an 18th-century mathematician, Daniel Bernoulli. We have encountered the zeroth-order Bessel function J_0.

Now, why does only J_0 appear? My coauthor, Kenji Seto, and I have investigated this problem and obtained the following result. In the multi-weight pendulum, we keep the assumption of the equal spacing but change the masses of the weights. What we have found is that if the masses of the weights are given by

$$m_k = \frac{\Gamma(\nu + k + 1)}{k!\,\Gamma(\nu + 1)} \cdot m_0,$$

where ν is a real constant such that $\nu + 1 > 0$, then the eigenvalue problem is describable by the associated Laguerre polynomial $L_k^{(\nu)}$, where

$$L_k^{(\nu)}(x) \equiv \sum_{r=0}^{k} \frac{(-1)^r \Gamma(\nu + k + 1)}{\Gamma(\nu + r + 1)(k - r)!\,r!} x^r.$$

In particular, $L_k^{(0)}(x) \equiv L_k(x)$. We then take the $n \to \infty$ limit, keeping the string length. In the obtained chain pendulum, the eigenvalue problem is describable by the Bessel function J_ν. This result follows from the fact that m_k/m_0 asymptotically behaves like k^ν as $k \to \infty$.

2.8 Partial Differential Equations

A partial differential equation is, of course, a differential equation which involves partial derivatives of unknown functions. So far, we have discussed the ordinary differential equation which contains only one independent variable, but if we wish to take into account both time and space, we have to deal with partial differential equations. Even for the problem independent of time, it is necessary to introduce more than one variable if the object is not a point but an extended one. More generally, if we wish to treat a field, the 3-dimensional coordinates must be regarded as independent variables.

The simplest example of a partial differential equation is

$$\frac{\partial \Phi(x,y)}{\partial x} = f(x,y), \tag{2.8.1}$$

where x and y are independent variables, and $\Phi(x,y)$ and $f(x,y)$ are an unknown function and a given one, respectively. The solution to (2.8.1) is, of course, given by

$$\Phi(x,y) = \int dx\, f(x,y). \tag{2.8.2}$$

Although there is no special symbol expressing "partial integration", the integration in (2.8.2) is to integrate *partially* with respect to x. Hence, the "integration constant" is now an arbitrary function of y. In general, the general solution to a partial differential equation of n independent variables contains arbitrary functions of $n-1$ independent variables. Cauchy–Riemann's differential equations, considered in (1.8.2) of Chapter 1, are examples of simultaneous partial differential equations. All holomorphic functions are the solutions.

The usual way of fixing arbitrariness contained in the general solution is, as done in the case of the ordinary differential equation, to impose initial conditions or to impose boundary conditions.

In the case of an initial-value problem, there is a special independent variable t, identified with time. Let the highest order of the derivative with respect to t be m. Then, given all values at $t = 0$ of unknown functions and their partial derivatives with respect to t of all orders lower than m, the problem is to find the solution to the given partial differential equation satisfying the above initial conditions. This problem is called the **Cauchy problem**, and the initial conditions are called the **Cauchy data**.

In the case of a boundary value problem, there arises a new situation qualitatively different from the case of ordinary differential equations, that is, even in the 2-dimensional case, the boundary curve of a domain can be quite complicated. Therefore, even if the partial differential equation is simple, the boundary value problem may become extremely difficult.

Another way of determining unknown functions is to impose partial differential equations more than the number of the unknown functions involved. The system of simultaneous partial differential equations, which determines the unknown functions just like the ordinary differential equations, is called the **maximally overdetermined system** or the **holonomic system**. This kind of system of equations may be used for defining special functions.

In considering the partial differential equation, it is very important not to forget that the independent variables are meaningful *not* as individual variables *but* as a system of variables. For simplicity, we consider the case of two independent variables (x, y). When we transform y only into $\hat{y}$, the new system of independent variables is *not* $(x, \hat{y})$ *but* $(\hat{x}, \hat{y})$. The transformation of partial derivatives is generally given by

$$\frac{\partial}{\partial x} = \frac{\partial \hat{x}}{\partial x}\frac{\partial}{\partial \hat{x}} + \frac{\partial \hat{y}}{\partial x}\frac{\partial}{\partial \hat{y}},$$
$$\frac{\partial}{\partial y} = \frac{\partial \hat{x}}{\partial y}\frac{\partial}{\partial \hat{x}} + \frac{\partial \hat{y}}{\partial y}\frac{\partial}{\partial \hat{y}}. \tag{2.8.3}$$

Therefore, even if $\hat{x} = x$, but as long as $\hat{y}$ depends on x, $\partial/\partial x = \partial/\partial \hat{x}$ does *not* hold.

2.8.1 *Second-order homogeneous linear partial differential equations*

Since nonlinear partial differential equations are usually very difficult to solve, we restrict our consideration to homogeneous linear partial differential equations. Furthermore, we consider the practically important case in which all coefficients of derivative terms are real constants.

For simplicity, we explicitly consider the case of two independent variables. The first-order partial differential equation is

$$a\frac{\partial \Phi}{\partial x} + b\frac{\partial \Phi}{\partial y} + f(x, y)\Phi = 0. \tag{2.8.4}$$

By the transformation $\hat{x} = x/a$, $\hat{y} = (x/a) - (y/b)$, (2.8.4) becomes

$$\frac{\partial \Phi}{\partial \hat{x}} + \hat{f}(\hat{x}, \hat{y})\Phi = 0. \tag{2.8.5}$$

Furthermore, setting $\Phi = \exp(-\hat{\Phi})$, we can reduce it to (2.8.1).

Likewise, the general second-order partial differential equation can be reduced to one of the three standard forms,

$$\frac{\partial^2 \Phi}{\partial x^2} + \frac{\partial^2 \Phi}{\partial y^2} + f(x,y)\Phi = 0,$$

$$\frac{\partial^2 \Phi}{\partial x^2} - \frac{\partial^2 \Phi}{\partial y^2} + f(x,y)\Phi = 0, \tag{2.8.6}$$

$$\frac{\partial^2 \Phi}{\partial x^2} - \frac{\partial \Phi}{\partial y} + f(x,y)\Phi = 0,$$

by an appropriate linear transformation. The necessary procedure is essentially the same as the reduction of the general quadratic equation into the canonical equations of an ellipse, a hyperbola and a parabola by means of translations and principal-axis transformations. Furthermore, by scale transformation, the remaining parameters have been eliminated.

The case of many independent variables can be treated similarly.

2.8.2 *Equations involving Laplacian*

The first formula of (2.8.6) with $f \equiv 0$ is called the 2-dimensional **Laplace equation**. We write it as

$$\triangle \Phi \equiv \left(\frac{\partial^2}{\partial x^2} + \frac{\partial^2}{\partial y^2} \right) \Phi = 0, \tag{2.8.7}$$

where the differential operator $\triangle$ is called the 2-dimensional **Laplacian**.

As is easily seen from Cauchy–Riemann's equations the real (or imaginary) part of an arbitrary holomorphic function satisfies Laplace equation (2.8.7).

The case in which $f(x,y)$ is non-trivial, that is,

$$\left(\triangle + f(x,y) \right)\Phi = 0, \tag{2.8.8}$$

is not solved generally, but in some special cases, it becomes solvable. For example, if $f(x,y) = g(x) + h(y)$, we can reduce (2.8.8) into ordinary differential equations by setting $\Phi(x,y) = \phi(x)\varphi(y)$. That is, (2.8.8) becomes

$$\frac{1}{\phi(x)} \left(\frac{\partial^2}{\partial x^2} + g(x) \right) \phi(x) = -\frac{1}{\varphi(y)} \left(\frac{\partial^2}{\partial y^2} + h(y) \right) \varphi(y). \tag{2.8.9}$$

Then, while the left-hand side is a function of x only, the right-hand side is a function of y only. In order for them to be equal to each other, both sides must be a constant. Writing it as $-\lambda$, we have

$$\left(\frac{d^2}{dx^2} + g(x) + \lambda\right)\phi(x) = 0,$$

$$\left(\frac{d^2}{dy^2} + h(y) - \lambda\right)\varphi(x) = 0.$$

(2.8.10)

If we can obtain the solution $\{\phi_\lambda(x), \varphi_\lambda(y)\}$, then the general solution to (2.8.8) is given by $\int d\lambda\, a(\lambda)\phi_\lambda(x)\varphi_\lambda(y)$, where $a(\lambda)$ is an arbitrary function of λ. (For the boundary value problem, the integration should be replaced by summation over eigenvalues of λ.)

2.8.3 Transformation of the 2-dimensional Laplacian into the polar coordinate system

The n-dimensional Laplacian is invariant under n-dimensional rotations. Therefore, it is convenient to express it in the polar coordinate system. In the 2-dimensional case, the polar coordinates (r, θ) are related to (x, y) by $x = r\cos\theta$ and $y = r\sin\theta$, or conversely, $r = \sqrt{x^2 + y^2}$ and $\theta = \arctan(y/x)$. The 2-dimensional Laplacian $\triangle$ in the polar coordinate system is shown to be

$$\triangle = \frac{\partial^2}{\partial r^2} + \frac{1}{r}\frac{\partial}{\partial r} + \frac{1}{r^2}\frac{\partial^2}{\partial\theta^2}.$$

(2.8.11)

The proof of (2.8.11) can be made straightforwardly by using formula (2.8.3). Since

$$\frac{\partial r}{\partial x} = \frac{x}{r} = \cos\theta, \quad \frac{\partial\theta}{\partial x} = \frac{-y}{r^2} = -\frac{\sin\theta}{r},$$

$$\frac{\partial r}{\partial y} = \frac{y}{r} = \sin\theta, \quad \frac{\partial\theta}{\partial y} = \frac{x}{r^2} = \frac{\cos\theta}{r},$$

(2.8.12)

we have

$$\frac{\partial}{\partial x} = \cos\theta\frac{\partial}{\partial r} - \frac{\sin\theta}{r}\frac{\partial}{\partial\theta},$$

$$\frac{\partial}{\partial y} = \sin\theta\frac{\partial}{\partial r} + \frac{\cos\theta}{r}\frac{\partial}{\partial\theta}.$$

(2.8.13)

Hence the Euler formula implies

$$\frac{\partial}{\partial x} \pm i\frac{\partial}{\partial y} = e^{\pm i\theta}\left(\frac{\partial}{\partial r} \pm \frac{i}{r}\frac{\partial}{\partial\theta}\right).$$

(2.8.14)

By using (2.8.14), we obtain

$$
\Delta = \left(\frac{\partial}{\partial x} - i \frac{\partial}{\partial y} \right) \left(\frac{\partial}{\partial x} + i \frac{\partial}{\partial y} \right)
$$

$$
= \left(\frac{\partial}{\partial r} - \frac{i}{r} \frac{\partial}{\partial \theta} \right) \left(\frac{\partial}{\partial r} + \frac{i}{r} \frac{\partial}{\partial \theta} \right) + \frac{1}{r} \left(\frac{\partial}{\partial r} + \frac{i}{r} \frac{\partial}{\partial \theta} \right). \qquad (2.8.15)
$$

Calculating the right-hand side of (2.8.15), we see that its imaginary part vanishes as it should, and that its real part equals the right-hand side of (2.8.11). $\qquad\qquad \square$

If $f(x, y)$ in (2.8.8) is dependent only on r, it is convenient to consider in the polar coordinate system. Setting $\Phi(x, y) = R_n(r) e^{in\theta}$, we can reduce the problem into the ordinary differential equation

$$
\left(\frac{d^2}{dr^2} + \frac{1}{r} \frac{d}{dr} - \frac{n^2}{r^2} + f(r) \right) R_n(r) = 0. \qquad (2.8.16)
$$

But the solution should be a one-valued function of (x, y). Therefore, n must be an integer. If $f(r)$ is a constant M^2, (2.8.16) becomes Bessel's differential equation (2.6.36) by a scale transformation $r \mapsto r/M$.

2.8.4 3-dimensional Laplacian in the polar coordinate system

We consider the 3-dimensional Laplacian in the polar coordinate system. In order to distinguish it from the 2-dimensional one, we denote the 3-dimensional Laplacian by a boldface Δ here only. That is, we write

$$
\Delta = \frac{\partial^2}{\partial x^2} + \frac{\partial^2}{\partial y^2} + \frac{\partial^2}{\partial z^2}. \qquad (2.8.17)
$$

The 3-dimensional polar coordinates are related to (x, y, z) by $x = r \sin\theta \cos\varphi$, $y = r \sin\theta \sin\varphi$, $z = r \cos\theta$. The 3-dimensional Laplacian in the polar coordinate system is given by

$$
\Delta = \frac{1}{r^2} \frac{\partial}{\partial r} \left(r^2 \frac{\partial}{\partial r} \right) + \frac{1}{r^2 \sin\theta} \frac{\partial}{\partial \theta} \left(\sin\theta \frac{\partial}{\partial \theta} \right) + \frac{1}{r^2 \sin^2\theta} \frac{\partial^2}{\partial \varphi^2}. \qquad (2.8.18)
$$

The proof of (2.8.18) is as follows. It is quite cumbersome if we try to prove (2.8.18) by means of the 3-dimensional transformation directly. It is wiser to make use of the cylindrical coordinates ($x = \rho \cos\varphi$, $y = \rho \sin\varphi$, $z = z$)

as the intermediate stage. From the result concerning the 2-dimensional Laplacian, we first rewrite (2.8.17) as

$$\Delta = \frac{\partial^2}{\partial\rho^2} + \frac{1}{\rho}\frac{\partial}{\partial\rho} + \frac{1}{\rho^2}\frac{\partial^2}{\partial\varphi^2} + \frac{\partial^2}{\partial z^2}. \tag{2.8.19}$$

Next, we make the transformation $\rho = r\sin\theta$, $z = r\cos\theta$. Since the sum of the first term and the last one of (2.8.19) can be regarded as the 2-dimensional Laplacian, we can again use (2.8.11). The second term of (2.8.19) can be calculated by using the second formula of (2.8.13); we obtain

$$\frac{1}{\rho}\frac{\partial}{\partial\rho} = \frac{1}{r\sin\theta}\left(\sin\theta\frac{\partial}{\partial r} + \frac{\cos\theta}{r}\frac{\partial}{\partial\theta}\right) = \frac{1}{r}\frac{\partial}{\partial r} + \frac{\cos\theta}{r^2\sin\theta}\frac{\partial}{\partial\theta}. \tag{2.8.20}$$

Therefore, (2.8.19) becomes

$$\Delta = \left(\frac{\partial^2}{\partial r^2} + \frac{1}{r}\frac{\partial}{\partial r} + \frac{1}{r^2}\frac{\partial^2}{\partial\theta^2}\right) + \left(\frac{1}{r}\frac{\partial}{\partial r} + \frac{\cos\theta}{r^2\sin\theta}\frac{\partial}{\partial\theta}\right) + \frac{1}{(r\sin\theta)^2}\frac{\partial^2}{\partial\varphi^2}. \tag{2.8.21}$$

It is easy to see that the right-hand side of (2.8.21) coincides with that of (2.8.18).[10] $\square$

In quantum mechanics, the time-independent Schrödinger equation for a one-particle system can be written in the form of

$$\big(\Delta + f(x,y,z)\big)\psi(x,y,z) = 0. \tag{2.8.22}$$

Therefore, this kind of differential equation is very important. In particular, if the force acting on the particle is rotationally invariant, then $f(x,y,z)$ depends only on $r = \sqrt{x^2+y^2+z^2}$. In this case, setting

$$\psi(x,y,z) = R_\ell(r)Y_\ell(\theta,\varphi), \tag{2.8.23}$$

from (2.8.18) we obtain

$$\left(\frac{d}{dr}r^2\frac{d}{dr} + r^2 f(r) - \ell(\ell+1)\right)R_\ell(r) = 0, \tag{2.8.24}$$

and

$$\left(\frac{1}{\sin\theta}\frac{\partial}{\partial\theta}\sin\theta\frac{\partial}{\partial\theta} + \frac{1}{\sin^2\theta}\frac{\partial^2}{\partial\varphi^2} + \ell(\ell+1)\right)Y_\ell(\theta,\varphi) = 0. \tag{2.8.25}$$

---

[10]Extending this method of proof, we can prove the general n-dimensional Laplacian in the n-dimensional polar coordinate system by means of mathematical induction with respect to n. In the case in which angular variables are unrelated, it reduces to $r^{-n+1}(\partial/\partial r)[r^{n-1}\partial/\partial r]$.

Here, owing to the one-valuedness of $\psi(x, y, z)$, ℓ must be an integer. (From the symmetry property, we may assume $\ell \geq 0$.) The function $Y_\ell(\theta, \varphi)$ is called the (3-dimensional) **spherical harmonic**.[11]

The dependence on φ of $Y_\ell(\theta, \varphi)$ is easily seen to be of the form $e^{im\varphi}$ ($m = -\ell, -\ell + 1, \ldots, \ell$). The dependence on θ of $Y_\ell(\theta, \varphi)$ is given by the **associated Legendre polynomial**, which is expressed in terms of the $|m|$-th-order derivative of the Legendre polynomial $P_\ell(z)$, ($z = \cos\theta$). For $m = 0$, (2.8.25) reduces to Legendre's differential equation (2.7.7) (with $\nu = \ell$) by setting $\cos\theta = z$ (then $1 - z^2 = \sin^2\theta$, $dz = -\sin\theta d\theta$). In the Schrödinger equation, (2.8.24) becomes the boundary value problem of Strum–Liouville and gives the eigenvalues (for the binding energies of the hydrogen atom, for instance).

2.8.5 *Wave equations*

The second and third equations of (2.8.6) are the differential equations which describe the time development of physical systems, where y is identified with a constant multiple of the time t. Indeed, in the second equation of (2.8.6), setting $f(x, y) \equiv 0$ and $y = vt$, v being a constant, we have

$$\left(\frac{\partial^2}{\partial x^2} - \frac{1}{v^2} \frac{\partial^2}{\partial t^2} \right) \Phi(x, t) = 0. \tag{2.8.26}$$

With two arbitrary functions $F(z)$ and $G(z)$, the general solution to (2.8.26) is given by

$$\Phi(x, t) = F(x - vt) + G(x + vt). \tag{2.8.27}$$

Each term of the right-hand side of (2.8.27) represents a 1-dimensional wave, propagating towards either the right or left direction with a velocity v.

In the 3-dimensional space, the wave equation becomes

$$\left(\Delta - \frac{1}{v^2} \frac{\partial^2}{\partial t^2} \right) \Phi(x, y, z, t) = 0. \tag{2.8.28}$$

This equation is called the **d'Alembert equation**. It is used to describe the propagation of electromagnetic waves (i.e., light waves in the broad sense). The differential operator $\Box \equiv \Delta - (\partial/\partial t)^2$ is called the **d'Alembertian**, where the light speed c is set equal to 1.

---

[11] According to group theory, the solution of a rotationally invariant differential equation forms a representation of the special orthogonal group $SO(3)$. Hence, we encounter the spherical harmonic $Y_\ell(\theta, \varphi)$.

The differential equation $(\Box - M^2)\Phi = 0$ is called the **Klein–Gordon equation**. It is a relativistic equation describing the quantum-theoretical wave associated with an elementary particle having a mass M (in natural unit). We have $M = 0$ for a photon, which is the quantum of electromagnetic wave.

The third equation of (2.8.6) with $f \equiv 0$ is called the **heat equation**, which describes heat conduction. In the 3-dimensional space, the heat equation is

$$\frac{\partial \Phi(x, y, z, t)}{\partial t} = \Delta \Phi(x, y, z, t). \tag{2.8.29}$$

The time-dependent Schrödinger equation for a one-particle system is

$$i\frac{\partial \psi(x, y, z, t)}{\partial t} = \left(-k\Delta + \hat{f}(x, y, z) \right) \psi(x, y, z, t), \tag{2.8.30}$$

where k is a constant inversely proportional to the mass of the particle. If the time-dependence of ψ is of the form $e^{-i\nu t}$, then (2.8.30) reduces to (2.8.22).

2.8.6 *Simultaneous partial differential equations*

In simultaneous partial differential equations, there arises a consistency problem. For example, consider

$$\frac{\partial \Phi(x, y)}{\partial x} = f(x, y),$$
$$\frac{\partial \Phi(x, y)}{\partial y} = g(x, y). \tag{2.8.31}$$

We then have

$$\frac{\partial}{\partial y}\left(\frac{\partial \Phi(x, y)}{\partial x} \right) = \frac{\partial f(x, y)}{\partial y},$$
$$\frac{\partial}{\partial x}\left(\frac{\partial \Phi(x, y)}{\partial y} \right) = \frac{\partial g(x, y)}{\partial x}. \tag{2.8.32}$$

Since partial differential operators are commutative, the left-hand sides of both formulas of (2.8.32) coincide. Hence the corresponding right-hand sides must also coincide, that is, the equality

$$\frac{\partial f(x, y)}{\partial y} = \frac{\partial g(x, y)}{\partial x}, \tag{2.8.33}$$

must hold. This condition is called the **integrability condition**. If the integrability condition is not satisfied at $(x, y) = (a, b)$, there exists no solution to the simultaneous differential equations extendable to (a, b).

In a complicated system of simultaneous partial differential equations, it is generally very difficult to confirm whether or not it satisfies the integrability condition. As the method for giving a consistent system of simultaneous partial differential equations, the **variational principle** is used in analytic mechanics and in quantum field theory. Given a functional $I[\Phi_A]$ of unknown functions Φ_A and their first derivatives, we take the minimal point of $I[\Phi_A]$ for the functional variation $\delta\Phi_A$ of Φ_A. We then obtain a consistent system of the simultaneous partial differential equations, which are called the **Euler–Lagrange equations** for Φ_A.

Differential Operators

The contents of this chapter are rather advanced. First, Mikusiński's operational calculus is presented. Second, the concept of the complex power of the differential operator is introduced. Finally, the constant-coefficient linear differential equation is extended to the case in which the coefficients are not commutative with the unknown function, and its solution under given initial conditions is found.

3.1 Operational Calculus

At the end of the 19th century, an electrical engineer, Oliver Heaviside, introduced **operational calculus** by treating the differential operator as if it were a number, in practical calculations. At the beginning, his method was unwelcome by mathematicians because his analysis was much too formal. But, as it worked very well, it has been widely used by applied mathematicians. Later, various theories have been proposed to make it rigorous by mathematicians. Finally, in the middle of the 20th century, Jan Mikusiński most elegantly gave mathematical justification to Heaviside's operational calculus.

3.1.1 *Functions of the differential operator*

As considered in §1.4 of Chapter 1, if the differential operator is regarded as an extended concept of the multiplication of a variable, we can carry out various calculations transparently. Throughout this chapter, we regard the differential operator d/dx as a mathematical object, and, hence, we denote it by D for simplicity. D is a derivation satisfying the rule (1.4.2) of Chapter 1, and the commutation relation

$$[D, f(x)] = f'(x) \tag{3.1.1}$$

holds, where $f(x)$ denotes the operator of the multiplication by a function $f(x)$ from the left.

Since we want to deal with D as an arbitrary quantity, hereafter we consider a function of D, denoted by $\varphi(D)$. Although we further wish to analyze a more general function $\varphi(x, D)$, we refrain from doing so because it is too difficult to present here.

When n is a positive integer, D^n is nothing but the operator of differentiating n times. D^{-1} is, of course, the operator of integration, that is,

$$D^{-1}F(x) \equiv \int dx\, F(x). \tag{3.1.2}$$

In order to avoid possible confusion, we usually use a lower-case letter for expressing the operator of multiplying a function, and a capital letter for a function appearing as the operand, which no more acts as an operator. For example, while $f(x)$ in (3.1.1) is of the former, $F(x)$ in (3.1.2) is of the latter.

Since the indefinite integral contains an integration constant, we replace (3.1.2) by the definite integral

$$D^{-1}F(x) \equiv \int_0^x dy\, F(y). \tag{3.1.3}$$

Then D^{-n} is to repeat (3.1.3) n times.

Now, we rewrite the Taylor expansion formula (1.6.8) of Chapter 1 into

$$F(x + a) = \sum_{n=0}^{\infty} \frac{a^n}{n!} F^{(n)}(x) = \left(\sum_{n=0}^{\infty} \frac{a^n D^n}{n!} \right) F(x), \tag{3.1.4}$$

by replacing x by $x+a$, a by x, and f by F. Noting the expansion formula, (1.6.13) of Chapter 1, of the exponential function, formally we have

$$F(x + a) = e^{aD} F(x). \tag{3.1.5}$$

From (3.1.5), we see that e^{aD} is the operator which induces the translation $x \mapsto x + a$.

As in (3.1.4), if $\varphi(z)$ is holomorphic at $z = 0$, we can formally define $\varphi(D)$ by

$$\varphi(D)F(x) = \sum_{n=0}^{\infty} \frac{\varphi^{(n)}(0)}{n!} D^n F(x). \qquad (3.1.6)$$

Furthermore, even if $\varphi(z)$ has a pole at $z = 0$, we can formally define $\varphi(D)$ by the Laurent expansion, (1.8.18) of Chapter 1,

$$\varphi(D) = \sum_{n=-N}^{\infty} c_n D^n. \qquad (3.1.7)$$

3.1.2 *Heaviside's operational calculus*

Let $\varphi(z)$ be a polynomial of degree N such that the coefficient of the highest degree term is normalized to 1. The constant coefficient linear ordinary differential equation can generally be written as

$$\varphi(D)\Phi(x) = F(x), \qquad (3.1.8)$$

where $F(x)$ and $\Phi(x)$ are a known function and an unknown function, respectively.

By means of Dirac's delta function considered in §1.10, we can write

$$F(x) = \int_{-\infty}^{\infty} dy\, \delta(x - y)F(y). \qquad (3.1.9)$$

Therefore, if we can solve

$$\varphi(D)\hat{\Phi}(x) = \delta(x), \qquad (3.1.10)$$

then the solution to (3.1.8) is given by the convolution[1]

$$\Phi(x) = \int_{-\infty}^{\infty} dy\, \hat{\Phi}(x - y)F(y). \qquad (3.1.11)$$

Now, (3.1.10) can formally be solved as

$$\hat{\Phi}(x) = \frac{1}{\varphi(D)}\delta(x) = \frac{D}{\varphi(D)}\theta(x). \qquad (3.1.12)$$

---

[1] This technique can be extended to the inhomogeneous linear partial differential equation. The solution to (3.1.10) (and its extension to the n-dimensional case) is called the **fundamental solution**.

Here, $\theta(x)$ denotes the Heaviside step function introduced in (1.10.2) of Chapter 1. We write the solutions to the algebraic equation $\varphi(z) = 0$ as $\alpha_1, \alpha_2, \ldots, \alpha_N$. Then, since we have

$$\varphi(D) = (D - \alpha_1)(D - \alpha_2) \cdots (D - \alpha_N) = \prod_{k=1}^{N}(D - \alpha_k), \qquad (3.1.13)$$

(3.1.12) becomes

$$\hat{\Phi}(x) = \frac{D}{\prod_{k=1}^{N}(D - \alpha_k)}\theta(x). \qquad (3.1.14)$$

Assuming that $\alpha_j \neq \alpha_k$ for any $j \neq k$, we decompose the right-hand side of (3.1.14) into partial fractions:

$$\hat{\Phi}(x) = \sum_{k=1}^{N} \frac{1}{\varphi'(\alpha_k)} \frac{D}{D - \alpha_k}\theta(x). \qquad (3.1.15)$$

Hence, we have only to calculate $D/(D - \alpha)\theta(x)$. Under the understanding of (3.1.3), we have

$$D^{-n}\theta(x) = \frac{x^n}{n!}\theta(x). \qquad (3.1.16)$$

Hence, we finally have

$$\frac{D}{D - \alpha}\theta(x) = \frac{1}{1 - \alpha D^{-1}}\theta(x) = \sum_{n=0}^{\infty} \alpha^n D^{-n}\theta(x)$$

$$= \sum_{n=0}^{\infty} \frac{\alpha^n x^n}{n!}\theta(x) = e^{\alpha x}\theta(x). \qquad (3.1.17)$$

Substituting (3.1.17) into (3.1.15), we obtain

$$\hat{\Phi}(x) = \sum_{k=1}^{N} \frac{e^{\alpha_k x}}{\varphi'(\alpha_k)}\theta(x). \qquad (3.1.18)$$

Corresponding to (3.1.17), we hereafter assume $F(x) \equiv 0$ for $x < 0$, that is, $F(x) \equiv F(x)\theta(x)$. Then substituting (3.1.18) into (3.1.11), we have

$$\Phi(x) = \int_{-\infty}^{\infty} dy \sum_{k=1}^{N} \frac{e^{\alpha_k(x-y)}}{\varphi'(\alpha_k)}\theta(x - y)F(y)\theta(y)$$

$$= \sum_{k=1}^{N} \frac{1}{\varphi'(\alpha_k)} \int_{0}^{x} dy\, e^{\alpha_k(x-y)} F(y). \qquad (3.1.19)$$

Thus we have found a solution to (3.1.8). This solution is a particular one under the initial conditions $\Phi(0) = \Phi'(0) = \cdots = \Phi^{(N-1)} = 0$.

For the general initial conditions in which $\Phi(0), \Phi'(0), \ldots, \Phi^{(N-1)}(0)$ are given, we have only to replace $F(x)$ by $F(x)$ plus

$$\varphi(D)\big(\Phi(x)\theta(x)\big) - \big(\varphi(D)\Phi(x)\big)\theta(x). \tag{3.1.20}$$

The reason for this is as follows. The first term of (3.1.20) is the left-hand side of (3.1.8) itself, while the second term of (3.1.20) (apart from the minus sign) is the part of it which is non-vanishing for $x > 0$. Therefore, (3.1.20) is nothing but the part of the delta-type singularities at $x = 0$ of (3.1.8), which are necessary to realize the given initial conditions. Later, we present a concrete example [in (3.1.31)] and the general explicit expression of (3.1.20) [in (3.3.49)].

When $\varphi(z) = 0$ has multiple solutions, we have only to take the limit of coincidence. Or, we have only to substitute the formula

$$\frac{D}{(D-\alpha)^n}\theta(x) = \frac{x^{n-1}}{(n-1)!}e^{\alpha x}\theta(x) \tag{3.1.21}$$

in the partial fraction decomposition formula. Here (3.1.21) is obtained by partially differentiating (3.1.17) with respect to α, $n-1$ times.

3.1.3 *Laplace transform*

The procedure of Heaviside's operational calculus may not be mathematically acceptable, but the final result (3.1.19) is correct. Examining this formula, we find that the essential point is the integral $\int_0^\infty dy\, e^{-\alpha y}F(y)$. Accordingly, it has been found that the **Laplace transform**

$$\mathcal{L}[F](s) \equiv \int_0^\infty dx\, e^{-sx} F(x) \tag{3.1.22}$$

is just the mathematical justification of Heaviside's operational calculus. The Laplace transform was introduced by a mathematician, Pierre-Simon Laplace, who lived a century earlier than Heaviside. If $F(x)$ is a continuous function asymptotically increasing at most as a power of x, then the integral (3.1.22) exists and gives a function of s, holomorphic in $\Re s > 0$.

Now, calculating the Laplace transform of $DF(x)$ by using integration by parts, we obtain

$$\mathcal{L}[DF](s) = \int_0^\infty dx\, e^{-sx} DF(x) = s\mathcal{L}[F](s) - F(0). \tag{3.1.23}$$

If $F(0) = 0$ (more precisely, $F(-0) = 0$), we see that D is represented by s in the Laplace transform. And if $F(x) \equiv F(x)\theta(x)$ is assumed as above, D^n corresponds to s^n in the Laplace transform. More generally, $\varphi(D)$ can be defined by $\varphi(s)$ in the Laplace transform. Thus the problem of solving the constant-coefficient linear differential equation is reduced to the problem of calculating the inverse Laplace transform $\mathcal{L}^{-1}$, which is given by a contour integral, called the **Bromwich integral**. But we omit further details.

3.1.4 *Mikusiński's operational calculus*

The differential equation is a *local* equation, that is, the behavior of the solution in a neighborhood of any point is determined only by the information in the neighborhood of that point. On the other hand, the Laplace transform is *not* local because it is defined by an integral over an interval extending to infinity. Therefore, the formulation based on the Laplace transform is an unnatural way of solving the differential equation. Indeed, the final form of (3.1.19) does not use information from values larger than x.

In Heaviside's method for finding a solution, the most important formula is

$$\frac{1}{D - \alpha} F(x) = \int_0^x dy\, e^{\alpha(x-y)} F(y). \qquad (3.1.24)$$

Hence, for the mathematical justification, we have only to establish (3.1.24) rigorously. This has been done by **Mikusiński's operational calculus**.

In general, given two continuous functions $F(x)$ and $G(x)$ for $x \geq 0$, the **convolution** of G and F, denoted by $G * F$, is a function of x defined by

$$(G * F)(x) \equiv \int_0^x dy\, G(x - y)F(y)$$

$$= \int_0^x dy\, G(y)F(x - y) = (F * G)(x). \qquad (3.1.25)$$

In the abstract sense, we regard the convolution as a kind of "multiplication". Then the totality of the continuous functions defined in $x \geq 0$ forms a commutative algebra. Furthermore, according to the **Titchmarsh theorem**, this "multiplication" has no zero divisor, that is, if $G * F = 0$, at least either $G = 0$ or $F = 0$ must hold. Therefore, we can define the "division" unambiguously.[2]

For Dirac's delta function, we have

$$\left(\delta * F\right)(x) = \int_0^x dy\, \delta(x - y)F(y) = F(x), \qquad (3.1.26)$$

that is, $\delta(x)$ is the unit element of the above "multiplication". Therefore, we have to extend our consideration to the distribution. To multiply $F(x)$ by a constant α in this "multiplication" is represented by the convolution

$$\alpha F(x) = \int_0^x dy\, \left(\alpha\delta(x - y)\right)F(y). \qquad (3.1.27)$$

Furthermore, the differential operator D is represented by the convolution

$$DF(x) = \int_0^x dy\, \delta'(x - y)F(y). \qquad (3.1.28)$$

Hence, $D - \alpha$ can be represented by the convolution. From the above consideration, its inverse $1/(D - \alpha)$ is uniquely definable and given by (3.1.24). Indeed, we see that

$$(D - \alpha)\int_0^x dy\, e^{\alpha(x-y)}F(y)$$

$$= \left(e^{\alpha(x-y)}F(y)\right)\Big|_{y=x} + \int_0^x dy\, \left(\frac{\partial}{\partial x} - \alpha\right)e^{\alpha(x-y)}F(y)$$

$$= F(x). \qquad (3.1.29)$$

The successive calculations are the same as in Heaviside's operational calculus.

3.1.5 *Concrete example*

By means of operational calculus, we solve a concrete example. The problem which we consider is as follows: "Let a and b be real numbers. Solve the differential equation

$$(D^2 + a^2)\Phi(x) = e^{bx}, \qquad (3.1.30)$$

under the initial conditions that $\Phi(0)$ and $\Phi'(0)$ are given real numbers."

---

[2] As a simple analogy, we consider the totality of integers, which forms a commutative algebra. For any two integers a and b, if $ab = 0$, then at least either a or b is zero, that is, there is no zero divisor. Hence we can introduce rational numbers unambiguously.

First, the formula (3.1.20) for the initial conditions becomes

$$(D^2 + a^2)\big(\Phi(x)\theta(x)\big) - \big((D^2 + a^2)\Phi(x)\big)\theta(x)$$
$$= 2\Phi'(x)\delta(x) + \Phi(x)\delta'(x)$$
$$= \Phi'(x)\delta(x) + D\big(\Phi(x)\delta(x)\big)$$
$$= \Phi'(0)\delta(x) + \Phi(0)\delta'(x). \tag{3.1.31}$$

Since the solution to $z^2 + a^2 = 0$ are $\alpha_1 = ia$ and $\alpha_2 = -ia$, (3.1.19) implies

$$\Phi(x) = \frac{1}{2ia} \int_{-\infty}^{x} dy\, e^{ia(x-y)} \left(e^{by}\theta(y) + \Phi'(0)\delta(y) + \Phi(0)\delta'(y) \right) + \text{c.c.},$$
$$\tag{3.1.32}$$

where c.c. means the complex conjugate of the precedent term. We calculate (3.1.32) and then find

$$\Phi(x) = \frac{1}{a^2 + b^2} \left(e^{bx} - \cos(ax) - \frac{b}{a}\sin(ax) \right)$$
$$+ \Phi(0)\cos(ax) + \frac{1}{a}\Phi'(0)\sin(ax). \tag{3.1.33}$$

3.2 Non-Integer-Order Derivatives

3.2.1 *Complex-order derivatives*

In the previous section, we have discussed how to treat the differential operator D as an ordinary quantity. By means of Mikusiński's method of using the convolution, we can define the **complex-order derivative** D^{ν}. (Although traditionally it is called the **fractional-order derivative**, this word is misleading because we are not restricted to discussing the order of rational numbers. Hence, we do not use this term.)

There may arise the question: "Why does one introduce such a curious concept?" The answer is: "There is a motto that any concept which is naturally defined for all positive integers should be extended to the one for complex numbers."[3] The (non-artificial) function of the complex variable is quite convenient because of its analyticity.

---

[3]For example, the dimension of a space is originally meaningful only for a positive integer. But it has been extended to the complex number; the theory of complex dimension was applied to the renormalization theory of quantum field theory and contributed to the establishment of the Standard Theory of elementary particle physics.

Let $F(x)$ be an arbitrary function such that it is identically zero for $x < 0$ and C^∞-class for $x \geq 0$. If $\nu = n$ is a positive integer, we know

$$D^n F(x) = \int_0^x dy\, \delta^{(n)}(x - y)F(y). \qquad (3.2.1)$$

If $\nu = -n$ is a negative integer, we have

$$D^{-n} F(x) = \int_0^x dy\, \frac{(x - y)^{n-1}}{(n - 1)!} F(y), \qquad (3.2.2)$$

as is evident by integrating by parts. In (3.2.2), if n is extended to a complex number after replacing $(n - 1)!$ by $\Gamma(n)$, it is called the **Riemann–Liouville integral**. Hence, remembering the Y distribution,

$$
\begin{aligned}
Y_\mu(x) &= \frac{x^{\mu-1}}{\Gamma(\mu)}\theta(x), \quad (\mu \neq 0, -1, -2, \ldots), \\
&= \delta^{(n)}(x), \quad (\mu = -n;\ n = 0, 1, 2, \ldots)
\end{aligned}
\qquad (3.2.3)
$$

introduced in §1.10, we define

$$D^\nu F(x) \equiv \int_0^x dy\, Y_{-\nu}(x - y)F(y) = \int_{-\infty}^\infty dy\, Y_{-\nu}(x - y)F(y), \qquad (3.2.4)$$

for ν complex. We emphasize that (1.10.30) of Chapter 1, assures the validity of

$$D^\mu D^\nu = D^{\mu+\nu}. \qquad (3.2.5)$$

We note that this property does not generally hold if the operand functions are not restricted to the functions which identically vanish for $x < 0$.

The linear differential equation involving D^ν becomes the **integral equation of Volterra type**.

3.2.2 *Logarithm-order derivatives*

We consider the **logarithm-order derivative**. Contrary to e^{aD}, $\log D$ cannot be defined by the Taylor expansion in powers of D. Hence, we define it by $\log D \equiv \lim_{\nu \to 0} \partial D^\nu / \partial \nu$. Partially differentiating the right-hand side of (3.2.4) with respect to ν, we obtain

$$\frac{\partial}{\partial \nu} \int_0^x dy\, Y_{-\nu}(x - y)F(y) = \int_0^x dy\, (\psi(-\nu) - \log(x - y)) Y_{-\nu}(x - y)F(y),$$

$$(3.2.6)$$

where $\psi(\xi)$ is the **digamma function**, defined by

$$\psi(\xi) \equiv \frac{d}{d\xi} \log \Gamma(\xi) = \frac{\Gamma'(\xi)}{\Gamma(\xi)}. \tag{3.2.7}$$

Therefore, we have $(d/d\xi)[1/\Gamma(\xi)] = -\psi(\xi)/\Gamma(\xi)$, which has been used in writing (3.2.6). The digamma function $\psi(\xi)$ has a simple pole at $\xi = 0$, and its Laurent expansion at $\xi = 0$ is

$$\psi(\xi) = -\frac{1}{\xi} - \gamma + O(\xi), \tag{3.2.8}$$

where $\gamma \equiv -\Gamma'(1)$ is **Euler's constant**,[4] and $O(\xi)$ represents some quantity such that $O(\xi)/\xi$ is bounded.

The proof of (3.2.8) is as follows. The function $\log\left[\Gamma(\xi+1)\right]$ is holomorphic at $\xi = 0$. Taking the logarithm of the identity $\Gamma(\xi) = \Gamma(\xi+1)/\xi$, we have $\log \Gamma(\xi) = \log \Gamma(\xi+1) - \log(\xi)$. Differentiating it, we obtain $\psi(\xi) = \psi(\xi+1) - \xi^{-1}$. Expanding $\psi(\xi+1)$ at $\xi = 0$, we obtain (3.2.8). $\square$

As is evident from its expression, the first term of (3.2.6) is infinite at $\nu = 0$. Moreover, its second term (after taking off the parenthesis) is $\lim_{\nu \to 0}\left[\log(x-y)Y_{-\nu}(x-y)\right]$, whose meaning is not manifestly clear. Hence we integrate the second term of (3.2.6) by parts; it becomes

$$\int_0^x dy\, Y_{-\nu}(x-y) \log(x-y) F(y)$$

$$= \int_0^x dy\, Y_{-\nu+1}(x-y)\left(-\frac{F(y)}{x-y} + \log(x-y)F'(y)\right). \tag{3.2.9}$$

We rewrite the first term of the right-hand side of (3.2.9) by using

$$\frac{Y_{-\nu+1}(x-y)}{x-y} = \frac{(x-y)^{-\nu-1}}{\Gamma(-\nu+1)}\theta(x-y) = \frac{1}{-\nu}Y_{-\nu}(x-y). \tag{3.2.10}$$

Then we find in (3.2.6) that the contribution from it just cancels the pole $1/\nu$ of the first term [cf. (3.2.8)].

---

[4]It is usually defined by $\gamma \equiv \lim_{n\to\infty}\left(\sum_{k=1}^{n} 1/k - \log n\right) = 0.5772\cdots$. It is believed to be an irrational number, but that is not yet proved.

From the above consideration, we see that (3.2.6) is rewritten as

$$\frac{\partial}{\partial \nu} D^\nu F(x) = \frac{\partial}{\partial \nu} \int_0^x dy \, Y_{-\nu}(x-y) F(y)$$

$$= \int_0^x dy \big((-\gamma + O(\nu)) Y_{-\nu}(x-y) F(y)$$

$$- \log(x-y) Y_{-\nu+1}(x-y) F'(y) \big). \qquad (3.2.11)$$

Taking the $\nu \to 0$ limit of (3.2.11), we obtain

$$(\log D) F(x) = -\gamma F(x) - \int_0^x dy \, \log(x-y) F'(y). \qquad (3.2.12)$$

Thus the logarithm-order derivative of $F(x)$ is given by (3.2.12). It should be noted that if $\lim_{y \to +0} F(y) = c \neq 0$, then $F'(y)$ on the right-hand side of (3.2.12) contains a term $c\delta(y)$.

We consider the difference of the following two formulas: one is obtained from (3.2.12) by replacing $F(x)$ by $f(x)F(x)$, while the other is obtained by multiplying both sides of (3.2.12) by $f(x)$. We obtain

$$(\log D)\big(f(x)F(x)\big) - f(x)(\log D) F(x)$$

$$= \int_0^x dy \, \log(x-y) \left(f(x) F'(y) - \big(f(y)F(y)\big)' \right). \qquad (3.2.13)$$

Integrating (3.2.13) by parts, we find an elegant formula

$$\big[\log D, \, f(x)\big] F(x) = \int_0^x dy \, \frac{f(x)-f(y)}{x-y} F(y). \qquad (3.2.14)$$

Comparing this with (3.1.1), we have here the difference quotient instead of the differential quotient.

3.2.3 *Derivative of a power-of-logarithm order*

In (3.2.12), by setting $F(x) = Y_{-\nu+1}(x)$, we have

$$(\log D) Y_{-\nu+1}(x) = -\gamma Y_{-\nu+1}(x) - \int_0^x dy \, \log(x-y) Y_{-\nu}(y). \qquad (3.2.15)$$

In order to calculate the right-hand side of (3.2.15), we make use of the beta function formula [cf. (1.9.12) and (1.9.13) of Chapter 1]

$$\int_0^1 dt\, t^{\mu-1}(1-t)^{\nu-1} = \frac{\Gamma(\mu)\Gamma(\nu)}{\Gamma(\mu+\nu)}. \tag{3.2.16}$$

We partially differentiate (3.2.16) with respect to ν, and then set $\nu = 1$. By using the definition (3.2.7) of $\psi(\xi)$, the recurrence formulas $\Gamma(\mu+1) = \mu\Gamma(\mu)$ and $\Gamma'(1) = -\gamma$, we obtain

$$\int_0^1 dt\, t^{\mu-1}\log(1-t) = -\frac{1}{\mu}\big(\gamma + \psi(\mu+1)\big). \tag{3.2.17}$$

We can calculate the integral appearing on the right-hand side of (3.2.15) by rewriting it as

$$\int_0^x dy\, \log(x-y)y^{-\nu-1}$$

$$= \log x \int_0^x dy\, y^{-\nu-1} + \int_0^x dy\, \log\left(1 - \frac{y}{x}\right)\cdot y^{-\nu-1}; \tag{3.2.18}$$

indeed, the first term is elementary integral and the integration of the second term can be carried out by setting $y/x = t$ and then using (3.2.17). Finally, dividing both sides by $\Gamma(-\nu)$, we have

$$\int_0^x dy\, \log(x-y)Y_{-\nu}(y) = \log x \cdot Y_{-\nu+1}(x) - \big(\gamma + \psi(-\nu+1)\big)Y_{-\nu+1}(x). \tag{3.2.19}$$

Substituting (3.2.19) into (3.2.15), we have

$$(\log D)Y_{-\nu+1}(x) = -\big(\log x - \psi(-\nu+1)\big)Y_{-\nu+1}(x). \tag{3.2.20}$$

The right-hand side of (3.2.20) is just the partial derivative of $Y_{-\nu+1}(x)$ with respect to ν [cf. (3.2.6)]. Thus we find an elegant formula

$$(\log D)Y_{-\nu+1}(x) = \frac{\partial}{\partial\nu}Y_{-\nu+1}(x). \tag{3.2.21}$$

By using (3.2.21) repeatedly, we obtain

$$(\log D)^n Y_{-\nu+1}(x) = \left(\frac{\partial}{\partial\nu}\right)^n Y_{-\nu+1}(x), \tag{3.2.22}$$

for $n = 0, 1, 2, \dots$. Dividing the right-hand side of (3.2.22) by $n!$ and summing up for $n \geq 0$, we obtain

$$\sum_{n=0}^{\infty} \frac{1}{n!} \left(\frac{\partial}{\partial \nu} \right)^n Y_{-\nu+1}(x) = \exp \left(\frac{\partial}{\partial \nu} \right) Y_{-\nu+1}(x). \qquad (3.2.23)$$

As presented in (3.1.5), the exponential function of the differential operator, $\exp(\partial/\partial \nu)$, is the translation operator $\nu \mapsto \nu + 1$, and therefore the right-hand side equals to $Y_{-\nu}(x)$. This must be equal to the quantity obtained by summing up the left-hand side of (3.2.22) after dividing it by $n!$. We thus find

$$\sum_{n=0}^{\infty} \frac{1}{n!} (\log D)^n Y_{-\nu+1}(x) = Y_{-\nu}(x). \qquad (3.2.24)$$

This formula is consistent with the formal equality

$$\sum_{n=0}^{\infty} \frac{1}{n!} (\log D)^n = \exp(\log D) = D, \qquad (3.2.25)$$

as expected by (1.10.28) of Chapter 1.

3.3 Operator-Coefficient Linear Differential Equations

3.3.1 *Heisenberg equation*

In §3.1, we have presented the method for solving the constant-coefficient linear ordinary differential equation by means of operator calculus. The coefficient of each term is, of course, a number, namely, a quantity commutative with any quantity.

If the coefficient of each term is a quantity such as a matrix or an operator, then it is not commutative with the unknown function and the given function. For example, the fundamental equation of quantum mechanics, the **Heisenberg equation**, is the following operator-coefficient differential equation. The **Hamiltonian operator**, denoted by H, is the operator which describes the total energy of a system. Then, for any physical quantity $X(t)$, the Heisenberg equation is

$$i \frac{d}{dt} X(t) + H X(t) - X(t) H = 0, \qquad (3.3.1)$$

(in natural unit). If H commutes with $X(t)$, (3.3.1) becomes a trivial equation. Since H is a constant operator noncommutative with $X(t)$, the

Heisenberg equation (3.3.1) is non-trivial. The solution to (3.3.1) for a given $X(0)$ is

$$X(t) = e^{iHt}X(0)e^{-iHt}. \qquad (3.3.2)$$

Indeed, we have

$$\frac{d}{dt}X(t) = iHe^{iHt}X(0)e^{-iHt} + e^{iHt}X(0)e^{-iHt}(-iH) = i\big(HX(t) - X(t)H\big).$$
$$(3.3.3)$$

This equation has been solved, because it is very simple. But if the given equation is more complicated, one cannot find the solution. The method for solving such an equation is not given in mathematical textbooks.

In this section, we present the general method for solving the constant-coefficient linear ordinary differential equation in which the coefficients depend only on one constant operator (or more than one constant operators which are all mutually commutative). We do not assume any special properties of noncommutative quantities, that is, all calculations are made without changing the ordering of operators.

3.3.2 *Operator-coefficient algebraic equations*

In order to get familiar with the treatment of noncommutative quantities, we first discuss the case of the operator-coefficient algebraic equation.

As the simplest example of noncommutative algebra, we take the 2×2 matrix algebra. Since the number of matrix elements of a 2×2 matrix is 4, an arbitrary 2×2 matrix is expressible as a linear combination of 4 basic matrices. In quantum mechanics, one adopts the unit matrix and the three Pauli matrices, σ_1, σ_2, σ_3 as the basic matrices.[5] What we need is the fact that they are mutually noncommutative and the square of each one is equal to the unit matrix. Since, in the following discussion, we want to avoid the concrete image as far as possible, we denote one of the Pauli matrix by A. Furthermore, the unit matrix is identified with 1.

Thus what we assume is that there is a quantity A such that $A^2 = 1$ and noncommutative with the unknown quantity X. The noncommutativity is realized by introducing another quantity B. Then the independent ones proportional to B are B, AB, BA, and ABA. Any quantity can be expressed as a linear combination of them.

---

[5]Their explicit expressions are $\sigma_1 = \left(\begin{smallmatrix} 0 & 1 \\ 1 & 0 \end{smallmatrix}\right)$, $\sigma_2 = \left(\begin{smallmatrix} 0 & -i \\ i & 0 \end{smallmatrix}\right)$, $\sigma_3 = \left(\begin{smallmatrix} 1 & 0 \\ 0 & -1 \end{smallmatrix}\right)$, but these are unnecessary here.

The operator-coefficient algebraic equation is

$$a_{00}X + a_{10}AX + a_{01}XA + a_{11}AXA = B, \qquad (3.3.4)$$

where a_{jk} is commutative with any quantity. By assumption, X can be expressed as

$$X = b_{00}B + b_{10}AB + b_{01}BA + b_{11}ABA, \qquad (3.3.5)$$

where b_{jk} is commutative with any quantity. Substituting (3.3.5) into (3.3.4), we obtain simultaneous linear algebraic equations for (b_{jk}):

$$
\begin{aligned}
a_{00}b_{00} + a_{10}b_{10} + a_{01}b_{01} + a_{11}b_{11} &= 1, \\
a_{00}b_{10} + a_{10}b_{00} + a_{01}b_{11} + a_{11}b_{01} &= 0, \\
a_{00}b_{01} + a_{10}b_{11} + a_{01}b_{00} + a_{11}b_{10} &= 0, \\
a_{00}b_{11} + a_{10}b_{01} + a_{01}b_{10} + a_{11}b_{00} &= 0.
\end{aligned}
\qquad (3.3.6)
$$

If we solve (3.3.6) by means of Cramer's method, and substitute the solution into (3.3.5), we obtain the solution to (3.3.4).

But this way is not interesting. We want to solve (3.3.4) in a smarter way by using the specialty of the problem. For this purpose, we employ the following trick. First, we introduce "left A", denoted by A_{L} and "right A", denoted by A_{R}. Both A_{L} and A_{R} are quantities commutative with any quantities and $A_{\mathrm{L}}^2 = A_{\mathrm{R}}^2 = 1$. Then, owing to the linear independence of $1, A_{\mathrm{L}}, A_{\mathrm{R}}, A_{\mathrm{L}}A_{\mathrm{R}}(= A_{\mathrm{R}}A_{\mathrm{L}})$, we can rewrite (3.3.6) as

$$\left(a_{00} + a_{10}A_{\mathrm{L}} + a_{01}A_{\mathrm{R}} + a_{11}A_{\mathrm{L}}A_{\mathrm{R}}\right)\left(b_{00} + b_{10}A_{\mathrm{L}} + b_{01}A_{\mathrm{R}} + b_{11}A_{\mathrm{L}}A_{\mathrm{R}}\right) = 1. \qquad (3.3.7)$$

We rewrite (3.3.4) by replacing A by A_{L} if A is placed at the left of B and by A_{R} if A is placed at the right of B. Then (3.3.4) becomes

$$\left(a_{00} + a_{10}A_{\mathrm{L}} + a_{01}A_{\mathrm{R}} + a_{11}A_{\mathrm{L}}A_{\mathrm{R}}\right)X = B. \qquad (3.3.8)$$

On the other hand, the solution to (3.3.4) can be expressed in the following way. In (3.3.5), we replace A by A_{L}, if its position is left of B, or by A_{R}, if its position is right of B. Then, since the solution to (3.3.4) is what is obtained by multiplying the second factor of the left-hand side of (3.3.7) by B, we can write it as

$$X = \frac{B}{a_{00} + a_{10}A_{\mathrm{L}} + a_{01}A_{\mathrm{R}} + a_{11}A_{\mathrm{L}}A_{\mathrm{R}}}. \qquad (3.3.9)$$

This is exactly the formula which is directly obtained from (3.3.8) by formal division.

If we want to rewrite (3.3.9) in terms of the original operator A, we must rationalize the denominator. First, we deform (3.3.9) into

$$X = \frac{\left(a_{00} + a_{11}A_{\mathrm{L}}A_{\mathrm{R}} - a_{10}A_{\mathrm{L}} - a_{01}A_{\mathrm{R}}\right)B}{\left(a_{00} + a_{11}A_{\mathrm{L}}A_{\mathrm{R}}\right)^2 - \left(a_{10}A_{\mathrm{L}} + a_{01}A_{\mathrm{R}}\right)^2}. \tag{3.3.10}$$

Then the denominator of (3.3.10) contains $A_{\mathrm{L}}A_{\mathrm{R}}$ but not A_{L} and A_{R}. We then again rationalize the denominator of (3.3.10) to eliminate $A_{\mathrm{L}}A_{\mathrm{R}}$. The final result is, of course, the same as what is obtained by substituting b_{jk} (solution of (3.3.6)) into (3.3.5).

The method presented here does not simplify the method for calculating the solution in the most general case. But what is important is that our method is quite elegant for the special case which we want to discuss. The most essential point of our method is that we can freely make the division in the formula involving operators. It is, therefore, quite convenient for applying to the operator calculus.

Now, we slightly extend the above consideration. This time, we assume $A^n = 1$, n being a positive integer, instead of $A^2 = 1$. Such an operator is realizable by a cyclic permutation of $\{1, 2, \ldots, n\}$. Its matrix representation is given by $(A)_{m\ell} = \delta_{m,\ell-1}$ $(m, \ell = 1, 2, \ldots, n;$ 0 should be replaced by $n)$.

The algebraic equation obtained by generalizing (3.3.4) is

$$\sum_{j,k=0}^{n-1} a_{jk} A^j X A^k = B. \tag{3.3.11}$$

Its solution corresponding to (3.3.5) is expressed as

$$X = \sum_{j,k=0}^{n-1} b_{jk} A^j B A^k. \tag{3.3.12}$$

Substituting (3.3.12) into (3.3.11) and comparing both sides, we obtain n^2 simultaneous algebraic equations,

$$\sum_{j,j',k,k'=0}^{n-1} \delta_{j+j',j''} \delta_{k+k',k''} a_{jk} b_{j'k'} = \delta_{j'',0} \delta_{k'',0}, \tag{3.3.13}$$

where subscripts should be understood in the modulo n, namely, a number greater than n should be subtracted by n.

As before, we introduce A_L and A_R, commutative with any quantity, such that $A_L{}^n = A_R{}^n = 1$. Then (3.3.13) is rewritten as

$$\left(\sum_{j,k=0}^{n-1} a_{jk} A_L{}^j A_R{}^k \right) \left(\sum_{j',k'=0}^{n-1} b_{j'k'} A_L{}^{j'} A_R{}^{k'} \right) = 1. \qquad (3.3.14)$$

On the other hand, (3.3.11) becomes

$$\left(\sum_{j,k=0}^{n-1} a_{jk} A_L{}^j A_R{}^k \right) X = B. \qquad (3.3.15)$$

From the above consideration, we see that the solution is obtained by formally dividing (3.3.15) by the first factor on its left-hand side. That is, we have

$$X = \frac{B}{\sum_{j,k=0}^{n-1} a_{jk} A_L{}^j A_R{}^k}. \qquad (3.3.16)$$

We present a concrete example, which will be used later. The solution to the equation

$$AX + XA = B, \qquad A^3 = 1 \qquad (3.3.17)$$

is given by

$$X = \frac{B}{A_L + A_R} = \frac{\left(A_L{}^2 - A_L A_R + A_R{}^2 \right) B}{A_L{}^3 + A_R{}^3} = \frac{1}{2} \left(A^2 B - ABA + BA^2 \right). \qquad (3.3.18)$$

In the above discussion, n is arbitrary; hence we can take the limit of $n \to \infty$ formally. In this case, the constraint $A^n = 1$ disappears (but the norm $\|A\| = 1$ is kept, see below.) The final solution is obtained by expanding the inverse of a polynomial in A_L and A_R into a formal power series. Therefore, if this series is convergent (under the assumption of $\|A\| = 1$), we have the true solution. Even in the case in which it is not convergent, we may still be able to obtain the true solution in the following way. We first replace A by zA. If for small $|z|$ the series becomes convergent, we analytically continue the obtained formal solution with respect to z to $z = 1$.

3.3.3 *Operator-coefficient linear ordinary differential equations*

Now, we return to considering the linear ordinary differential equation involving operators in its coefficients.

As seen above, it is convenient to replace the element operator A by either of two commutative quantities A_{L} and A_{R}, corresponding to the place of A. For example, the Heisenberg equation (3.3.1) becomes

$$\big(iD + H_{\mathrm{L}} - H_{\mathrm{R}}\big)X(t) = 0. \tag{3.3.19}$$

(If we want to include the initial condition, the right-hand side should be replaced by $X(0)\delta(t)$.)

In general, let $\Phi(x)$ and $F(x)$ be an unknown operator function and a given one, respectively. The differential equation which we discuss here is the one which can be written as

$$D^N\Phi(x) + \sum_{k=1}^{N}\sum_{j} f_{kj}(A)D^{N-k}\Phi(x)g_{kj}(A) = F(x). \tag{3.3.20}$$

Here, $f_{kj}(A)$ and $g_{kj}(A)$ are arbitrary functions of A, and the summation over j is a finite one. Although it may look somewhat complicated, (3.3.20) is a normal-type N-th-order linear ordinary differential equation for $\Phi(x)$ such that each-order derivative of $\Phi(x)$ has the coefficient which contains powers of A acting on $\Phi(x)$ from both sides in various ways.

Rewriting (3.3.20) by using A_{L} and A_{R}, we simply have

$$\varphi(D)\Phi(x) = F(x), \quad \varphi(D) \equiv \sum_{k=0}^{N} E_k D^{N-k}, \tag{3.3.21}$$

where

$$E_0 \equiv 1, \quad E_k \equiv \sum_{j} f_{kj}(A_{\mathrm{L}})g_{kj}(A_{\mathrm{R}}), \quad (k = 1, 2, \ldots, N). \tag{3.3.22}$$

The formal solution to (3.3.21) is

$$\Phi(x) = \frac{1}{\varphi(D)}F(x). \tag{3.3.23}$$

We calculate the right-hand side of (3.3.23) by Heaviside–Mikusiński's method.

Since the algebraic equation $\varphi(z) = 0$ contains "mysterious" objects, A_{L} and A_{R}, we cannot simply apply the fundamental theorem of algebra to it. But, in the heuristic sense, we assume that $\varphi(z) = 0$ has N solutions $z = G_k$ $(k = 1, 2, \ldots, N)$. Furthermore, we assume $G_k \neq G_\ell$ for $k \neq \ell$.

Then the partial-fraction decomposition of $1/\varphi(z)$ is written as

$$\frac{1}{\varphi(z)} = \sum_{k=1}^{N} \frac{H_k}{z - G_k}, \tag{3.3.24}$$

where $H_k \equiv 1/\varphi'(G_k)$. Substituting (3.3.24) (z is replaced by D) into (3.3.23), we have

$$\Phi(x) = \sum_{k=1}^{N} \frac{H_k}{D - G_k} F(x). \tag{3.3.25}$$

We assume that Mikusiński's operator calculus can be applied to (3.3.25). From (3.1.24), we have

$$\frac{1}{D - G} F(x) = \int_0^x dy \, e^{G(x-y)} F(y). \tag{3.3.26}$$

Hence, (3.3.25) becomes

$$\Phi(x) = \int_0^x dy \sum_{k=1}^{N} H_k e^{G_k(x-y)} F(y). \tag{3.3.27}$$

For illustration, we present a concrete example. What we consider is a second-order differential equation

$$D^2 \Phi(x) - \left(A^2 \Phi(x) + 2A\Phi(x)A + \Phi(x)A^2 \right) = F(x), \quad A^3 = 1. \tag{3.3.28}$$

Using A_{L} and A_{R}, we rewrite (3.3.28) as

$$\left(D^2 - (A_{\mathrm{L}} + A_{\mathrm{R}})^2 \right) \Phi(x) = F(x), \quad A_{\mathrm{L}}^3 = A_{\mathrm{R}}^3 = 1. \tag{3.3.29}$$

Therefore, from (3.3.25) and (3.3.27), we obtain

$$\Phi(x) = \frac{1}{2(A_{\mathrm{L}} + A_{\mathrm{R}})} \left(\frac{1}{D - A_{\mathrm{L}} - A_{\mathrm{R}}} - \frac{1}{D + A_{\mathrm{L}} + A_{\mathrm{R}}} \right) F(x)$$

$$= \frac{1}{2(A_{\mathrm{L}} + A_{\mathrm{R}})} \int_0^x dy \left(e^{(A_{\mathrm{L}} + A_{\mathrm{R}})(x-y)} - e^{-(A_{\mathrm{L}} + A_{\mathrm{R}})(x-y)} \right) F(y). \tag{3.3.30}$$

Remembering (3.3.18), we find the solution given by

$$\Phi(x) = \frac{1}{4} \left(A^2 \Psi(x) - A\Psi(x)A + \Psi(x)A^2 \right), \tag{3.3.31}$$

where

$$\Psi(x) = \int_0^x dy \left(e^{A(x-y)} F(y) e^{A(x-y)} - e^{-A(x-y)} F(y) e^{-A(x-y)} \right). \tag{3.3.32}$$

This example is a lucky case. In general, G_k and H_k can be complicated irrational functions. Therefore, we do not know how strange the G_k and H_k as the functions of A_L and A_R are, so that the expressions in (3.3.27) are not meaningful. That is, it is a very important problem whether or not A_L and A_R can be separated in the final result. In what follows, we prove this separability problem explicitly.

3.3.4 *Proof of separability*

First, we define the concept of **separability**. We wish to consider a finite sum $\sum_{k,\ell} A^k \Phi_{k\ell} A^\ell$, where $\Phi_{k\ell}$ is a quantity noncommutative with A. Introducing A_L and A_R, we replace it by $\sum_{k,\ell} A_L^k A_R^\ell \Phi_{k\ell}$. If a given quantity $\Phi(A_L, A_R)$ can be written as $\sum_{k,\ell} A_L^k A_R^\ell \Phi_{k\ell}$, we say that $\Phi(A_L, A_R)$ satisfies the separability condition. Such a quantity straightforwardly reproduces the expression in terms of A. Therefore, what we should prove is that the solution can be written in terms of the quantities satisfying the separability condition.

If we expand the inverse of the polynomial

$$\varphi(z) = \sum_{k=0}^{N} E_k z^{N-k}, \tag{3.3.33}$$

into the *formal* power series of $1/z$, we have[6]

$$\frac{1}{\varphi(z)} = \sum_{n=0}^{\infty} \frac{C_n}{z^{N+n}}. \tag{3.3.34}$$

Multiplying both sides of (3.3.34) by $\varphi(z)$ and substituting (3.3.33) into the right-hand side, we obtain

$$1 = \sum_{m=0}^{\infty} \sum_{k=0}^{N} \frac{E_k C_m}{z^{k+m}} = \sum_{n=0}^{\infty} \frac{1}{z^n} \sum_{k=0}^{\min(n,N)} E_k C_{n-k}. \tag{3.3.35}$$

Since (3.3.35) is an identity, the coefficient of $1/z^n$ ($n = 1, 2, \ldots$) must vanish. That is, we have

$$E_0 C_0 = 1, \quad \sum_{k=0}^{\min(n,N)} E_k C_{n-k} = 0, \quad (n = 1, 2, \ldots). \tag{3.3.36}$$

---

[6]It does not affect the following reasoning whether or not it is convergent.

Since the coefficient of C_n is always $E_0 = 1$, every formula can be written in the form $C_n = \cdots$. From (3.3.22), E_k satisfies the separability condition. Therefore, C_n also satisfies the separability condition, as is easily proved by mathematical induction.

On the other hand, expanding the right-hand side of (3.3.24) into the power series of $1/z$, we have

$$\frac{1}{\varphi(z)} = \sum_{k=1}^{N} \sum_{m=1}^{\infty} \frac{H_k G_k^{m-1}}{z^m} = \sum_{n=-N+1}^{\infty} \frac{\sum_{k=1}^{N} H_k G_k^{n+N-1}}{z^{N+n}}. \qquad (3.3.37)$$

Comparing (3.3.37) with (3.3.34), we obtain the relation

$$\sum_{k=1}^{N} H_k G_k^{n+N-1} = 0, \quad (n < 0) \qquad (3.3.38)$$

$$= C_n, \quad (n \geq 0).$$

In the solution (3.3.27), we formally expand the exponential function into a power series. Then we have

$$\Phi(x) = \int_0^x dy \sum_{k=1}^{N} H_k \sum_{m=0}^{\infty} \frac{G_k^m (x-y)^m}{m!} F(y)$$

$$= \int_0^x dy \sum_{n=-N+1}^{\infty} \frac{\sum_{k=1}^{N} H_k G_k^{n+N-1}}{(n+N-1)!} (x-y)^{n+N-1} F(y). \qquad (3.3.39)$$

Hence (3.3.38) implies

$$\Phi(x) = \int_0^x dy \sum_{n=0}^{\infty} \frac{C_n}{(n+N-1)!} (x-y)^{n+N-1} F(y). \qquad (3.3.40)$$

As we have already shown, C_n satisfies the separability condition. Hence each order of the expansion for $\Phi(x)$ satisfies the separability condition. Thus we can restore it in the expression in terms of A. □

From the continuity of the coefficients, the solution (3.3.40) remains correct even if $\varphi(z) = 0$ has multiple solutions.

Since the solution $\Phi(x)$ is presented in the form of a series expansion, it is important to prove the convergence of the series. For this purpose, it is necessary to give the norms of the operators E_k. We assume that the norm, $\|E_k\| (> 0)$, of E_k is defined in some way. We set

$$M \equiv \max\{1, \|E_k\|^{1/k}, \ (k = 1, 2, \ldots, N)\}. \qquad (3.3.41)$$

Then, owing to the recurrence formula (3.3.36) for C_n, we can show that $\|C_n\| \le (2M)^n$ by mathematical induction with respect to n. First, for $n = 0$, since $C_0 = 1$, it is all right. Hence, under the assumption of the induction, we have

$$\|C_n\| \le \sum_{k=1}^{\min(n,N)} \|E_k\| \cdot \|C_{n-k}\| \le \sum_{k=1}^{n} M^k (2M)^{n-k}$$

$$= \left(\sum_{k=1}^{n} 2^{n-k} \right) M^n \le (2M)^n. \tag{3.3.42}$$

Therefore, the series in (3.3.40) has a non-zero convergence radius. $\quad\square$

3.3.5 *Solution to the operator-coefficient differential equation*

In order to confirm that (3.3.40) is, indeed, the solution to the original differential equation (3.3.21) by means of term-by-term differentiation, we rewrite (3.3.40) in terms of the Y distribution. That is, we want to confirm that

$$\Phi(x) = \int_0^x dy \sum_{n=0}^{\infty} C_n Y_{n+N}(x-y) F(y) \tag{3.3.43}$$

satisfies (3.3.21).

Before discussing the general case, we present a concrete example. What we consider here is an operator-coefficient $N = 2$ differential equation,

$$D^2 \Phi(x) - \big(A\Phi(x) + \Phi(x)A \big) = F(x). \tag{3.3.44}$$

Since

$$\varphi(z) = z^2 - (A_{\mathrm{L}} + A_{\mathrm{R}}), \tag{3.3.45}$$

we have

$$\frac{1}{\varphi(z)} = \sum_{k=0}^{\infty} \frac{(A_{\mathrm{L}} + A_{\mathrm{R}})^k}{z^{2k+2}}. \tag{3.3.46}$$

Therefore, by comparing this with the $N = 2$ case of (3.3.34), C_n is 0 for n odd and $C_{2k} = (A_{\mathrm{L}} + A_{\mathrm{R}})^k$ for $n = 2k$. Hence, in the present example, the solution (3.3.43) becomes

$$\Phi(x) = \int_0^x dy \sum_{k=0}^{\infty} Y_{2k+2}(x-y)(A_{\mathrm{L}} + A_{\mathrm{R}})^k F(y). \tag{3.3.47}$$

It is easy to confirm that (3.3.47), indeed, satisfies (3.3.44) in the following way.

The $k = 0$ term of $D^2\Phi(x)$ reproduces the right-hand side of (3.3.44), namely, $F(x)$, as is shown by means of $D^2 Y_2(x-y) = Y_0(x-y) = \delta(x-y)$. The $k > 0$ terms of $D^2\Phi(x)$ is equal to $A\Phi(x) + \Phi(x)A$, as is shown by replacing k by $k + 1$ and using $(A_{\mathrm{L}} + A_{\mathrm{R}})^{k+1} = (A_{\mathrm{L}} + A_{\mathrm{R}})(A_{\mathrm{L}} + A_{\mathrm{R}})^k$, as the result, which cancels the remainder terms in the left-hand side of (3.3.44).

Now we return to the general case. We assume that $F(x)$ is given in the form of a power series of x. In this case, the integration over y in (3.3.43) can be explicitly carried out. Since $F(x) \equiv F(x)\theta(x)$, we may write

$$F(x) = \sum_{k=-N}^{\infty} F_k Y_{k+1}(x). \tag{3.3.48}$$

Here the negative k part corresponds to the initial conditions because $Y_{-|k|+1}(x) = \delta^{(|k|-1)}(x)$. Calculating (3.1.20), we find

$$F_k = \sum_{j=0}^{N-|k|} E_{N-|k|-j}\Phi^{(j)}(0), \quad \text{for } k < 0. \tag{3.3.49}$$

The proof of (3.3.49) is somewhat complicated. We present it in §3.3.6 later.

Substituting (3.3.48) into (3.3.43), we obtain

$$\Phi(x) = \int_0^x dy \sum_{m=0}^{\infty} C_m Y_{m+N}(x-y) \sum_{k=-N}^{\infty} F_k Y_{k+1}(y). \tag{3.3.50}$$

Using the convolution formula (1.10.30) of Chapter 1, we carry out the integration over y; then

$$\Phi(x) = \sum_{k=-N}^{\infty} \sum_{m=0}^{\infty} C_m F_k Y_{m+N+k+1}(x). \tag{3.3.51}$$

Setting $m = n - j$ and $k = j - N$, we rewrite (3.3.51) in the form of a power series. Then we have a beautiful result,

$$\Phi(x) = \sum_{n=0}^{\infty} \left(\sum_{j=0}^{n} C_{n-j} F_{j-N} \right) Y_{n+1}(x). \tag{3.3.52}$$

Finally, we confirm that (3.3.52) satisfies (3.3.21). Substituting (3.3.52) in $\varphi(D)\Phi(x)$, we have

$$\varphi(D)\Phi(x) = \sum_{k=0}^{N} E_k \sum_{n=0}^{\infty} \sum_{j=0}^{n} C_{n-j} F_{j-N} Y_{n-N+k+1}(x). \tag{3.3.53}$$

With $m = n - N + k$, namely, $n = m + N - k$, (3.3.53) becomes[7]

$$\varphi(D)\Phi(x) = \sum_{m=-N}^{\infty} \sum_{j=0}^{m+N} \left(\sum_{k=0}^{\min(m+N-j,N)} E_k C_{(m+N-j)-k} \right) F_{j-N} Y_{m+1}(x). \tag{3.3.54}$$

The quantity in the large parentheses is equal to 1 only for $m + N - j = 0$ and to 0 otherwise, owing to (3.3.36). Therefore, only the part of $j = m + N$ survives. Thus we have

$$\varphi(D)\Phi(x) = \sum_{m=-N}^{\infty} F_m Y_{m+1}(x) = F(x), \tag{3.3.55}$$

which is nothing but (3.3.21). $\qquad\qquad\square$

3.3.6 Proof of (3.3.49)

We present the proof of the initial-condition formula (3.3.49). As preparation, we prove the following lemma.

Lemma 3.3.1 For two arbitrary functions F and G, the following identity holds:

$$D^n(FG) - (D^n F)G = \sum_{m=0}^{n-1} D^m \big((D^{n-m-1} F) DG \big). \tag{3.3.56}$$

---

[7]From (3.3.53) we have $0 \le k \le N$ and $0 \le j \le n$. With $m = n - N + k$, we check the values taken by the three summation indices k, j, and m in this order. As for k, $0 \le k \le N$ and $k = m + N - n \le m + N - j$; therefore $0 \le k \le \min(m+N-j, N)$. As for j, $0 \le j \le n = m + N - k \le m + N$. As for m, $m = n - N + k \ge -N$.

The proof of (3.3.56) is made by mathematical induction with respect to n. For $n = 1$, both sides are equal to FDG. Hence we assume the validity of (3.3.56) as the induction assumption. Let D operate on both sides of (3.3.56). Then, its left-hand side becomes

$$D\big(D^n(FG) - (D^nF)G\big) = D^{n+1}(FG) - (D^{n+1}F)G - (D^nF)DG,$$
(3.3.57)

and its right-hand side of (3.3.56) becomes

$$D\sum_{m=0}^{n-1} D^m\big((D^{n-m-1}F)DG\big) = \sum_{m=1}^{n} D^m\big((D^{n-m}F)DG\big)$$

$$= \sum_{m=0}^{n} D^m\big((D^{n-m}F)DG\big) - (D^nF)DG.$$
(3.3.58)

In the formula that (3.3.57) equals (3.3.58), both last terms cancel and the remainders just give the formula which is obtained from (3.3.56) by replacing n by $n+1$. □

Now, what we want to do is the generalization of the calculation of (3.1.31). From (3.1.20) and (3.3.21), the part of $F(x)$ which corresponds to the initial condition is

$$F(x)_{\text{initial}} \equiv \varphi(D)\big(\Phi(x)\theta(x)\big) - \big(\varphi(D)\Phi(x)\big)\theta(x)$$

$$= \sum_{k=0}^{N-1} E_k \big(D^{N-k}(\Phi(x)\theta(x)) - (D^{N-k}\Phi(x))\theta(x)\big). \quad (3.3.59)$$

Since the coefficient of E_N is 0, the summation over k ends at $k = N - 1$. The coefficient of E_k is what is obtained from the left-hand side of (3.3.56) by setting $F = \Phi$, $G = \theta$ and $n = N - k$. Noting that $D\theta(x) = \delta(x)$ and $\phi(x)\delta(x) = \phi(0)\delta(x)$ for any function $\phi(x)$, we have

$$F(x)_{\text{initial}} = \sum_{k=0}^{N-1} E_k \sum_{m=0}^{N-k-1} D^m\big(D^{N-k-m-1}\Phi(x) \cdot \delta(x)\big)$$

$$= \sum_{k=0}^{N-1} E_k \sum_{m=0}^{N-k-1} \Phi^{(N-k-m-1)}(0)\delta^{(m)}(x)$$

$$= \sum_{m=0}^{N-1} \left(\sum_{k=0}^{N-m-1} E_k \Phi^{(N-k-m-1)}(0) \right) \delta^{(m)}(x). \quad (3.3.60)$$

On the other hand, the negative k part of (3.3.48) is written as

$$F(x)_{\text{initial}} = \sum_{k=-N}^{-1} F_k \delta^{(-k-1)}(x) = \sum_{m=0}^{N-1} F_{-m-1} \delta^{(m)}(x). \qquad (3.3.61)$$

Comparing (3.3.60), where k is replaced by n, with the coefficient of $\delta^{(m)}(x)$ in (3.3.61), we find

$$F_{-m-1} = \sum_{n=0}^{N-m-1} E_n \Phi^{(N-n-m-1)}(0). \qquad (3.3.62)$$

That is, we obtain

$$F_k = \sum_{n=0}^{N+k} E_n \Phi^{(N+k-n)}(0) = \sum_{j=0}^{N+k} E_{N+k-j} \Phi^{(j)}(0). \qquad (3.3.63)$$

Since k is negative, (3.3.63) is nothing but (3.3.49). $\qquad\qquad\qquad\square$

Position Information of the Operator

In western languages, sentences are always written in the linear ordering from the left to the right. Mathematical expressions are also usually written in the same way as in the sentences. Furthermore, it is a custom that the operator acts on the operand from the left. For example, the differential operator D acts from the left, that is, fDg means fg' but not $f'g$. When we need the differential operator to act from the right, we use an over-arrow, that is, $f\overleftarrow{D}g$ means $f'g$. In this case, we denote fg' by $f\overrightarrow{D}g$ for clarity.

In §3.3, we introduced the two symbols, a subscript L and a subscript R, in such a way that A_{L} and A_{R} indicate the left and right positions of an operator A with respect to another operator B. If A_{L} and A_{R} are placed in the original position, the subscript symbols are of the same meaning as the over-arrow symbols. But $\overrightarrow{A}$ and $\overleftarrow{A}$ cannot change their places. On the other hand, A_{L} and A_{R} are freely movable. The subscript symbols are the memories of the positions where they have been located initially. We emphasize that the quantities A_{L} and A_{R} are the operators which contain the position information as a part of operator properties.

This way of treating the position information brings us great convenience for operator calculus, because the division can be made freely as seen in the text. Apart from division, calculations may be simplified. For example, if we write (3.3.47) in terms of A by using the binomial expansion, then it becomes much more cumbersome to prove that (3.3.47) satisfies (3.3.44).

In quantum field theory, one encounters the following problem: "Let $f(z)$ be an analytic function of z, and A and Φ be noncommutative operators. Then express the commutator $[f(A), \Phi]$ in terms of A and $[A, \Phi]$." The usual approach to this problem is to expand $f(A)$ into the Taylor series and calculate $[A^n, \Phi]$. If $f(z)$ has a singular point at $z = 0$, this method does not work. If we use A_L and A_R, the solution to the above problem is given by

$$[f(A), \Phi] \Rightarrow \frac{f(A_L) - f(A_R)}{A_L - A_R}[A, \Phi].$$

The formal proof can be made by using the Cauchy integral representation (1.8.13) of Chapter 1.

Extension to the multiple (n-ple) commutator

$$[\cdots [[f(A), \Phi]_1, \Phi]_2 \cdots]_n$$

is also possible. In this case, we have to introduce $n + 1$ mutually commutative operators $A_{(0)}, A_{(1)}, \ldots, A_{(n)}$, where $(0), (1), \ldots, (n)$ are operator ordering indices.

––––––––––

For details, see the following paper: M. Abe, N. Ikeda and N. Nakanishi, "Operator ordering index method for multiple commutators and Suzuki's quantum analysis", *Journal of Mathematical Physics*, **38**(2) (1997), 547–555.

Epilogue to PART I

Differential equations are wise.
The Einstein equation is wiser than Einstein!
The Schrödinger equation is wiser than Schrödinger!
The Dirac equation is wiser than Dirac!

Albert Einstein believed that the universe would be eternally static and immortal. But his equation correctly described the expanding universe. Erwin Schrödinger believed that Schrödinger's wave function would describe actual real waves. But it was related to the observational only in the sense of the probability. Paul Dirac believed that the positively-charged particle which was predicted by the Dirac equation for the electron would be identified with the proton. But, soon after, the positron, which is positively charged and has the same mass as the electron, was discovered experimentally.

The Einstein equation, Schrödinger equation and Dirac equation are all systems of partial differential equations. They are the representatives of the three greatest achievements of modern fundamental physics, namely, relativity, quantum mechanics and relativistic quantum theory, respectively.

The Einstein equation is a system of 10 simultaneous nonlinear partial differential equations. The gravitational field is identified with the space-time metric tensor, that is, gravity is characterized as the deformation of the space-time structure. The Schrödinger equation is a linear partial differential equation. It makes clear that even the invisible world of atoms is quantitatively describable by mathematics, though it is a strange world in which our daily knowledge no longer holds. The Dirac equation makes the remarkable prediction that every Dirac particle must have its own antiparticle, which has the same mass but the opposite charge. This fact has been confirmed experimentally.

Nowadays, the most successful fundamental theory of elementary particle physics is the **Standard Theory** (or often called the **Standard Model**). This theory is formulated on the basis of **quantum field theory**. Since elementary particles are the things which can be created and annihilated, each of them cannot be regarded as a theoretically fundamental object. The quantum field is, mathematically, an operator-valued distribution of the space-time coordinates and, physically, the object which dominates the quantum nature of each elementary particle. A particular model of quantum field theory is determined by giving a 4-dimensional space-time integral, called the **action integral**, which is a functional of various quantum fields. Each quantum field satisfies its partial differential equations, which follow the action integral through the variational principle. The operator properties of each quantum field is also derived from the action integral through the standard procedure, called the **canonical quantization.**

The Standard Theory is not yet perfect. Its action integral is not beautiful enough to be regarded as the ultimate theory. Many attempts at modifying the Standard Theory have been done, but there has been no success.

Even apart from the aesthetic aspects, the Standard Theory itself is incomplete because it does not describe gravity. The gravitational force, the most time-honored force discovered by Isaac Newton, is now completely describable by Einstein's **general relativity**, or more aptly called **Einstein gravity**. It is the most successful theory, both mathematically and experimentally. Attempts at extending Einstein gravity have all failed.

Einstein gravity is, however, still incomplete in the sense that it is a classical theory, that is, it does not describe any quantum effects in spite of the fact that gravitational force acts on elementary particles. From the consistency requirement, Einstein gravity should be quantized. Although it was a long-standing problem to quantize Einstein gravity satisfactorily, it has been successfully resolved, that is, Einstein gravity has been canonically quantized in closed form, without violating manifest relativistic covariance, and it couples with the generally covariantized Standard Theory of elementary particles. This theory is called **quantum Einstein gravity**, and correspondingly, Einstein gravity is called **classical Einstein gravity**.

Quantum Einstein gravity is a beautiful theory,[1] so perhaps it could be a part of the ultimate theory.

In spite of this fact, the wrong traditional belief that Einstein gravity cannot be quantized is widespread. The origin of this criticism against quantum Einstein gravity is as follows.

It is extremely difficult to solve the equations of quantum field theory except for those of toy models. Therefore, one usually employs an approximation method, called the **covariant perturbation theory**. The starting point of this approach is to decompose the action integral into the free part and the interaction part artificially. Then one adopts the **interaction picture**, in which the free part is eliminated by a simple transformation, and the theory is reduced to the one based on the modified action consisting of the interaction Lagrangian density. This theory can be solved by expanding the relevant quantity into the power series of the parameter, called the coupling constant. Each order is explicitly calculable at least in principle on the basis of the routine technique of the Feynman diagram. Unfortunately, however, in this calculation, one often encounters **divergence difficulty**, that is, the calculation becomes infinite. But, in the lucky cases, owing to the indefiniteness of the parameters involved, the indefiniteness caused by the removal of infinity can be absorbed into them. Such a theory is called the **renormalizable theory**. The Standard Theory is renormalizable. Therefore, as long as the parameters are adjusted to be equal to the observed values, one obtains finite values to all orders of perturbation theory. In this sense, the Standard Theory has the physical predictability.

On the other hand, quantum Einstein gravity is *not* renormalizable. Hence it has been claimed that quantum Einstein gravity has no physical predictability. This is the basis of the traditional objection to quantum Einstein gravity. This reasoning has been repeatedly utilized by the theoreticians who wish to propose or advocate an alternative quantum theory of gravity.

The above criticism is evidently logically absurd because the divergence difficulty is based on the perturbative approach. It is unclear whether or not the divergence difficulty is the inherent nature of quantum Einstein gravity itself. More vitally, we can show that the basis of the above criticism is *wrong*, that is, the interaction picture on which the perturbative

---

[1]Its formulation is presented in N. Nakanishi and I. Ojima, *Covariant Operator Formalism of Gauge Theories and Quantum Gravity*, (1990) World Scientific Publishing, Chapter 5.

approach is based cannot be constructed in quantum Einstein gravity. This is because, in order to set up the interaction picture, one must *assume* that the gravitational field tends to a classical space-time metric, such as the Minkowski metric, as the gravitational constant (denoted by κ (Einstein) or by G (Newton)) goes to zero. But this assumption is incorrect; indeed, we can prove that the $\kappa \to 0$ limit of the gravitational field is still an operator, but not a classical quantity. Any *a priori* space-time does not exist. This fact is quite natural if we suppose that quantum Einstein gravity is a part of the ultimate theory. The action integral of Einstein gravity nowhere contains any particular space-time metric. The Minkowski metric of the Standard Theory of elementary particles arises as a consequence of the mechanism of **spontaneous symmetry breakdown**.[2]

Quantum Einstein gravity must be solved without employing the interaction picture. That is, it should be solved in the **Heisenberg picture**, in which we have a system of the partial differential equations for the gravitational field coupled with the other elementary particle fields, without making any artificial procedure. The method for solving quantum field theory in the Heisenberg picture has been developed by Mitsuo Abe (my research associate) and me.[3]

The general method for solving quantum field theory in the Heisenberg picture consists of two steps. In the first step, we should solve a system of partial differential equations. But in quantum field theory, quantum fields are operators. That is, we must solve the operator version of the Cauchy problem, in which the initial conditions are given by the equal-time commutation relations determined by the canonical quantization. Except for simple models, it is impossible to solve it exactly. We have to employ some approximation method. In quantum Einstein gravity, we expand its system in powers of κ. This power series is different from that of the usual perturbative approach, in which the expansion parameter is $\sqrt{\kappa}$. In the Heisenberg picture, the zeroth-order of quantum Einstein gravity is exactly solvable, though the partial differential equations are nonlinear. In all subsequent higher-order approximation, the partial differential equations which

---

[2]The mechanism of spontaneous symmetry breakdown was first discovered in the theory of superconductors, known as the Bardeen–Cooper–Schrieffer (BCS) theory. Yoichiro Nambu succeeded in formulating this concept in quantum field theory. Steven Weinberg and Abdus Salam's electro-weak theory is based on the spontaneous symmetry breakdown. The electro-weak theory is an important part of the Standard Theory.

[3]For a review, see N. Nakanishi, "Method for solving quantum field theory in the Heisenberg picture", *Progress of Theoretical Physics*, **111**(3) (2004) 301–337.

we encounter are linear. The coefficients of their homogeneous parts are some operators written in terms of the zeroth-order solution. Their qualitative feature can be seen by the operator-coefficient ordinary differential equation discussed in §3.3. That is, if the analysis done there can be extended to the partial differential equation, we can solve the operator version of the Cauchy problem. Unfortunately, the higher-order approximation to quantum Einstein gravity is very complicated; the calculations which we encounter are extremely cumbersome.

In the second step, our task is to construct a linear representation of the operator solution obtained in the first step in terms of an infinite-dimensional complex vector space. This space contains the unique vacuum state, which is translationally invariant. The spontaneous symmetry breakdown stated above takes place as a consequence of the uniqueness of the vacuum state. The representation of the zeroth-order approximation to quantum Einstein gravity has been explicitly constructed. We omit more details because it does not relate to the differential equation.

Anyway I believe that the ultimate theory will be formulated with some extension of the partial differential equation.

PART II

Applications

Prologue to PART II

Let me introduce myself. My name is Kenji Seto. I am a theoretical techno-physicist. I started my academic carrier as an elementary-particle physicist. After getting my degree of doctor of science, I worked in collaboration with Noboru Nakanishi, the coauthor of this book. In my middle age, however, I left elementary-particle physics because of a change in my job. I have studied various fields of natural science, and my wide knowledge has helped me in the writing of PART II of this book.

What I present here is a collection of the papers which I had previously written in a communication circular distributed on the Internet. Each chapter is self-contained, so the reader can follow without reading the preceding chapters. However, the reader is expected to be familiar with the physics to some extent. My writing may seem disorganized, but my intent is to collect as many original models as possible that are not covered in other books. And I describe the calculation process in as much detail as possible. Therefore, I hope that the reader will find the chapters informative and interesting.

$$************************$$

PART II consists of the following four chapters.

Chapter 4 presents some of the simplest examples of ordinary differential equations. Specifically they are the tracking-line model, the traction-line model, the model of dropping an object from an airplane to hit a target, and a little fantastic model of a worm crawling along a stretched rubber rod. In addition, we deal with the SIR model for virus infection. Finally, we examine the possibility of building a space elevator. All of these models are reduced to simple differential equations. Therefore, they will be of interest to beginners.

Chapter 5 presents partial differential equations. First, we look at the simplest string vibration model. Next, we deal with the string vibration when it has both tension and rigidity. Finally, we examine mathematical models of wind ripples, sand dunes and wave phenomena on the highway as examples of wave phenomenon that occur in nature.

Chapter 6 presents problems involving the Bessel function. First, we explain Kepler's equation which was first solved by Friedrich Wilhelm Bessel using the Bessel function. Next we deal with the buckling problem of an elastic rod in which the Airy function plays an important role. Furthermore, we explain some problems in which the Bessel function works well, for example, the vibration of a one-dimensional lattice and the vibration of a keyboard percussion instrument.

Chapter 7 presents potential problems in quantum mechanics. First, we discuss the Aharonov–Bohm effect in quantum mechanics in which we can see how the Bessel function works. Next, we deal with a slightly difficult potential problem called the Rosen–Morse potential. Finally, we deal with two periodic potential problems in quantum mechanics. One is where the delta functions are arranged periodically and the other is the serrated (sawtoothed) shape potential problem.

****

I am most grateful to my teacher Noboru Nakanishi for his careful reading of my manuscript and for all his helpful comments. I also thank him for correcting my poor English.

Ordinary Differential Equations

In this chapter, we discuss some of the simplest examples of ordinary differential equations. We begin with the tracking-line model, the traction-line model, the model of dropping an object from an airplane to hit a target, and a little fantastic model of a worm crawling along a stretched rubber rod. In addition, we deal with the SIR model for virus infection that can be applied to the COVID-19 pandemic. Finally, we deal with the question, "Is it possible to build a space elevator?" All of these are reduced to simple differential equations, but they may be of interest to beginners.

4.1 Tracking-Line and Traction-Line

4.1.1 *Tracking-line*

We consider the model of an object A which is moving freely in two-dimensional space and is tracking an object B which is moving on a straight line. Generally speaking, we can think of A as an airplane and B as a ship which is moving in a straight line. This is called the **tracking-line** problem. Here, for the sake of convenience later, the y-axis and x-axis are taken in the horizontal direction and in the vertical upward direction, respectively. The ship shall travel in the positive direction on the y-axis at a constant velocity v_s, and the airplane chasing the ship shall fly at a constant speed of v_p. Let the position of the airplane when the ship passes the origin be (x_p, y_p). Tracking means that the ship is always located on the tangent to the locus (trajectory) along which the airplane flies. Let the expression of the trajectory be given by $y = y(x)$, where the coordinates (x, y) denote the position of the airplane when time t has elapsed after the ship has passed the origin $(0,0)$. Then $v_p t$ is equal to the distance by which the airplane traveled within t. Therefore, we have

$$t = \frac{1}{v_p} \int_x^{x_p} \sqrt{1 + y'^2} dx. \tag{4.1.1}$$

Note that the upper limit of this integral is x_p and the lower limit is x, because the airplane always moves in the direction of the decreasing value of x. The equation for the tangent to $y = y(x)$ at the point (x, y) is

$$Y - y = y'(X - x), \tag{4.1.2}$$

where (X, Y) denotes the coordinates of an arbitrary point lying on the tangent. Therefore, for the airplane to track the ship, this tangent should pass through $(0, v_s t)$ which is the position of the ship, so we have

$$v_s t - y = -xy'. \tag{4.1.3}$$

Eliminating the variable t by using (4.1.1), we obtain the differential-integral equation for the trajectory of the airplane

$$\alpha \int_x^{x_p} \sqrt{1 + y'^2} dx - y = -xy', \tag{4.1.4}$$

where we set

$$\alpha = \frac{v_s}{v_p}. \tag{4.1.5}$$

Hereafter, we use the notation,

$$p = y'. \tag{4.1.6}$$

By differentiating (4.1.4) with respect to x, we obtain the nonlinear differential equation,

$$\alpha \sqrt{1 + p^2} = x \frac{dp}{dx}. \tag{4.1.7}$$

This expression can be separated into variables immediately,

$$\frac{dp}{\sqrt{1 + p^2}} = \alpha \frac{dx}{x}, \tag{4.1.8}$$

and by integrating both sides, we obtain

$$\log \left(p + \sqrt{1 + p^2} \right) = \alpha \log \left(\frac{x}{x_0} \right), \tag{4.1.9}$$

where x_0 is a constant of integration. We deform (4.1.9) into

$$p + \sqrt{1 + p^2} = \left(\frac{x}{x_0}\right)^\alpha,$$

(4.1.10)

and solving this equation with respect to p, we have

$$p = \frac{1}{2}\left[\left(\frac{x}{x_0}\right)^\alpha - \left(\frac{x}{x_0}\right)^{-\alpha}\right].$$

(4.1.11)

Finally, integrating this with x, we obtain

$$y = \begin{cases} \dfrac{x_0}{2}\left[\dfrac{(x/x_0)^{1+\alpha}}{1+\alpha} - \dfrac{(x/x_0)^{1-\alpha} - 1}{1-\alpha}\right] - y_0, & \alpha \neq 1, \\[4mm] \dfrac{x_0}{2}\left[\dfrac{(x/x_0)^2}{2} - \log\left(\dfrac{x}{x_0}\right)\right] - y_0, & \alpha = 1, \end{cases}$$

(4.1.12)

where y_0 is a constant of integration.

The constants of integration x_0, y_0 contained in this solution should be determined by the position (x_p, y_p) of the airplane when the ship passes the origin, which is called the initial condition. That is, it must be $x = x_p$, $y = y_p$ for $t = 0$ in expression (4.1.3). Therefore, we have

$$\frac{y_p}{x_p} = p(x_p).$$

(4.1.13)

This means that the slope of the straight line connecting the origin and the position (x_p, y_p) of the airplane at that time is equal to the inclination of the airplane in the direction of flight. Substituting expression (4.1.11) into p on the right-hand side of this expression

$$\frac{y_p}{x_p} = \frac{1}{2}\left[\left(\frac{x_p}{x_0}\right)^\alpha - \left(\frac{x_p}{x_0}\right)^{-\alpha}\right].$$

(4.1.14)

By solving x_0 from this, we have

$$x_0 = \frac{x_p}{\left[(y_p/x_p) + \sqrt{1 + (y_p/x_p)^2}\right]^{1/\alpha}},$$

(4.1.15)

where only the positive root was adopted when solving the quadratic equation. To determine y_0, we can use $y = y_p$ when $x = x_p$ is set in (4.1.12). That is

$$y_0 = \begin{cases} \dfrac{x_0}{2}\left[\dfrac{(x_p/x_0)^{1+\alpha}}{1+\alpha} - \dfrac{(x_p/x_0)^{1-\alpha} - 1}{1-\alpha}\right] - y_p, & \alpha \neq 1, \\[4mm] \dfrac{x_0}{2}\left[\dfrac{(x_p/x_0)^2}{2} - \log\left(\dfrac{x_p}{x_0}\right)\right] - y_p, & \alpha = 1. \end{cases}$$

(4.1.16)

Normally, the speed of an airplane v_p will be greater than the speed of a ship v_s, therefore, α defined by (4.1.5) is supposed to be less than 1. Under this condition, in the upper formula of (4.1.12), we can set $x = 0$, so $y = \dfrac{x_0}{2(1-\alpha)} - y_0$, and this value of y is determined by (4.1.16) as

$$y = y_p - \frac{x_0}{2}\left[\frac{(x_p/x_0)^{1+\alpha}}{1+\alpha} - \frac{(x_p/x_0)^{1-\alpha}}{1-\alpha}\right]. \qquad (4.1.17)$$

This allows the airplane to catch up with the ship.

If $\alpha \geq 1$, that is, if the ship is faster than the airplane, y will diverge at $x = 0$. In this case the airplane will not catch up with the ship as a matter of course.

The tracking-lines when $x_p = 1$, $y_p = 1$ are shown in Figs. 4.1 to 4.3. The value of α is 0.5, 1 and 1.5 in Figs. 4.1, 4.2 and 4.3, respectively. In Fig. 4.1 with $\alpha = 0.5$, the airplane is catching up to the ship at about $y = 0.609 \cdots$.

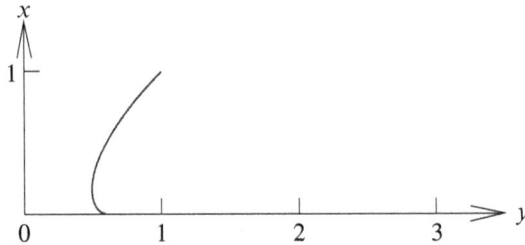

Fig. 4.1   Tracking-line $\alpha = 0.5$

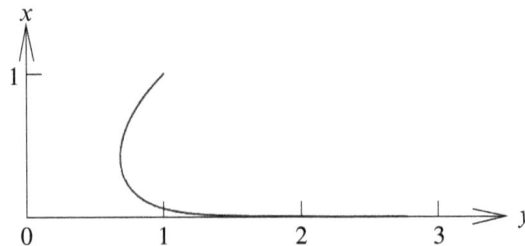

Fig. 4.2   Tracking-line $\alpha = 1$

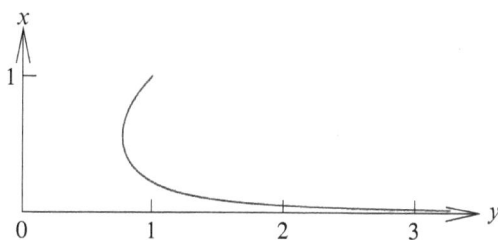

Fig. 4.3 Tracking-line $\alpha = 1.5$

4.1.2 *Traction-line*

Incidentally, there is a **traction-line** or **tractrix** which looks very similar to the tracking-line introduced in the previous subsection. These two curves are often confused, probably because they are quite analogous. However, in reality, they are completely different, so we will present here the model of the traction-line so as to avoid misunderstanding. The traction-line is also known as the **dog-curve**. This is because it refers to the curve along which a dog on a leash moves as it is led by its owner walking beside it. This dog is very obedient and does not walk off on its own. The owner starts from the origin $(0, 0)$, and moves along the y-axis. His speed may vary along the way, but as long as the dog follows, there is no problem with the speed.

It is assumed that the dog is initially at the point $(\ell, 0)$ on the x-axis, where ℓ denotes the length of the leash. Let the curve that the dog "draws" be $y = y(x)$. Because the dog is always connected to the owner, the owner's position lies on the tangent of the curve that the dog follows, keeping the length of the leash ℓ. Let the coordinates of an arbitrary point of the tangent be (X, Y), then the tangent equation at point (x, y), is given by, similarly to (4.1.2),

$$Y - y = y'(X - x). \tag{4.1.18}$$

The Y intercept of this tangent is $Y = y - xy'$ which gives the owner's position as $(0, y - xy')$. The length connecting this point to the dog's position (x, y) must be the length of the leash ℓ. So, we have

$$x^2 + x^2 y'^2 = \ell^2. \tag{4.1.19}$$

This is a nonlinear differential equation. However, in this case, it can be linearized immediately. From now on, we find y' and note that its sign is negative,

$$y' = -\frac{\sqrt{\ell^2 - x^2}}{x}. \tag{4.1.20}$$

We integrate this under the initial condition $y = 0$ when $x = \ell$,

$$y = -\int_\ell^x \frac{\sqrt{\ell^2 - x^2}}{x} dx. \tag{4.1.21}$$

By using the transformation of the variable of integration

$$x = \ell \sin \theta, \tag{4.1.22}$$

we obtain

$$y = -\ell \int_{\pi/2}^\theta \left(\frac{1}{\sin \theta} - \sin \theta \right) d\theta = -\ell \left[\log \left(\tan \frac{\theta}{2} \right) + \cos \theta \right]. \tag{4.1.23}$$

Here, we set θ to the original x, by using the expression

$$\ell \cos \theta = \sqrt{\ell^2 - x^2}, \quad \tan \frac{\theta}{2} = \frac{\ell - \sqrt{\ell^2 - x^2}}{x} = \frac{x}{\ell + \sqrt{\ell^2 - x^2}}, \tag{4.1.24}$$

and finally, we obtain the result,

$$y = \ell \log \left(\frac{\ell + \sqrt{\ell^2 - x^2}}{x} \right) - \sqrt{\ell^2 - x^2}. \tag{4.1.25}$$

Equations (4.1.22) and (4.1.23) express the relationship of x and y with θ as a parameter. If we put here, $\theta = \phi + \frac{\pi}{2}$, we can express the result by using the **inverse-Gudermann function** gd^{-1}, as

$$x = \ell \cos(\phi),$$

$$y = -\ell \left[\log \tan \left(\frac{\phi}{2} + \frac{\pi}{4} \right) - \sin \phi \right] = -\ell \left[\mathrm{gd}^{-1}(\phi) - \sin \phi \right]. \tag{4.1.26}$$

Figure 4.4 illustrates the traction line when $\ell = 1$.

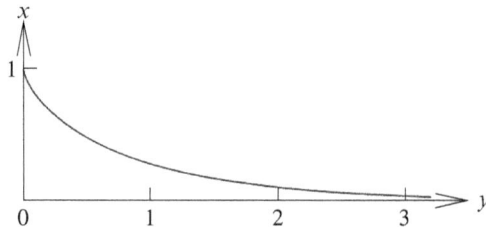

Fig. 4.4   Traction-line $\ell = 1$

4.2 Dropping an Object from an Airplane

4.2.1 *Introduction to the equation*

We consider here the model of dropping an object from an airplane to hit a target. Thankfully not done in peace time, it can be thought of as dropping a bomb from an airplane. When dropping a bomb, how should the airplane fly to hit the target point no matter when the bomb is dropped from the airplane? The problem is to find the trajectory that the airplane should fly. For simplicity, the speed of the airplane is always constant, and the initial speed of the bomb is zero when viewed from the airplane. In other words, in the case of a stationary system, its speed is the same as the flight speed of the airplane, and its direction is also the same as that of the airplane. Also, for simplicity, the air resistance acting on the bomb is ignored.

To find the trajectory of the airplane, we take the x-axis horizontally and the y-axis vertically above. Let the flight trajectory of the airplane be $y = y(x)$, and let its speed be a constant value denoted by v. Now if we drop a bomb from the airplane at the point (x, y), and let θ be the upward angle of the airplane at this point, we have the relation,

$$y' = \tan \theta. \tag{4.2.1}$$

Since the bomb dropped at this point has a horizontal velocity $v \cos \theta$ and a vertical velocity $v \sin \theta$, the coordinates (X, Y) of the bomb at the time t after it was dropped is written as

$$X = v \cos(\theta)t + x, \quad Y = v \sin(\theta)t - \frac{1}{2}gt^2 + y, \tag{4.2.2}$$

where g is the gravitational acceleration. Let the target point for the bomb be the origin $(0, 0)$; we set $X = 0$, $Y = 0$ at a certain time t, so that

$$v \cos(\theta)t + x = 0, \quad v \sin(\theta)t - \frac{1}{2}gt^2 + y = 0. \tag{4.2.3}$$

Eliminating t from (4.2.3) and substituting (4.2.1), we can write the differential equation,

$$-xy' - \frac{1}{2}\frac{g}{v^2}x^2(1 + y'^2) + y = 0. \tag{4.2.4}$$

This equation looks difficult at first glance, because this is a nonlinear differential equation containing the square of y'. But, using the following method, it is possible to find the exact solution. Before that, to simplify the formula, we introduce a constant ℓ with length dimension,

$$\frac{g}{v^2} = \frac{1}{\ell},$$

(4.2.5)

and we put

$$y' = p;$$

(4.2.6)

then equation (4.2.4) is rewritten as

$$y = xp + \frac{x^2}{2\ell}(1 + p^2).$$

(4.2.7)

4.2.2 *Method 1*

By differentiating (4.2.7) with respect to x, we have

$$\frac{dp}{dx}\left(1 + \frac{x}{\ell}p\right) + \frac{1}{\ell}(1 + p^2) = 0.$$

(4.2.8)

We change here our viewpoint and consider x as a function of p, that is, $x = x(p)$ and formulate this equation,

$$\frac{dx}{dp} + \frac{p}{1 + p^2}x = -\frac{\ell}{1 + p^2}.$$

(4.2.9)

It is possible to find an exact solution for this equation by using the **variation of constant**, because it is an **inhomogeneous linear differential equation** for x. First, if we solve with the right-hand side equal to zero,

$$x = \frac{C}{\sqrt{1 + p^2}},$$

(4.2.10)

where C is a constant of integration. Next, let C be regarded as a variable, and substituting (4.2.10) into equation (4.2.9), we have

$$\frac{dC}{dp} = -\frac{\ell}{\sqrt{1 + p^2}}.$$

(4.2.11)

Hereafter, let x_0 be the true constant of integration, then

$$C = x_0 - \ell \log\left(p + \sqrt{1 + p^2}\right).$$

(4.2.12)

Returning to equation (4.2.10), we have

$$x = \frac{x_0 - \ell \log\left(p + \sqrt{1 + p^2}\right)}{\sqrt{1 + p^2}}.$$

(4.2.13)

From this equation, x_0 is the value of x when $p = 0$. Substituting this x into (4.2.7), we obtain y as a function of p,

$$y = \frac{\left[x_0 - \ell \log\left(p + \sqrt{1 + p^2}\right)\right]p}{\sqrt{1 + p^2}} + \frac{1}{2\ell}\left[x_0 - \ell \log\left(p + \sqrt{1 + p^2}\right)\right]^2.$$

$$(4.2.14)$$

Now the relationship between x and y is obtained through a parameter p.

This gives us an exact solution, but expressions (4.2.13) and (4.2.14) are too complicated, so it is not easy to see what the flight trajectory of the airplane will be. The only thing to be found is that when $x = 0$, y is also zero. Therefore, the trajectory is always a curve that passes through the origin. This happens when $p = \sinh(x_0/\ell)$. We cannot see anything more than that. As a test, the curve when $x_0 = 0$ is calculated numerically and drawn in Fig. 4.5.

This figure is drawn with $x_0 = 0$ and the value of p varied from -6 to $+6$. The top point on the right side indicated by the thick solid line corresponds to $p = -6$. As the value of p increases from there the graph goes down, and when p is about -1.5, it goes back suddenly and becomes a **cusp**. At this point, p which is the first derivative of y is continuous, but the second derivative dp/dx is discontinuous. Going up gently from there and passing through the origin when $p = 0$, then going down from here, it becomes a cusp again at about $p = 1.5$. After that, it rises and reaches the highest point on the left side when $p = 6$. As we can see in this figure, the function $y = y(x)$ is a double-valued one, that is, two values of y correspond to one value of x. Therefore, there are two possible flight trajectories for an airplane. It seems that the point at which the curve becomes a cusp corresponds to the point where the curvature becomes infinite; therefore,

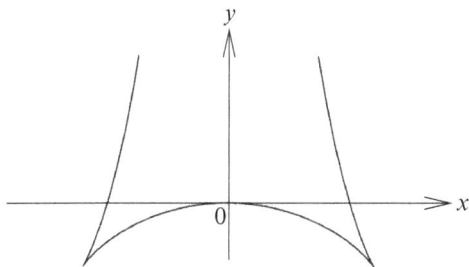

Fig. 4.5 $x_0 = 0$, $-6 \leq p \leq +6$

considering equation (4.2.8), we have

$$1 + \frac{x}{\ell}p = 0. \tag{4.2.15}$$

By substituting x of (4.2.13) with $x_0 = 0$ into this equation, and solving the value of p numerically, it becomes approximately $p = \pm 1.5088 \cdots$. This value coincides with the graph with good accuracy.

What happens in the case of non-zero x_0? At this time, the local maximum position, that is, the point corresponding to $p = 0$, we have from (4.2.13) and (4.2.14),

$$x = x_0, \quad y = \frac{x_0^2}{2\ell}. \tag{4.2.16}$$

From these equations, we see that when x_0 becomes negative and increases, the maximum position shifts to the upper left. Also, when x_0 becomes positive and large, the maximum position shifts to the upper right.

Let us find the trajectory of the airplane using actual values. For the value of ℓ which is defined as $\ell = v^2/g$ in (4.2.5), the airplane speed v is required. Here, assuming a small airplane, we set this value temporarily as $v = 250\,\text{km/h}$, and $g = 9.8\,\text{m/s}^2$, that is

$$v = \frac{250\,000}{3600}\,\text{m/s} \cong 69.444\,\text{m/s}, \quad \ell = \frac{v^2}{g} \cong 492.095\,\text{m}. \tag{4.2.17}$$

First, we give an example of the airplane rising near the target point. Figure 4.6 shows the trajectory of the airplane with a thick solid line, with the parameters $x_0 = -500\,\text{m}$, and p changing from 0.6 to 4.

For comparison, we have also drawn a graph with p from -30 to 0.6 (light gray line). The trajectories of a bomb dropped when $p = 1.5$ and $p = 2.5$, are shown by the dotted lines. No matter when the bomb is dropped, it can be seen that it reaches the origin, i.e. the target point.

Another trajectory of the airplane is shown by the light gray line in Fig. 4.6. It is redrawn again in Fig. 4.7, with the value of $x_0 = -500\,\text{m}$, which is the same as in Fig. 4.6, and the value of p in the range of 0 to -1.2. In this case, the trajectory of the airplane and the trajectory of the bomb almost overlap, so the trajectory of the bomb is not drawn. As can be seen, if the airplane flies as it does, it will also collide with the target point of the bomb.

4.2.3 *Method 2*

Although (4.2.4) is a nonlinear differential equation, it was easily solved. So, it might seem that this equation is a very special case. This equation is,

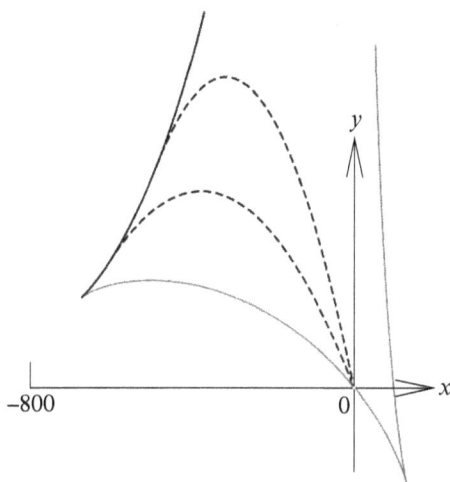

Fig. 4.6 $x_0 = -500\,\text{m}$, $0.6 \le p \le 4$

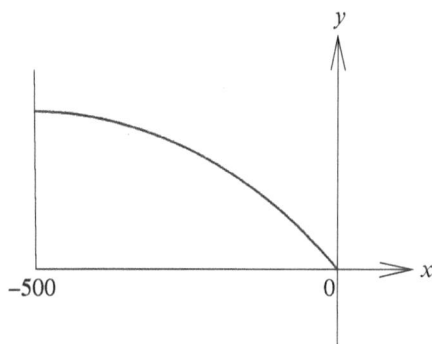

Fig. 4.7 $x_0 = -500\,\text{m}$, $-1.2 \le p \le 0$

however, one of the general ones that can be solved by using the **Legendre transformation**. The method based on the Legendre transformation is briefly described below. To begin with, we define the transformation from the (x, y) to (X, Y) system in this case

$$X = y', \quad Y = y - xy'. \tag{4.2.18}$$

Differentiating Y with respect to X,

$$\frac{dY}{dX} = \frac{dY/dx}{dX/dx} = \frac{-xdX/dx}{dX/dx} = -x. \tag{4.2.19}$$

By using these equations, we rewrite equation (4.2.7) to the (X, Y) system, with p as X,

$$2\ell Y = \left(\frac{dY}{dX}\right)^2 (1 + X^2).$$ (4.2.20)

Here, assuming $dY/dX (= -x) > 0$, we get

$$\sqrt{2\ell Y} = \frac{dY}{dX}\sqrt{1 + X^2}.$$ (4.2.21)

This can be separated immediately,

$$\frac{1}{\sqrt{1 + X^2}}dX = \frac{1}{\sqrt{2\ell Y}}dY.$$ (4.2.22)

Integrating and multiplying both sides by ℓ, we have

$$\ell \log\left(X + \sqrt{1 + X^2}\right) - x_0 = \sqrt{2\ell Y},$$ (4.2.23)

where x_0 is a constant of integration. Substituting this right-hand side into (4.2.21), and using (4.2.19) with $dY/dX = -x$, we obtain x as a function of X,

$$x = \frac{x_0 - \ell \log(X + \sqrt{1 + X^2})}{\sqrt{1 + X^2}}.$$ (4.2.24)

This is the same as (4.2.13), because $X = p$ from (4.2.6). Also, $y = Y + xX$ from the second expression of (4.2.18), and we obtain y by using (4.2.23) and (4.2.24)

$$y = \frac{[x_0 - \ell \log(X + \sqrt{1 + X^2})]X}{\sqrt{1 + X^2}} + \frac{1}{2\ell}\left[x_0 - \ell \log(X + \sqrt{1 + X^2})\right]^2.$$

(4.2.25)

This is nothing but equation (4.2.14).

4.3 Worm Crawling along a Stretched Rubber Rod

4.3.1 *Introduction to the problem*

When I took out an old mechanics textbook from my bookshelf, I found a piece of paper inside on which I had written, a long time ago, "The problem of a worm crawling along a stretched rubber rod." I probably wrote it when I was a student. Since the source was not written, I do not know where I had gotten it. Although it is a simple differential equation problem, it is an interesting one, so I will present it here. The problem is as follows, "From one end of a rubber rod, initially of length $1\,\mathrm{m}$, a worm crawls along at a speed of $1 \, \text{m/h}$. It is assumed that this rubber rod grows at a speed of $1 \, \text{km/h}$ from the time when the worm begins to crawl. Can this worm reach the other end of the rod?" This seems impossible at first glance. But, in fact, it can be found that it is possible in principle by calculation, but it will take a long, long, long time. In the following, we will deal with this problem and other slightly-modified problems.

4.3.2 *Analysis 1*

The problem is as mentioned above. Here, the initial length of the rod is ℓ_0. Let the fixed end of the rod be A and the other end that extends be B. Let the speed at which the worm crawls on the rod be v_0, and the speed at which the other end B of the rod grows be V_0. And let the length of the rod be $\ell(t)$ at time t after the worm starts at the point A. That is,

$$\ell(t) = \ell_0 + V_0 t. \tag{4.3.1}$$

Also, we set the coordinate of the position of the worm at time t as $x(t)$ measured from point A. The velocity of the growth of the rod at this point $x(t)$ is considered to be $\left[x(t)/\ell(t)\right] V_0$ taking into account the linearity of the rod. Since the worm's velocity dx/dt seen from the rest system is the sum of the worm's own velocity v_0 and the velocity of the growth of this rod, so we have

$$\frac{dx}{dt} = v_0 + \frac{x}{\ell(t)} V_0. \tag{4.3.2}$$

This can be solved by using the **variation of constant**, because this is an **inhomogeneous linear differential equation** for the unknown variable $x(t)$. By taking the homogeneous part of this equation, and substituting $\ell(t)$ of (4.3.1), we have

$$\frac{dx}{dt} = \frac{x}{\ell_0 + V_0 t} V_0. \tag{4.3.3}$$

This equation is easily integrated by using the **separation of variables**, and it can be solved with the constant of integration C,

$$x = C(\ell_0 + V_0 t). \tag{4.3.4}$$

Now, we regard C as a function of time t, and substitute this into (4.3.2),

$$\frac{dC}{dt} = \frac{v_0}{\ell_0 + V_0 t}; \tag{4.3.5}$$

by integrating this, we obtain

$$C = \frac{v_0}{V_0} \log \left(\frac{\ell_0 + V_0 t}{\ell_0} \right). \tag{4.3.6}$$

It is noticed that C becomes zero when $t = 0$. Returning this to expression (4.3.4), we obtain the final solution

$$x = \frac{v_0}{V_0} (\ell_0 + V_0 t) \log \left(\frac{\ell_0 + V_0 t}{\ell_0} \right). \tag{4.3.7}$$

To find the time t that the worm takes to reach point B, the other end of the rod, setting $x = \ell_0 + V_0 t$, we have

$$\frac{v_0}{V_0} \log \left(\frac{\ell_0 + V_0 t}{\ell_0} \right) = 1. \tag{4.3.8}$$

Hence, the arrival time t is solved as

$$t = \frac{\ell_0}{V_0} \left(e^{V_0/v_0} - 1 \right). \tag{4.3.9}$$

If we put $V_0 = 0$, as a check, this formula becomes $0/0$. Hence, taking the limit of $V_0 \to 0$, it becomes $t \to \ell_0/v_0$. This is the self-evident solution when the rod does not stretch at all.

Let us apply expression (4.3.9) to the numerical calculation. We put $\ell_0 = 1\,\mathrm{m}$, $v_0 = 1\,\mathrm{m/h}$, $V_0 = 1000\,\mathrm{m/h}$ to get

$$t = \frac{1}{1000} \left(e^{1000} - 1 \right) \text{ hours} \cong 2.248 \times 10^{427} \text{ years}. \tag{4.3.10}$$

Mathematically, the worm can reach the end of the rod, but it will take a tremendous number of years, far beyond the age of the universe ($=1.38 \times 10^{10}$ years).

It does sound strange that the worm can reach the other end. But over time, the worm gains velocity over the growth of the rod. From (4.3.2), when $x(t)$ becomes larger than $[(V_0 - v_0)/V_0]\ell(t)$, the velocity of the worm dx/dt exceeds the velocity of the growth of the rod V_0. As a result, the worm can reach the other end.

4.3.3 *Analysis 2*

In the previous problem, the worm's crawling speed v_0 remains unchanged. But we assume that the rod grows at an accelerating rate A_0, and its initial velocity is zero. With this assumption, the velocity at time t at end B is $A_0 t$. Since the extended length is $\frac{1}{2} A_0 t^2$, the length of the rod at time t is

$$\ell(t) = \ell_0 + \frac{1}{2} A_0 t^2. \tag{4.3.11}$$

As before, we set the position of the worm at time t as $x(t)$, and the velocity of the growth of the rod at that point is $[x(t)/\ell(t)] A_0 t$, then the worm's velocity dx/dt seen in the rest system is

$$\frac{dx}{dt} = v_0 + \frac{x}{\ell_0 + \frac{1}{2} A_0 t^2} A_0 t. \tag{4.3.12}$$

In the previous analysis, the equation was solved by using the **variation of constant**. But this is troublesome. Instead, we divide this expression by $(\ell_0 + \frac{1}{2} A_0 t^2)$, arrange the differentiation to get

$$\frac{d}{dt}\left(\frac{x}{\ell_0 + \frac{1}{2} A_0 t^2}\right) = \frac{v_0}{\ell_0 + \frac{1}{2} A_0 t^2}, \tag{4.3.13}$$

and integrate this expression

$$\frac{x}{\ell_0 + \frac{1}{2} A_0 t^2} = v_0 \sqrt{\frac{2}{A_0 \ell_0}} \, \mathrm{Tan}^{-1}\left(\sqrt{\frac{A_0}{2\ell_0}} t\right), \tag{4.3.14}$$

where Tan^{-1} is the principal value of the arctangent function. When $t = 0$, x is equal to zero. From this expression, we set $x = \ell_0 + \frac{1}{2} A_0 t^2$, in order for the worm to reach point B, then, the left-hand side of (4.3.14) becomes 1,

$$\mathrm{Tan}^{-1}\left(\sqrt{\frac{A_0}{2\ell_0}} t\right) = \frac{1}{v_0} \sqrt{\frac{A_0 \ell_0}{2}}. \tag{4.3.15}$$

Hence, the arrival time t is solved

$$t = \sqrt{\frac{2\ell_0}{A_0}} \tan\left(\frac{1}{v_0} \sqrt{\frac{A_0 \ell_0}{2}}\right). \tag{4.3.16}$$

However, caution must be taken here. Because Tan^{-1} of the left-hand side of (4.3.15) cannot take a value larger than $\pi/2$, we have a restriction

$$\frac{1}{v_0} \sqrt{\frac{A_0 \ell_0}{2}} < \frac{\pi}{2}. \tag{4.3.17}$$

For example, in the case of $\ell_0 = 1\,\mathrm{m}$, $v_0 = 1\,\mathrm{m/h}$, we have a condition of inequality for the worm to reach the other end B,

$$A_0 < \frac{\pi^2 v_0^2}{2\ell_0} = 4.9348\,\mathrm{m/h^2}. \tag{4.3.18}$$

4.3.4 *Analysis 3*

Here, we will generalize this problem. We assume that the speed at which the worm crawls and the speed at which the rod grows are arbitrary functions of time t, namely, we put these speeds as $v(t)$ and $V(t)$, respectively. In this case, the rod length $\ell(t)$ at time t is

$$\ell(t) = \ell_0 + \int_0^t V(t')dt'. \tag{4.3.19}$$

Let the position of the worm be $x(t)$ at this time t, and its speed is

$$\frac{dx}{dt} = v(t) + \frac{x}{\ell(t)}V(t). \tag{4.3.20}$$

Dividing both sides by $\ell(t)$ and rearranging,

$$\frac{d}{dt}\left(\frac{x}{\ell(t)}\right) = \frac{v(t)}{\ell(t)}. \tag{4.3.21}$$

Integrating, we obtain

$$\frac{x}{\ell(t)} = \int_0^t \frac{v(t')}{\ell(t')}dt'. \tag{4.3.22}$$

For the worm to reach the other end B, we set $x = \ell(t)$

$$\int_0^t \frac{v(t')}{\ell(t')}dt' = 1. \tag{4.3.23}$$

We can solve the arrival time from this integral equation.

The easiest way to perform this integration is by having the numerator in the same form as the derivative of the denominator in the integrand as before. This is done by using an arbitrary function $f(t)$, to which both $v(t)$ and $V(t)$ are proportional,

$$v(t) = v_0 f(t), \quad V(t) = V_0 f(t). \tag{4.3.24}$$

In this case, the integration of (4.3.23) is easily carried out,

$$\log\left[1 + \frac{V_0}{\ell_0}\int_0^t f(t')dt'\right] = \frac{V_0}{v_0}. \tag{4.3.25}$$

For example, we consider here the function $f(t)$ as an exponential function

$$f(t) = e^{\omega t}. \tag{4.3.26}$$

In this case, (4.3.25) becomes

$$\log\left[1 + \frac{V_0}{\omega \ell_0}(e^{\omega t} - 1)\right] = \frac{V_0}{v_0}, \tag{4.3.27}$$

and from this, the arrival time t is solved as

$$t = \frac{1}{\omega} \log \left[1 + \frac{\omega \ell_0}{V_0} \left(e^{V_0/v_0} - 1 \right) \right].$$ (4.3.28)

If the limit of $\omega \to 0$ is taken in this expression, then it becomes the same as expression (4.3.9).

Furthermore, if we set

$$f(t) = \cos(\omega t),$$ (4.3.29)

(4.3.25) becomes

$$\log \left[1 + \frac{V_0}{\omega \ell_0} \sin(\omega t) \right] = \frac{V_0}{v_0}.$$ (4.3.30)

We obtain from this

$$\sin(\omega t) = \frac{\omega \ell_0}{V_0} \left(e^{V_0/v_0} - 1 \right).$$ (4.3.31)

Here, if the right-hand side is less than 1, a solution exists, and then we obtain the arrival time t

$$t = \frac{1}{\omega} \mathrm{Sin}^{-1} \left[\frac{\omega \ell_0}{V_0} \left(e^{V_0/v_0} - 1 \right) \right],$$ (4.3.32)

where Sin^{-1} is the principal value of the arcsine function. In this expression as well, the limit of $\omega \to 0$ coincides with expression (4.3.9).

Finally, we consider the case where both the rod and the worm move at an accelerated rate. We put $f(t)$ of (4.3.24) as

$$f(t) = \omega t.$$ (4.3.33)

This means that the acceleration of the worm is $v_0 \omega$, and that of the rod is $V_0 \omega$, but both initial velocities are zero. In this case, (4.3.25) becomes

$$\log \left[1 + \frac{V_0 \omega}{2 \ell_0} t^2 \right] = \frac{V_0}{v_0}.$$ (4.3.34)

We can solve the arrival time t,

$$t = \sqrt{\frac{2 \ell_0}{V_0 \omega} \left(e^{V_0/v_0} - 1 \right)}.$$ (4.3.35)

If we set $\ell_0 = 1\,\mathrm{m}$, $v_0 = 1\,\mathrm{m/h}$, $V_0 = 1000\,\mathrm{m/h}$, $\omega = 1\,\mathrm{h}^{-1}$, the numerical value is

$$t = \sqrt{\frac{2}{1000} \left(e^{1000} - 1 \right)}\ \mathrm{hours} \cong 7.16 \times 10^{211}\ \mathrm{years}.$$ (4.3.36)

It will take a tremendous amount of time, though not as much as expression (4.3.10).

4.4 SIR Model of Virus Infection

4.4.1 *Introduction to the problem*

COVID-19 (COronaVIrus Disease 2019) caused by the SARS-CoV-2 virus was first identified in Wuhan, China, at the end of 2019. This disease is highly infectious and quickly spread to Japan as early as February 2020. By March, 2020, it had spread to Europe, the United States and the rest of the world.

The **SIR model** analyses how infectious diseases spread. This model of viral infection was proposed by William Ogilvy Kermack and Anderson Gray McKendrick in 1927 as a model to analyze how the number of infected persons varies over time. The acronym stands for **S**usceptible, **I**nfectious, and **R**ecovered.[1] The term "Susceptible" may be difficult to understand; in the following, we will assume it to mean a healthy person who has not been infected yet.

When we set the values (number of people) for each of the variables S, I and R at time t as $S(t)$, $I(t)$, and $R(t)$, respectively, this model is described as follows

$$\frac{dS(t)}{dt} = -\beta S(t)I(t),$$

$$\frac{dI(t)}{dt} = \beta S(t)I(t) - \gamma I(t), \qquad (4.4.1)$$

$$\frac{dR(t)}{dt} = \gamma I(t).$$

The first equation means that the number of healthy persons $S(t)$ decreases in proportion to the product of the number of healthy people and the number of infected ones. We put here the constant of proportion as β. The second equation means that the number of infected people $I(t)$ increases by the amount that healthy people become infected and it decreases by the amount that the infected recover. We set the coefficient for recovery of the infected person as γ. The last equation means that the number of recovered persons $R(t)$ increases as infected persons recover. However, as we can see from these equations, it is the first two equations that determine $S(t)$, $I(t)$, and the third equation is no more than describing that $R(t)$ is determined only by $I(t)$.

---

[1]This R refers to a person who has stopped infecting others; strictly speaking, it would include those who have recovered as well as those who have died.

In the following, we make the variables $S(t)$, $I(t)$, $R(t)$ dimensionless, to make later calculations simpler. By using the initial value $S(0)$ of the variable $S(t)$, we rewrite $S(t)$, $I(t)$, $R(t)$ and β as

$$\frac{S(t)}{S(0)} \to S(t), \quad \frac{I(t)}{S(0)} \to I(t), \quad \frac{R(t)}{S(0)} \to R(t), \quad \beta S(0) \to \beta. \qquad (4.4.2)$$

Now the initial value of $S(t)$ becomes 1. Also, β and γ have the same dimension of the reciprocal of time. In this sense, β and γ are called the "infection rate" and the "recovery rate", respectively. Note that the form of the equations remains unchanged for this transformation.

Adding all three equations, in each side

$$\frac{d}{dt}(S + I + R) = 0, \quad \text{therefore,} \quad S + I + R = \text{const.} \qquad (4.4.3)$$

The initial values when infection has not started yet are $S(0) = 1$, $I(0) = R(0) = 0$, so the constant of integration in this expression is equal to 1.

4.4.2 *Analytical solution*

$S(t)$ and $I(t)$ are determined by the first two equations in (4.4.1), as mentioned above. The purpose is to solve the equations analytically, as far as possible. The first two equations are separated into variables,

$$\frac{dS}{S} = -\beta I dt, \quad \frac{dI}{I} = (\beta S - \gamma)dt, \qquad (4.4.4)$$

and integrating these equations, we have

$$\log S = -\beta \int I dt, \quad \log I = \int (\beta S - \gamma)dt. \qquad (4.4.5)$$

The first equation of (4.4.5) will be used later. By finding I from the second equation and substituting it into the first one of (4.4.1), we have an equation which contains S only

$$\dot{S} = -\beta S e^{\int(\beta S - \gamma)dt}. \qquad (4.4.6)$$

Hereafter, dS/dt will be referred to as $\dot{S}$ in the Newton style. We rewrite this equation as

$$\log\left(\frac{\dot{S}}{-\beta S}\right) = \int (\beta S - \gamma)dt. \qquad (4.4.7)$$

Further differentiating with respect to t, we have

$$\frac{S\ddot{S} - \dot{S}^2}{S\dot{S}} = \beta S - \gamma. \qquad (4.4.8)$$

Here, we put

$$\dot{S} = P, \tag{4.4.9}$$

and differentiating this, we have

$$\ddot{S} = \frac{dP}{dt} = \frac{dS}{dt}\frac{dP}{dS} = P\frac{dP}{dS}. \tag{4.4.10}$$

Because of this, (4.4.8) becomes

$$\frac{dP}{dS} - \frac{1}{S}P = \beta S - \gamma. \tag{4.4.11}$$

This is an **inhomogeneous linear differential equation** with respect to P. To solve this equation, we consider the corresponding homogeneous equation,

$$\frac{dP}{dS} - \frac{1}{S}P = 0. \tag{4.4.12}$$

This has a solution,

$$P = CS, \tag{4.4.13}$$

where C is a constant of integration. Then we follow the **variation of constant**. Substituting this into (4.4.11), C as a variable of S,

$$\frac{dC}{dS} = \beta - \frac{\gamma}{S}. \tag{4.4.14}$$

We integrate this equation,

$$C = \beta S - \gamma \log S + D, \tag{4.4.15}$$

where D is a constant of integration. Returning to (4.4.13) with this result, and using (4.4.9), we obtain

$$\dot{S} = (\beta S - \gamma \log S + D)S. \tag{4.4.16}$$

To determine the constant D, we use the initial condition; when $t = 0$,

$$S(0) = 1, \quad \dot{S}(0) = 0. \tag{4.4.17}$$

Then, we have $D = -\beta$, and therefore, (4.4.16) becomes

$$\dot{S} = F(S), \quad \text{where} \quad F(S) \equiv \left[\beta(S-1) - \gamma \log S\right]S. \tag{4.4.18}$$

Separating the variables of this equation, and making it into a definite integral form,

$$t = \int_1^S \frac{dS'}{F(S')}. \tag{4.4.19}$$

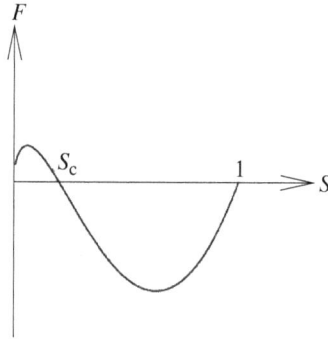

Fig. 4.8   Graph of $F(S)$, case of $\beta = 0.2/\text{day}$, $\gamma = 0.1/\text{day}$

It will be almost impossible to perform this integral analytically. Note that the number of infected people I increases over time, therefore, the number of healthy people S gradually decreases from the first value 1. As a result, the range actually applied is the one where $F(s)$ is negative. As is seen from this integral form, in order to be $t > 0$ where $S < 1$, $F(S)$ must be negative. The values of this function $F(S)$ and its derivative at $S = 1$ are

$$F(1) = 0, \quad \frac{dF(S)}{dS}\bigg|_{S=1} = \beta - \gamma. \qquad (4.4.20)$$

In order to be $F(S) < 0$ on the side of $S < 1$, the condition

$$\beta > \gamma \qquad (4.4.21)$$

is required. That is, the infection rate β must be greater than the recovery rate γ. Figure 4.8 shows a graph of the function $F(S)$ when $\beta = 0.2/\text{day}$, $\gamma = 0.1/\text{day}$ under this condition.

As is seen from this graph, when $S = 1$, we have $F(S) = 0$. In addition, there is another zero point in the range of $(0, 1)$. Let this point be S_c. The range that the variable S can take is $(S_c, 1)$ where $F(S)$ has a negative value. In other words, the value of S cannot be less than the value S_c. Because the equation for determining S_c is a transcendental equation, no exact solution can be obtained.

4.4.3 *Numerical trial*

It seems impossible to execute the integration of (4.4.19) analytically, as mentioned above. We calculate it numerically in the following. First, we find the number of healthy people $S(t)$. Let $\dot{S}$ in equation (4.4.18) be $\Delta S/\Delta t$, so we have

$$\Delta t = \frac{\Delta S}{F(S)}. \tag{4.4.22}$$

We set here $\Delta S = 0.001$, and seek Δt. At this time, if the initial value of S is set to 1, the denominator on the right-hand side of this equation becomes zero. So we find $(\Delta t)_1$ with the initial value of S as $1 - \Delta S$. This determines the point of $(S, t) = (1 - \Delta S, (\Delta t)_1)$. Next, with the value of S as $1 - 2\Delta S$, find $(\Delta t)_2$, and the point $(S, t) = (1 - 2\Delta S, (\Delta t)_1 + (\Delta t)_2)$ is determined. And so on, we seek the point $(S, t) = \left(1 - n\Delta S, \sum_{k=1}^{n}(\Delta t)_k\right)$. The solution curve shown in Fig. 4.9 is obtained by connecting these points with a smooth line. It can be seen from this figure that the value of $S(t)$ converges to the S_c at $t \to \infty$.

Next, we find the number of infected people $I(t)$. We substitute $\dot{S}$ of (4.4.18) into the first equation of (4.4.1) to get

$$I(t) = -\frac{F(S(t))}{\beta S(t)} = 1 - S(t) + \frac{\gamma}{\beta}\log S(t), \tag{4.4.23}$$

Thus, $I(t)$ is represented by using the $S(t)$ which is already found numerically. Since the initial value of $S(t)$ is 1, the initial value of $I(t)$ is now zero. And the value of $I(t)$ converges to zero at $t \to \infty$, because this equation is equivalent to the one for finding S_c, and the value of $S(t)$ becomes S_c at $t \to \infty$. That is, the infected person will eventually disappear. Figure 4.9 shows graphs of these variables $S(t)$ and $I(t)$ where the parameters are $\beta = 0.2/\text{day}$, $\gamma = 0.1/\text{day}$, the same as before.

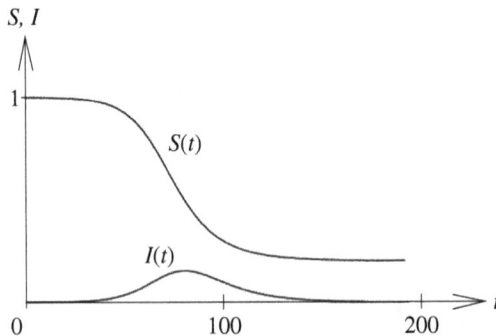

Fig. 4.9 Graph of $S(t)$, $I(t)$, case of $\beta = 0.2/\text{day}$, $\gamma = 0.1/\text{day}$

4.4.4 *The curve $I(t)$*

The curve $I(t)$ is a mountain-style curve as shown in Fig. 4.9. To obtain the maximum point of this curve, we set $dI/dt = 0$ in the second equation of (4.4.1), so that

$$S(t) = \frac{\gamma}{\beta}; \tag{4.4.24}$$

then, we obtain the maximum value from (4.4.23),

$$I_{\max} = 1 - \frac{\gamma}{\beta} + \frac{\gamma}{\beta} \log\left(\frac{\gamma}{\beta}\right). \tag{4.4.25}$$

This value is determined only by the value of γ/β. From now on, if the value of β is much larger than the value of γ, which is kept constant, the $I_{\max}$ becomes closer to 1. Conversely, if the value of β is closer to γ, the $I_{\max}$ becomes closer to zero.

The larger the maximum value $I_{\max}$ becomes, the narrower the width of the peak of this curve becomes. Therefore, it is expected that the integrated value of $I(t)$ will not vary so much. In fact, the integrated value of $I(t)$ from 0 to ∞ can be obtained from the first equation of (4.4.5)

$$\int_0^\infty I(t)dt = -\frac{1}{\beta} \log S_c = \frac{1}{\gamma}(1 - S_c), \tag{4.4.26}$$

where the S_c is the zero point of $F(S)$ defined in (4.4.18), and this value is less than 1. This last expression is obtained by using $F(S_c) = 0$. From this, the integral value is the final total number of infected people $1 - S_c$ divided by γ. Since this S_c is a quantity determined by γ/β, the integral value is a quantity that depends on both β and γ, and therefore, this value cannot be easily said to be a constant value.

Final Comments on the SIR Model

There are many papers concerning the SIR model which seek to solve numerically. So, the main focus here is to solve analytically as much as possible. As a result, we obtained the existence of the convergent value S_c of $S(t)$ in $t \to \infty$, and obtained the formula for the maximum value of $I(t)$. These are major achievements. Of course, the analysis here does not directly apply to the real world. Because humans take appropriate countermeasures against virus infection, the infection rate β varies from time to time. The infections that should have subsided have reappeared as second and third waves.

The method described here applies to only one of those waves. Therefore this model can only be used as a rough prediction of what will happen in the real world.

The SIR model dealt with here is the simplest of the virus infection models. There are several extensions to this model. For example, there is an incubation period during which the infected are not yet infectious. Or some infected people are immunized and therefore not infectious. Models that take into account these circumstances can be considered. Those who are interested are encouraged to read up on these models as well.

COVID-19 spread rapidly across the globe and the World Health Organization (WHO) declared it a pandemic in March 2020. As of August 2021, there were 211 million confirmed cases and 4.4 million COVID-19-related deaths. With the rising number of cases and the rapid spread of the variants, humanity seemed doomed.

Yet, if we follow the analysis here, no matter how widespread the infection, people with only S_c will not get infected, and can survive. This shows that there is hope yet ... especially since vaccines have been developed and rolled out as of early 2021.

4.5 Is it Possible to Construct a Space Elevator?

The idea of an elevator that connects earth to a **geostationary satellite** has captured people's attention. You can find many discussions on **space elevators** on the Internet, as well as on television. Some may think that such space elevators will become a reality soon. How convenient would that be? There would be no need to spend a lot of money on rockets. Besides, rockets release toxic gases into our atmosphere, and consumes enormous amounts of energy. With a space elevator, it is calculated that energy consumption can be reduced from $1/100$ to $1/1000$ compared to rockets.

Here, let us consider the preliminary and initial stages of making such a device, mainly from a mechanical standpoint. The biggest problem is whether the cable that supports such an elevator can withstand the weight.

4.5.1 *Basic analysis*

First, for simplicity, we will analyze based on the following three assumptions. (1) The cable used here shall have a uniform thickness and shall not stretch or shrink. (2) The mass of the elevator cage (climber) is ignored compared to the mass of the cable. (3) The position of the geostation-ary satellite shall not vary, when the elevator is attached. Although these assumptions are ambitious, they are essential for an initial analysis.

Let the cross-sectional area, the mass density and the tension of the cable be S, ρ and T, respectively. However, the tension varies from place to place. We let the tension at the point x be $T(x)$, where the coordinate x is taken to the direction of the geostationary satellite from the center of the earth as the origin. We set the radius and the mass of the earth as R_e and M_e, respectively. Furthermore, we set the acceleration of gravity on the ground, the gravitational constant and the angular velocity of earth's rotation as g, G and ω, respectively.

Now consider the balance of forces acting between the points x and $x + dx$ on the cable. The upward forces are the tension $T(x + dx)$ and the centrifugal force $Sdx\rho\, x\omega^2$, while the downward forces are the tension $T(x)$ and the gravity $GM_e Sdx\rho/x^2 = R_e^2 gSdx\rho/x^2$. Therefore, the equation of the balance of forces is

$$T(x + dx) + Sdx\rho\, x\omega^2 = T(x) + \frac{R_e^2 gSdx\rho}{x^2}. \tag{4.5.1}$$

We obtain a differential equation from this,

$$\frac{1}{S}\frac{dT}{dx} = \frac{R_e^2 g\rho}{x^2} - \rho\omega^2 x. \tag{4.5.2}$$

In the following, this equation is rewritten by using the stress $\sigma(x) = T(x)/S$ which is the tension per unit cross-sectional area,

$$\frac{d\sigma}{dx} = \frac{R_e^2 g\rho}{x^2} - \rho\omega^2 x. \tag{4.5.3}$$

Integrating this, we have

$$\sigma(x) = -\frac{R_e^2 g\rho}{x} - \frac{1}{2}\rho\omega^2 x^2 + C. \tag{4.5.4}$$

The constant of integration C is selected so that the stress becomes a con-stant σ_e on the ground ($x = R_e$), then, the stress at coordinate x is rewrit-ten as

$$\sigma(x) = R_e^2 g\rho\left(\frac{1}{R_e} - \frac{1}{x}\right) + \frac{1}{2}\rho\,\omega^2(R_e^2 - x^2) + \sigma_e. \tag{4.5.5}$$

To make the x dependency of this expression a little clearer, let us exam-ine the relationship with the geostationary satellite. Let the mass of the geostationary satellite be M_s and its x-coordinate be R_s. The relation for the balance between gravity and centrifugal force acting on a geostationary satellite is

$$G\frac{M_e M_s}{R_s^2} = M_s R_s \omega^2, \quad \text{therefore,} \quad R_e^2 g = R_s^3 \omega^2. \tag{4.5.6}$$

Using this relation, we rewrite (4.5.5) into the formula in which ω is eliminated,

$$\frac{\sigma(x)}{\rho} = R_e^2 g \left[\frac{1}{R_e} - \frac{1}{x} + \frac{1}{2R_s^3}(R_e^2 - x^2)\right] + \frac{\sigma_e}{\rho}. \tag{4.5.7}$$

To find the x dependency of this relation, we calculate the derivative of $\sigma(x)$,

$$\frac{1}{\rho}\frac{d\sigma}{dx} = \frac{R_e^2 g}{R_s^3} \cdot \frac{R_s^3 - x^3}{x^2}. \tag{4.5.8}$$

It can be seen, from this expression, that the stress will increase with height in the range of $R_e \le x \le R_s$, and it reaches its maximum value at the position of the geostationary satellite. In the following, let us find the maximum value of the stress. Let the stress σ at the point of the geostationary satellite be σ_s. We obtain from expression (4.5.7)

$$\frac{\sigma_s}{\rho} = \frac{R_e g}{2R_s^3}(R_s - R_e)^2(2R_s + R_e) + \frac{\sigma_e}{\rho}. \tag{4.5.9}$$

In what follows, we discuss the results so far by substituting actual numerical values:

$$R_e = 6370 \times 10^3 \,\text{m}, \quad g = 9.8 \,\text{m/s}^2,$$

$$\omega = \frac{2\pi}{(24 \times 60 \times 60)}/\text{s} = 7.2722 \times 10^{-5}/\text{s}. \tag{4.5.10}$$

First, we must find the distance R_s of the geostationary satellite from the center of the earth, using these values and relation (4.5.6). We have

$$R_s = \left(\frac{R_e^2 g}{\omega^2}\right)^{1/3} = 42\,207.647 \times 10^3 \,\text{m} \approx 42\,200 \,\text{km}. \tag{4.5.11}$$

Substituting this value into relation (4.5.9), we obtain

$$\frac{\sigma_s}{\rho} \approx 48\,400 \,\text{kN} \cdot \text{m/kg} + \frac{\sigma_e}{\rho}. \tag{4.5.12}$$

We consider, in the following, what this value means. The stress σ when a substance is pulled and broken is called the **tensile strength** and is denoted by σ_y. Also, this tensile strength σ_y divided by the mass density ρ,

Table 4.1 The tensile strength, density, specific strength and breaking
length of some typical substances

Material	Tensile strength σ_y (MPa)	Density ρ (g/cm^3)	Specific strength σ_y/ρ (kN·m/kg)	Breaking length $\sigma_y/(\rho g)$ (km)
Steel wire	5500	7.87	698	71.2
Carbon fiber	4300	1.75	2457	250
Kevlar	3620	1.44	2514	256
Spectra	3510	0.97	3619	369
CNT	62 000	1.34	46 268	4716
CCT	6900	0.116	59 483	6066

(Kevlar and Spectra are the product names of the aromatic polyamide resin
and the ultrahigh molecular weight polyethylene, respectively.)

that is σ_y/ρ, is called **specific strength**. The lighter and stronger a substance is, the larger the specific strength. The tensile strength, density,
specific strength and breaking length of 6 kinds of typical substances are
listed in Table 4.1. The **breaking length** is the specific strength divided
by the gravitational acceleration g, having a dimension of length, and represents the maximum length until the substance breaks when it is hung like
a string.

The feasibility of a space elevator was largely due to the appearance
of the **Carbon NanoTube (CNT)**. The CNT was discovered in 1991 by
S. Iijima. It attracted the world's attention as a substance with amazing
tensile strength. However, as seen in (4.5.12), the specific strength of the
CNT is not sufficient, even if we ignore the value of σ_e/ρ on the ground.
Of course, in the case of the CNT, if we can remove the impurities in it
and increase its purity in future, it may be possible to increase its specific
strength. But, it cannot be expected to increase by an order of magnitude.

In 2008, the **Colossal Carbon Tube (CCT)** was discovered which has
a higher specific strength than the CNT. As is seen in Table 4.1, the specific
strength of the CCT exceeds the value obtained in (4.5.12). Moreover, the
breaking length of the CCT is almost equal to the radius of the earth. The
definition of the breaking length itself is that the gravitational acceleration
g is a constant value on the ground, and something like a centrifugal force
is not, of course, contained.

When it comes to actually making the space elevator, we notice that
there are many problems to be solved. To avoid the effects of the wind, the
cable needs to be under considerable tension. The weight of the elevator
cage (climber) itself must also be considered, and so on. As for safety, to

ride with confidence, the specific strength must be at least 10 times more than what is obtained here. Besides, the CCT and CNT are still in the experimental stage, and the length is at most a few centimeters. When it comes to actually making this elevator, even if we have a relay point along the way, it is necessary to make a cable with a length of at least tens of thousands of kilometers seamlessly. How many years will it take to be able to make this kind of space elevator?

4.5.2 *Analysis when the thickness of the cable is varied*

If we want to make the space elevator a reality, we must give up the idea of using a cable of uniform thickness because, currently, there is nothing better than CNT or CCT. The only way to do this is to use a cable that becomes thicker as it goes up. The carbon bond state in the CNT is usually made of 6-membered rings. If a structure of 8-membered rings is incorporated, the tube can be branched. Therefore, the thickness of the cable can be varied.

We set the cross-sectional area, the mass density and the tension of the cable as S, ρ and T, respectively, as before. In the present case, however, not only the tension T, but also the cross-sectional area S vary depending on the location of the cable. When the coordinate x is taken toward the geostationary satellite with the center of the earth as the origin, we set the cross-sectional area and the tension of the cable at the point x as $S(x)$ and $T(x)$, respectively.

At this time, equation (4.5.2) of the tension shown in the previous subsection

$$\frac{1}{S}\frac{dT}{dx} = \frac{R_e^2 g \rho}{x^2} - \rho \omega^2 x \qquad (4.5.13)$$

is established as it is. In the present case, we assume that the stress generated in the cable should be a constant value σ_0 regardless of the height, that is,

$$\frac{T(x)}{S(x)} \equiv \sigma_0. \qquad (4.5.14)$$

Then, we analyze what kind of x-dependency the cross-sectional area $S(x)$ has. By using this expression, equation (4.5.13) becomes

$$\frac{1}{S}\frac{dS}{dx} = \frac{\rho}{\sigma_0}\left(\frac{R_e^2 g}{x^2} - \omega^2 x\right). \qquad (4.5.15)$$

From this equation, we eliminate ω by using (4.5.6) which expresses the balance between gravity and the centrifugal force acting on the geostationary satellite,

$$\frac{1}{S}\frac{dS}{dx} = \frac{\rho R_e^2 g}{\sigma_0}\left(\frac{1}{x^2} - \frac{x}{R_s^3}\right). \tag{4.5.16}$$

Here, R_s is the distance from the center of the earth to the geostationary satellite. Integrating this to find the cross-sectional area $S(x)$, we obtain

$$S(x) = S_e \exp\left[\frac{\rho R_e^2 g}{\sigma_0}\left(\frac{1}{R_e} - \frac{1}{x} + \frac{R_e^2 - x^2}{2R_s^3}\right)\right]. \tag{4.5.17}$$

Here, S_e is a constant of integration, which means the cross-sectional area on the ground. It can be seen from (4.5.16) and (4.5.17), the cross-sectional area increases as the height increases, and it reaches its maximum value at the location of the geostationary satellite.

In the following, we perform a numerical analysis based on this expression. Let the cross-sectional area at the geostationary satellite be $S_s = S(R_s)$. We find the ratio of S_s to the cross-sectional area S_e on the ground,

$$\frac{S_s}{S_e} = \exp\left[\frac{R_e g}{2(\sigma_0/\rho)R_s^3}(R_s - R_e)^2(2R_s + R_e)\right]. \tag{4.5.18}$$

The value of this ratio is determined by the value of σ_0/ρ. Here, CNT is assumed as the material of the cable and we write the specific strength of CNT as $(\sigma_y/\rho)_{\rm C}$. We estimate (4.5.18) numerically, in the case of $\sigma_0/\rho = (1, 1/2, 1/5$ and $1/10)(\sigma_y/\rho)_{\rm C}$. The result of these 4 cases are

$$\frac{S_s}{S_e} = \begin{cases} 2.8463, & \sigma_0/\rho = (\sigma_y/\rho)_{\rm C} \\ 8.1019, & \sigma_0/\rho = (\sigma_y/\rho)_{\rm C}/2 \\ 186.8416, & \sigma_0/\rho = (\sigma_y/\rho)_{\rm C}/5 \\ 34\,909.7924, & \sigma_0/\rho = (\sigma_y/\rho)_{\rm C}/10. \end{cases} \tag{4.5.19}$$

As the value of σ_0/ρ is set smaller, the ratio of cross-sectional areas increases exponentially. From the viewpoint of cable safety, the value of σ_0/ρ should be kept not more than $1/10$ of the specific strength of CNT. The cross-sectional area ratio in this case is about $35\,000$ times. This does not seem real at all.

Final Comments on the Space Elevator

There are too many problems with the construction of the space elevator. With regard to condition (3) mentioned at the beginning (the position of the geostationary satellite shall not vary, when the elevator is attached), since the geostationary satellite is subjected to a very large tension from the cable, if nothing is done, the satellite itself will fall to earth. To solve this problem, the cable must be extended from the satellite to a point in the sky further in the direction opposite to earth, so that the centrifugal force applied to it cancels out the tension in the direction of the earth. The length to be extended in this case can be obtained by setting expression (4.5.7) to be zero. For this reason, the length of the cable becomes longer. However, there is another solution to this problem: move the satellite to a position higher than the geosynchronous orbit and utilize the centrifugal force applied to the satellite body. Which of these solutions is better depends on how easily they can be constructed.

Even if such an elevator can be constructed, it will take a long time to reach the satellite. If the elevator cage can run at the same speed as a high-speed train at $300\,\mathrm{km/h}$, it will take as long as five days to travel the distance between the ground and the satellite $R_s - R_e \approx 35\,830\,\mathrm{km}$. In addition, near the earth, how do you prevent the cable from deterioration and vibration due to rain and wind? There are many other problems that need to be solved, such as lightning strikes, deterioration due to sunlight, acts of terrorism, and collisions with aircrafts, etc.

In any case, the material that makes the cable would not be possible except for carbon compound. New carbon-based materials similar to CNT and CCT will continue to be discovered. But it would not be expected that the specific strength increases dramatically. Considering that diamond which is the hardest substance in the world, is made of carbon, any other element would not do.

Sure, the space elevator is an interesting dream subject. However, in my opinion, unless there is a real revolutionary technological breakthrough in the next 100 years, the feasibility of constructing a space elevator is almost zero. The dream of constructing a bridge between earth and the moon will remain a dream.

Partial Differential Equations

In this chapter, we discuss some problems which can be formulated in terms of partial differential equations. At first, we deal with the simplest vibrating string problem. We analyze here the question of how a string vibrates when we pull up the center or at another point of the string and then release it. This problem seems to be an interesting one, because it is a little unexpected even for those who are familiar with physics. Next, we analyze what happens in the case of a string which has a certain degree of rigidity, for example the string in an actual instrument such as a piano. Finally, we deal with wave phenomena that occur in nature, using mathematical interpretations of wind ripples, sand dunes and wave phenomena on the highway.

5.1 Surprising Results in Vibrating String Problems

To begin with, we pose a question: suppose that there is a string whose both end points are fixed with some tension. If we pull up the middle point of this string and then release it, how will the string vibrate? Choose your answer from Figs. 5.1–5.3.

In each of the Figs. 5.1–5.3, it is assumed that time elapses in the order of (1), (2), (3) and (4). Figure 5.1 shows that the string retains the shape of a triangle even as its height decreases. Figure 5.2 shows that the height decreases just like in Fig. 5.1, but the string goes down in the form of a curve. Figure 5.3 shows that the height decreases and the string flattens from the top. However, the strings mentioned here do not have rigidity. So we think of them as threads or chains in reality. Those who can answer this question immediately will be those who have a very good physical sense.

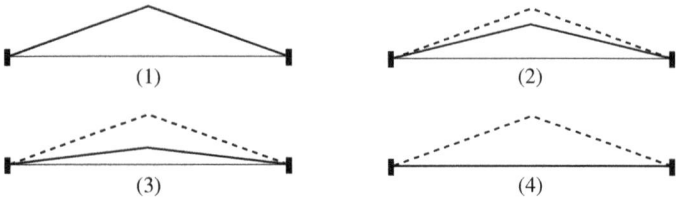

Fig. 5.1

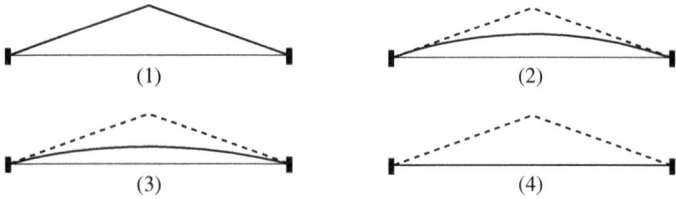

Fig. 5.2

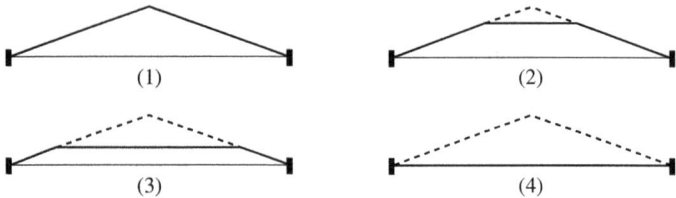

Fig. 5.3

The correct answer is Fig. 5.3. However, most people chose Fig. 5.1 or Fig. 5.2, while none chose Fig. 5.3.

It is easy to understand that Fig. 5.1 and Fig. 5.2 are incorrect by thinking as follows: when the string is pulled up the tension cancels out in the straight parts, and as a result, no lateral acceleration occurs in the string. In other words, even if there is a gradient, the second derivative is zero in the straight part, and therefore, considering the wave equation, the acceleration is also zero. That is, if it was stationary at first, this straight part does not start moving immediately. The first part that starts moving is only the part where it is released. This movement gradually propagates to the other parts over time and moves as a whole.

I noticed this problem when I made computer simulations of various physical phenomena to show my students. The result that this Fig. 5.3 was the correct answer was surprising even for myself.

As mentioned above, a string in physics means something that does not have rigidity. A real string in a musical instrument such as the violin, harp, or piano has a certain degree of rigidity, therefore, its vibration does not completely look like in Fig. 5.3. This will be discussed in a later section.

5.1.1 *Attempt to solve the equation*

Let us consider this problem by solving the **wave equation**. The vertical displacement of the string at the coordinate x and time t is denoted by $U(x, t)$. As is well known, the equation for the string vibration is written as

$$\frac{\partial^2 U(x, t)}{\partial t^2} = c^2 \frac{\partial^2 U(x, t)}{\partial x^2}, \quad c = \sqrt{\frac{T}{\rho}}. \tag{5.1.1}$$

Here, c is the **wave propagation velocity** which is defined by the second expression, where ρ is the linear density of the string and T is the strength of the tension. Furthermore, we assume that the center of the string is the coordinate origin, and we take the x-axis to the right along the string. Also, the string length is 2ℓ, therefore, x is assumed to be within the range of $[-\ell, \ell]$.

We solve this displacement $U(x, t)$ in a **separation of variables** with respect to x and t. Let us put it as

$$U(x, t) = X(x)P(t). \tag{5.1.2}$$

For this problem, the x-dependent part $X(x)$ must be an even function obviously, and the boundary condition $X(\pm \ell) = 0$ must be satisfied. Therefore the possible function of $X(x)$ is $\cos\left((k + \frac{1}{2})\pi x/\ell\right)$ where k is a non-negative integer. Substituting this into equation (5.1.1), the time-dependent part $P(t)$ is of the form $\cos\left((k + \frac{1}{2})\pi ct/\ell\right)$. Here, we have excluded a possible additional term of sine function, because the initial velocity is zero. The shape of the actual displacement $U(x, t)$ is represented by the **superposition** of the product of these functions

$$U(x, t) = \sum_{k=0}^{\infty} C_k \cos\left(\left(k + \frac{1}{2}\right)\frac{\pi x}{\ell}\right) \cos\left(\left(k + \frac{1}{2}\right)\frac{\pi ct}{\ell}\right). \tag{5.1.3}$$

The coefficient C_k is determined from the initial value of the string. Before carrying out this procedure, let the **vibration period** of each term of this expression be t_k, and its reciprocal be the **frequency** ν_k. Then we have

$$t_k = \frac{2\ell}{(k + \frac{1}{2})c}, \quad \nu_k = \frac{(k + \frac{1}{2})c}{2\ell}. \tag{5.1.4}$$

In particular, the vibration of $k = 0$ is called **fundamental vibration**, then, the period t_0 and the frequency ν_0 are

$$t_0 = \frac{4\ell}{c}, \quad \nu_0 = \frac{c}{4\ell}. \tag{5.1.5}$$

The vibrations of $k \geq 1$ are generally called **overtone vibrations** or **harmonic vibrations**, and their frequency is

$$\nu_k = (2k + 1)\nu_0, \quad (k = 1, 2, 3, \ldots). \tag{5.1.6}$$

It can be seen that this frequency is an odd-number multiple of the fundamental frequency. There is no even-number multiple vibration because $X(x)$ is restricted to even functions.

Next, let us determine the coefficient C_k. First, we assume that the middle point of the string is pulled up by a distance U_0, and the initial displacement $U(x, 0)$ is set as

$$F(x) = U_0 \frac{\ell - |x|}{\ell}. \tag{5.1.7}$$

Applying this to expression (5.1.3), we have

$$F(x) = \sum_{k=0}^{\infty} C_k \cos\left(\left(k + \frac{1}{2}\right)\frac{\pi x}{\ell}\right). \tag{5.1.8}$$

Multiplying both sides of this expression by $\cos\left((k' + \frac{1}{2})\pi x/\ell\right)$, integrating from $-\ell$ to ℓ, and using the orthogonality relation of the cosine function

$$\int_{-\ell}^{\ell} \cos\left(\left(k + \frac{1}{2}\right)\frac{\pi x}{\ell}\right) \cos\left(\left(k' + \frac{1}{2}\right)\frac{\pi x}{\ell}\right) dx = \ell\, \delta_{k,k'}, \tag{5.1.9}$$

the coefficient C_k is determined as

$$C_k = \frac{1}{\ell} \int_{-\ell}^{\ell} F(x) \cos\left(\left(k + \frac{1}{2}\right)\frac{\pi x}{\ell}\right) dx = \frac{2U_0}{\pi^2 (k + \frac{1}{2})^2}. \tag{5.1.10}$$

Returning this to expression (5.1.3), we obtain the solution

$$U(x, t) = \frac{2U_0}{\pi^2} \sum_{k=0}^{\infty} \frac{\cos\left((k + \frac{1}{2})\pi x/\ell\right) \cos\left((k + \frac{1}{2})\pi ct/\ell\right)}{(k + \frac{1}{2})^2}. \tag{5.1.11}$$

However, we cannot imagine how the strings actually move at first glance. By using here the product-to-sum formula for trigonometric functions, this solution is deformed into

$$U(x,t) = \frac{U_0}{\pi^2} \sum_{k=0}^{\infty} \frac{\cos\left((k+\frac{1}{2})\pi(x-ct)/\ell\right) + \cos\left((k+\frac{1}{2})\pi(x+ct)/\ell\right)}{(k+\frac{1}{2})^2}.$$

(5.1.12)

This means that it is separated into a right-propagating wave and a left-propagating wave. Here, we substitute C_k of (5.1.10) into (5.1.8), and replace the variable x with $y = x \pm ct$

$$F(y) = \frac{2U_0}{\pi^2} \sum_{k=0}^{\infty} \frac{\cos\left((k+\frac{1}{2})\pi y/\ell\right)}{(k+\frac{1}{2})^2}.$$

(5.1.13)

If we use this relation, $U(x,t)$ of (5.1.12) is rewritten as

$$U(x,t) = \frac{1}{2}\Big(F(x-ct) + F(x+ct)\Big).$$

(5.1.14)

When we compare this with the initial value (5.1.7), it shows that half of the initial displacement becomes the right-propagating wave and the other half becomes a left-propagating wave. This formula can be derived directly from the general solution of the wave equation without using the Fourier series solution described above.

Here, caution must be taken. If we put $t \to \infty$ in expression (5.1.14), both $F(x-ct)$ and $F(x+ct)$ become $-\infty$ from the definition of $F(x)$ (5.1.7). That is, something strange happens because it does not match the actual situation. Originally, $F(x)$ defined in (5.1.7) is applied only in the range $[-\ell, \ell]$. Therefore, (5.1.13) is an expression that holds in the same range $[-\ell, \ell]$ of the y from the process of its formation. However, the right-hand side of this expression is the periodic function of period 4ℓ with respect to y, and it is recognized that the expression is correct in the range $[-2\ell, 2\ell]$ of y. In fact, in numerically confirming this formula, both sides match within this range, and does not hold in the region beyond this range. Therefore, when $x - ct$ or $x + ct$ in (5.1.14) exceeds the range $[-2\ell, 2\ell]$, this formula cannot be used as it is.

In order to resolve this, we extend the function $F(y)$ to a periodic function. That is, for the range $[4n\ell - 2\ell, 4n\ell + 2\ell]$ of y, n being any integer, we put

$$F(y) \equiv F(y - 4n\ell).$$

(5.1.15)

This is the extension of $F(y)$ to a periodic function. Now we can use expression (5.1.14) for all time t.

Although it is the same thing, we may describe the above situation in the following way. We restrict the domain of the function $F(y)$ to $[-2\ell, 2\ell]$ by replacing y by $\langle y \rangle$:

$$y \mapsto y - 4\ell \left[\frac{y + 2\ell}{4\ell} \right] \overset{\text{def}}{=} \langle y \rangle, \qquad (5.1.16)$$

where [ ] denotes the Gauss symbol. As a result, expression (5.1.14) is modified as

$$U(x,t) = \frac{1}{2} \left[F(\langle x - ct \rangle) + F(\langle x + ct \rangle) \right]. \qquad (5.1.17)$$

Thus, the expression for $U(x,t)$ becomes elegant if it is written in terms of $\langle \ \rangle$. However, it will be cumbersome to classify the cases following this expression. Here, we make a program according to this formula and try numerical calculation. From this analysis, the shape shown in Fig. 5.3 is reproduced. The continuation of Fig. 5.3 is shown in Fig. 5.4. The speed at which the horizontal part of this figure descends is a constant value $U_0\, c/\ell$. Half of one cycle is executed up to this point, the direction is changed, and the original state is restored at the same speed.

Anyone who has learned physics in university should know that the vibrating string problem can be solved using the Fourier series. However, few people have tried to solve how strings vibrate by analyzing the initial value problem. This problem is rarely dealt with in ordinary textbooks, so we attempt to do it here. Again we realize there are many unexpected unknowns.

Here, we pull up the middle point of the string and release it. This problem is a relatively simple calculation, because the situation is very symmetric. However, if we pull up a part off the middle point of the string, the calculation is not so simple. In the next section, we will deal with such cases.

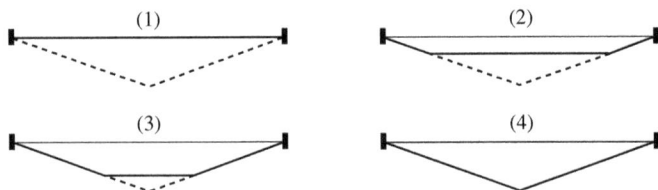

Fig. 5.4

5.2 Another Vibrating String Problem

First, we consider the following problem. We have a string, with both ends fixed and tensioned. If we pull up at a point $1/4$ length from the left end and then quietly release it without initial velocity, how will the string vibrate? Figure 5.5(1) shows the initial state, while Fig. 5.5(2)–(4) show the progress in the middle of the vibration.

Which of these, A, B, C or D is the correct vibration pattern?

A: by repeating (1) (2), (1) (2), ...
B: by repeating (1) (2) (3) (4), (1) (2) (3) (4), ...
C: by repeating (1) (2) (4) (3), (1) (2) (4) (3), ...
D: by repeating (1) (4), (1) (4), ...

Even a person who is familiar with physics may be perplexed. The correct answer is D: (1) (4), (1) (4), ...

5.2.1 *Attempt to solve the equation*

Let us consider this problem by solving the wave equation. As mentioned in the previous section, when we set the displacement in the direction perpendicular to the string as $U(x,t)$ at coordinate x and time t, the equation for string vibration is

$$\frac{\partial^2 U(x,t)}{\partial t^2} = c^2 \frac{\partial^2 U(x,t)}{\partial x^2}. \tag{5.2.1}$$

Here, we select the left end of the string as the coordinate origin O, and the x-axis is taken to the right along the string. The length of the string was 2ℓ in the previous case, but in the present case, we set it to be ℓ. Therefore, it is assumed that the range of the variable x is $[0, \ell]$.

The method to solve the equation is the same as in the previous section, but since the setting situation has changed, we show the steps again. First, we solve this lateral displacement $U(x,t)$ in the separation of variables with

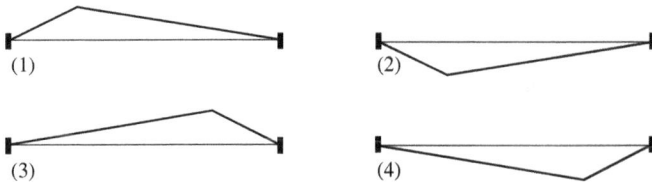

(1)

(2)

(3)

(4)

Fig. 5.5

respect to x and t,

$$U(x,t) = X(x)P(t). \tag{5.2.2}$$

5.2.2 *Case of both ends being fixed*

We consider the case in which both ends of the string are fixed. Then, the boundary condition of the x-dependent part $X(x)$ is given by

$$X(0) = 0, \quad X(\ell) = 0. \tag{5.2.3}$$

In this case, the function that can be a candidate of $X(x)$ is $\sin(k\pi x/\ell)$, with k as a natural number. Substituting this into the equation, the time-dependent part $P(t)$ is in the form $\cos(k\pi ct/\ell)$. Then, the sine function is excluded because the initial velocity is zero. The actual displacement $U(x,t)$ is represented by the superposition of the product of these functions

$$U(x,t) = \sum_{k=1}^{\infty} C_k \sin\left(\frac{k\pi x}{\ell}\right) \cos\left(\frac{k\pi ct}{\ell}\right). \tag{5.2.4}$$

The coefficient C_k is determined from the initial value of the string. Before that, let the vibration period of the k term of this expression be t_k, and its reciprocal be the vibration frequency ν_k. Thus we have

$$t_k = \frac{2\ell}{kc}, \quad \nu_k = \frac{kc}{2\ell}. \tag{5.2.5}$$

In particular, the vibration of $k = 1$ is called **fundamental vibration**. Then, the period t_1 and the frequency ν_1 are

$$t_1 = \frac{2\ell}{c}, \quad \nu_1 = \frac{c}{2\ell}. \tag{5.2.6}$$

The vibrations with $k \geq 2$ are generally called **overtone vibrations**, and the frequency at that time is given by

$$\nu_k = k\nu_1, \quad k = 2, 3, \ldots. \tag{5.2.7}$$

This is an integral multiple of the fundamental frequency.

Next, we determine the coefficient C_k. First, we define the function $F(x)$ which represents the initial shape of the string when the string is pulled up at the point x_0 by the distance U_0 as

$$F(x) = \begin{cases} \dfrac{U_0}{x_0}x, & 0 \leq x \leq x_0, \\[2mm] \dfrac{U_0}{\ell - x_0}(\ell - x), & x_0 \leq x \leq \ell. \end{cases} \tag{5.2.8}$$

Since this gives the initial value $U(x, 0)$ of the displacement, from expression (5.2.4), we have

$$F(x) = \sum_{k=1}^{\infty} C_k \sin\left(\frac{k\pi x}{\ell}\right).$$
(5.2.9)

Multiplying both sides by $\sin(k'\pi x/\ell)$, integrating with respect to x, and using the orthogonality of the sine function

$$\int_0^\ell \sin\left(\frac{k\pi x}{\ell}\right) \sin\left(\frac{k'\pi x}{\ell}\right) dx = \frac{\ell}{2}\delta_{k,k'},$$
(5.2.10)

the coefficient C_k is determined as

$$C_k = \frac{2}{\ell} \int_0^\ell F(x) \sin\left(\frac{k\pi x}{\ell}\right) dx = \frac{2\ell^2 U_0}{\pi^2 x_0(\ell - x_0)} \frac{\sin(k\pi x_0/\ell)}{k^2}.$$
(5.2.11)

Returning this to (5.2.4), we see that the solution $U(x, t)$ is

$$U(x, t) = \frac{2\ell^2 U_0}{\pi^2 x_0(\ell - x_0)} \sum_{k=1}^{\infty} \frac{\sin(k\pi x_0/\ell)}{k^2} \sin\left(\frac{k\pi x}{\ell}\right) \cos\left(\frac{k\pi ct}{\ell}\right).$$
(5.2.12)

Even if we look at this solution, we cannot imagine how the strings actually move. By using, here, the product-to-sum formula for trigonometric functions, this solution is deformed into

$$U(x, t) = \frac{\ell^2 U_0}{\pi^2 x_0(\ell - x_0)} \sum_{k=1}^{\infty} \frac{\sin(k\pi x_0/\ell)}{k^2}$$
$$\times \left[\sin\left(\frac{k\pi(x - ct)}{\ell}\right) + \sin\left(\frac{k\pi(x + ct)}{\ell}\right)\right].$$
(5.2.13)

This means that the displacement $U(x, t)$ is separated into a right-propagating wave and a left-propagating wave. Here, substituting C_k of (5.2.11) into (5.2.9), and replacing the variable x by $y = x \pm ct$,

$$F(y) = \frac{2\ell^2 U_0}{\pi^2 x_0(\ell - x_0)} \sum_{k=1}^{\infty} \frac{\sin(k\pi x_0/\ell)}{k^2} \sin\left(\frac{k\pi y}{\ell}\right).$$
(5.2.14)

By using this expression, the displacement $U(x, t)$ of (5.2.13) is rewritten as

$$U(x, t) = \frac{1}{2}\left[F(x - ct) + F(x + ct)\right].$$
(5.2.15)

When we compare this with the initial value (5.2.8), it shows that half of the initial displacement becomes the right-propagating wave, and the other half becomes a left-propagating wave. This expression can be directly derived from the general solution of the wave equation.

Caution is required here. The original function $F(x)$ of (5.2.8) is defined only in the range $[0, \ell]$. Therefore, if we replace x by $y = x \pm ct$, as in expression (5.2.15), it deviates from the original domain depending on the value of t. To solve this, we take advantage of the fact that the right-hand side of expression (5.2.14) is an odd function of y and is a periodic function with a period 2ℓ, and we expand the domain of the function $F(y)$ of the left-hand side to adapt to this situation. That is, for y in the range $[-\ell, 0]$, we put

$$F(y) \equiv -F(-y), \qquad (5.2.16)$$

and from this, the domain of y becomes $[-\ell, \ell]$. Furthermore, in order to give this function periodicity, we assume that the expression

$$F(y) \equiv F(y - 2n\ell), \qquad (5.2.17)$$

holds for the range $[2n\ell - \ell,\ 2n\ell + \ell]$ of y where n is an arbitrary integer. Now we can use expression (5.2.15) for y over all areas. Or, by leaving the domain of the odd function $F(y)$ as $[-\ell,\ \ell]$, we replace formally the independent variable y, by using the Gauss symbol [ ]

$$y \mapsto y - 2\ell\left[\frac{y + \ell}{2\ell}\right] \overset{\text{def}}{=} \langle y \rangle. \qquad (5.2.18)$$

With this definition, expression (5.2.15) is replaced by

$$U(x, t) = \frac{1}{2}\Big[F(\langle x - ct\rangle) + F(\langle x + ct\rangle)\Big]. \qquad (5.2.19)$$

Figures 5.6 and 5.7 show the numerical values of the displacement $U(x, t)$ according to expression (5.2.19). In these figures, let the coordinate x-axis be taken horizontally to the right, the time t-axis be diagonally upward and the displacement $U(x, t)$ be in the upward direction; these are drawn three-dimensionally. Figure 5.6 shows the case of $x_0 = \ell/2$, and Fig. 5.7 shows the case of $x_0 = \ell/4$. Figure 5.6 makes it easier to see how the values change over time. On the other hand, the change in Fig. 5.7 is a little difficult to understand. But the change is better understood by comparing Figs. 5.6 and 5.7 closely.

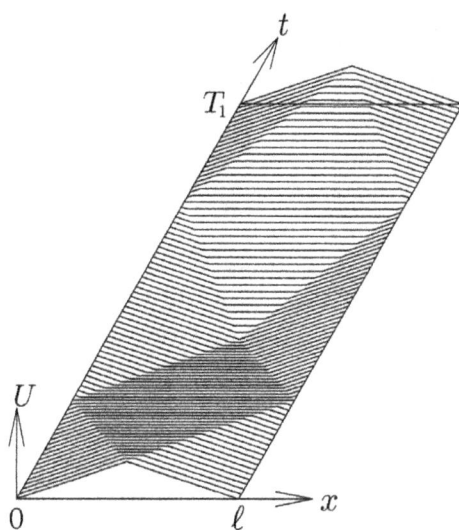

Fig. 5.6   Case of $x_0 = \ell/2$

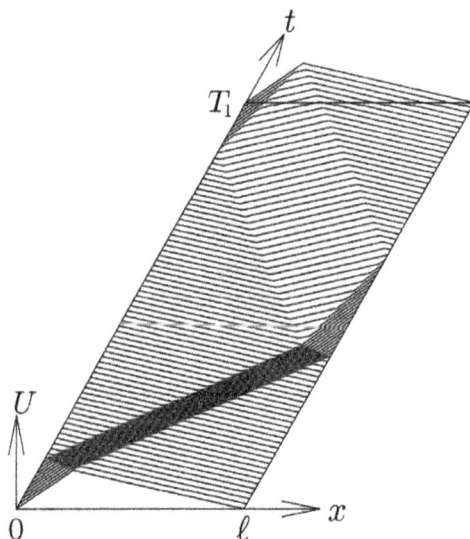

Fig. 5.7   Case of $x_0 = \ell/4$

5.2.3 *Case of left end being free*

We consider here the boundary condition where the left end of the string is free and the right end is fixed.[1] In this case, the boundary condition of the x-dependent part $X(x)$ is

$$\frac{dX(x)}{dx}\bigg|_{x=0} = 0, \quad X(\ell) = 0. \tag{5.2.20}$$

At this time, the function that can be a candidate for $X(x)$ is $\cos\left((k + \frac{1}{2})\pi x/\ell\right)$ with k as a non-negative integer. Substituting this into equation (5.2.1), the time-dependent part $P(t)$ becomes $\cos\left((k + \frac{1}{2})\pi ct/\ell\right)$. The displacement $U(x, t)$ is the superposition of the products of these functions, with D_k as an arbitrary constant,

$$U(x, t) = \sum_{k=0}^{\infty} D_k \cos\left(\left(k + \frac{1}{2}\right)\frac{\pi x}{\ell}\right) \cos\left(\left(k + \frac{1}{2}\right)\frac{\pi ct}{\ell}\right). \tag{5.2.21}$$

At this time, the period t_k and the frequency ν_k of the term k are

$$t_k = \frac{2\ell}{(k + \frac{1}{2})c}, \quad \nu_k = \frac{(k + \frac{1}{2})c}{2\ell}. \tag{5.2.22}$$

The **fundamental period** t_0 and the **fundamental frequency** ν_0 are

$$t_0 = \frac{4\ell}{c}, \quad \nu_0 = \frac{c}{4\ell}. \tag{5.2.23}$$

Furthermore, the **overtone frequency** ν_k with $k \geq 1$ is

$$\nu_k = (2k + 1)\nu_0, \quad k = 1, 2, \dots . \tag{5.2.24}$$

This is the odd multiple of the fundamental frequency ν_0.

The coefficient D_k in (5.2.21) is determined by the initial value of the string. We assume here that the point of $x = x_0$ is raised by the length U_0, and the initial value $U(x, 0)$ is set,

$$G(x) = \begin{cases} U_0, & 0 \leq x \leq x_0 \\ \dfrac{U_0}{(\ell - x_0)}(\ell - x), & x_0 \leq x \leq \ell. \end{cases} \tag{5.2.25}$$

---

[1] A physical way to make one end of the string a free end is to stand the rod in a direction perpendicular to the length of the string and to pass a ring with negligible mass through this rod and connect the end of the string to this ring. However, this ring must be able to move along the rod without friction. This can give tension to the strings, but their vertical component is zero. As a result, the derivative of the displacement becomes zero at this point.

The method is the same as in the previous section, but the form of the equation will change, so we will repeat the steps. Applying this to expression (5.2.21), we have

$$G(x) = \sum_{k=0}^{\infty} D_k \cos\left(\left(k + \frac{1}{2}\right)\frac{\pi x}{\ell}\right). \tag{5.2.26}$$

Multiplying both sides by $\cos\left((k' + \frac{1}{2})\pi x/\ell\right)$, integrating with x, and using the orthogonal relation of the cosine function

$$\int_0^\ell \cos\left(\left(k + \frac{1}{2}\right)\frac{\pi x}{\ell}\right)\cos\left(\left(k' + \frac{1}{2}\right)\frac{\pi x}{\ell}\right)dx = \frac{\ell}{2}\delta_{k,k'}, \tag{5.2.27}$$

we have the coefficient D_k,

$$D_k = \frac{2}{\ell}\int_0^\ell G(x)\cos\left(\left(k + \frac{1}{2}\right)\frac{\pi x}{\ell}\right)dx = \frac{2\ell U_0}{\pi^2(\ell - x_0)}\frac{\cos\left((k + \frac{1}{2})\pi x_0/\ell\right)}{(k + \frac{1}{2})^2}. \tag{5.2.28}$$

Substituting this into (5.2.21), we obtain the displacement $U(x,t)$ as

$$U(x,t) = \frac{2\ell U_0}{\pi^2(\ell - x_0)}\sum_{k=0}^{\infty}\frac{\cos\left((k + \frac{1}{2})\pi x_0/\ell\right)}{(k + \frac{1}{2})^2}$$

$$\times \cos\left(\left(k + \frac{1}{2}\right)\frac{\pi x}{\ell}\right)\cos\left(\left(k + \frac{1}{2}\right)\frac{\pi ct}{\ell}\right). \tag{5.2.29}$$

Here as before, using the product-to-sum formula of trigonometric functions, we rewrite the displacement $U(x,t)$ as

$$U(x,t) = \frac{\ell U_0}{\pi^2(\ell - x_0)}\sum_{k=0}^{\infty}\frac{\cos\left((k + \frac{1}{2})\pi x_0/\ell\right)}{(k + \frac{1}{2})^2}$$

$$\times\left[\cos\left(\left(k + \frac{1}{2}\right)\frac{\pi(x - ct)}{\ell}\right) + \cos\left(\left(k + \frac{1}{2}\right)\frac{\pi(x + ct)}{\ell}\right)\right]. \tag{5.2.30}$$

This is a form decomposed into a right-propagating wave and a left-propagating wave. Furthermore, by substituting D_k of (5.2.28) into (5.2.26), and changing the variable x to $y = x \pm ct$

$$G(y) = \frac{2\ell U_0}{\pi^2(\ell - x_0)}\sum_{k=0}^{\infty}\frac{\cos\left((k + \frac{1}{2})\pi x_0/\ell\right)}{(k + \frac{1}{2})^2}\cos\left(\left(k + \frac{1}{2}\right)\frac{\pi y}{\ell}\right). \tag{5.2.31}$$

Using this function G, expression (5.2.30) is rewritten as

$$U(x,t) = \frac{1}{2}\big[G(x - ct) + G(x + ct)\big]. \tag{5.2.32}$$

Here we extend the domain of the function G as before. In this case, noting that the right-hand side of (5.2.31) is antisymmetric with respect to the point of $y = \ell$, we define for y

$$G(y) = -G(2\ell - y), \tag{5.2.33}$$

in the range $[\ell, 2\ell]$. This expression extends the domain of the function G to $[0,\ 2\ell]$. Then, using the fact that the right-hand side of (5.2.31) is an even function, we set for y in the range $[-2\ell,\ 0]$,

$$G(y) = G(-y). \tag{5.2.34}$$

From this, we can expand the domain of G to $[-2\ell,\ 2\ell]$. Finally, since the right-hand side of (5.2.31) is a periodic function with a period of 4ℓ, we can also expand the function G to the periodic one. That is, we set for y in the range $[4n\ell - 2\ell,\ 4n\ell + 2\ell]$ with an arbitrary integer n,

$$G(y) = G(y - 4n\ell). \tag{5.2.35}$$

As a result, the domain is expanded to the entire interval of y, and now, expression (5.2.32) can be used for all time t.

Or, as we did in the previous expression (5.2.18), we keep the domain of the function G as $[-2\ell,\ 2\ell]$, we replace the variable y, using the Gauss symbol $[\ ]$,

$$y \mapsto y - 4\ell\left[\frac{y + 2\ell}{4\ell}\right] \stackrel{\text{def}}{=} \langle y\rangle. \tag{5.2.36}$$

Therefore, expression (5.2.32) becomes

$$U(x,t) = \frac{1}{2}\big[G(\langle x - ct\rangle) + G(\langle x + ct\rangle)\big]. \tag{5.2.37}$$

This is an expression that can be used for all time t. The numerical values which are obtained based on this expression are shown in Figs. 5.8 and 5.9 below. These figures correspond to the case of $x_0 = \ell/2$ and the case of $x_0 = \ell/4$, respectively. There are some *moiré* patterns that make the picture difficult to see clearly. It is better to enlarge the picture to see it in detail.

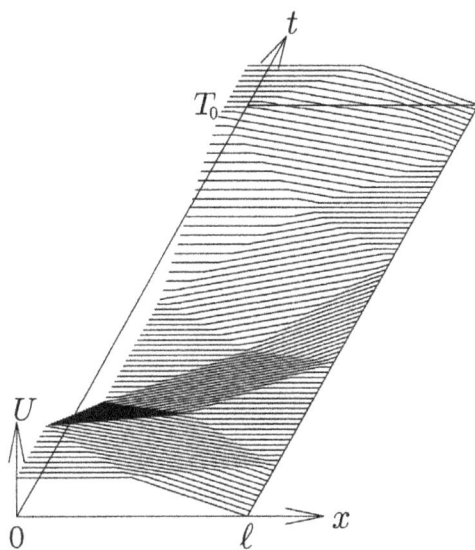

Fig. 5.8   Case of $x_0 = \ell/2$

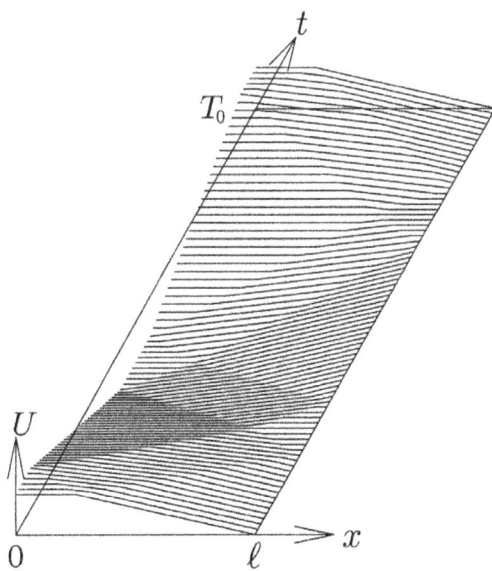

Fig. 5.9   Case of $x_0 = \ell/4$

Saxophone and Clarinet

When I retired, I learned to play the saxophone to pass time. My teacher teaches both the saxophone and the clarinet. When I play the saxophone, my teacher plays along with me on his clarinet. However, this is a very difficult thing to do because the saxophone and clarinet have completely different pitches even if the same [Do] sound is played. I am always impressed by how my teacher is able to transpose the score in his head in order to produce the same pitch on his clarinet to match that of my saxophone.

Here, we describe the physical difference between a saxophone and a clarinet, both wind instruments. These instruments have a "reed" shaped like a spatula attached to the mouthpiece. By vibrating the reed, the air inside the tube resonates and makes a sound. Both are reed instruments, but their sound mechanisms are completely different. The reeds on the mouthpieces of both the saxophone and the clarinet stay closed most of the time, even when they are vibrating. Therefore, in this sense, the mouthpieces are closed ends (fixed ends in terms of string vibration). However, there is a great difference between these two instruments. The saxophone has a tube that expands in a conical shape from the mouthpiece to the opposite side; on the other hand, the clarinet has an almost straight tube.

In the saxophone, the other end of the tube becomes a vibration node because the air vibration diminishes there. That is, although the tube is open, the effect is the same as the closed end. In other words, the saxophone is an instrument with a closed mouthpiece and a closed other end. However, the saxophone is classified as an "open end tube". This is because the overtone outputs are the same as other instruments with an open mouthpiece and an open other end, such as a trumpet or a trombone.

On the other hand, in the clarinet, the opposite end of the tube is the antinode of air vibration. It literally becomes the open end (the free end in terms of string vibration). In other words, the clarinet has a closed mouthpiece and an open other end. Such an instrument is called a "closed end tube". The only wind instrument of Western music which is classified as a closed end tube is the clarinet.

Generally in music, the interval between a sound with a certain frequency and a sound with twice the frequency is called an octave. The saxophone has the "octave key". For example, if the octave key is pressed while playing the [Do] sound, the sound becomes a [Do] sound which is an octave higher, having twice the frequency of the original one. In this sense, the overtone at this time follows expression (5.2.7). On the other hand, this is not the case in the clarinet, because the clarinet is a closed and open end instrument. The clarinet has no octave key; instead it has a "register key". By pressing this key, the sound becomes an overtone. However, this overtone is not one octave higher, but a sound with a frequency that is tripled according to expression (5.2.24). For example, if the register key is pressed while playing the [Do] sound, the triple overtone sound becomes a [So] sound one octave higher. The reason is that "3" is represented by using exponents as $2^{\log 3/\log 2}$, and this value is approximately equal to the value $2^{1+7/12}$ which is the frequency of the [So] sound one octave higher. In fact, the values of these exponents become $\log 3/\log 2 = 1.5849\cdots$ and $1+7/12 = 1.5833\cdots$ (see the description below).

There is another difference between a saxophone and a clarinet. For the saxophone, the fundamental frequency is $c/(2\ell)$ as shown by (5.2.6), and for the clarinet, it is $c/(4\ell)$ as shown by (5.2.23). This means that the clarinet produces a sound that is twice as low as the saxophone, if they have the same tube length. In fact, since a soprano saxophone (a higher-register variety of the saxophone) and a clarinet have almost the same length, the clarinet can produce a sound twice as low as the soprano saxophone.

Let me describe this a little more. Before learning music, I thought that the frequency of the scale of music, the so-called [Do, Re, Mi, Fa, So, La, Ti, Do] would rise uniformly. But this is not correct. If the frequency of [Do] is ν and writing this as [Do] $= \nu$, the frequency of each note, in the currently used "equal temperament" system, is

$$[\text{Do}] = \nu, \quad [\text{Re}] = 2^{2/12}\nu, \quad [\text{Mi}] = 2^{4/12}\nu, \quad [\text{Fa}] = 2^{5/12}\nu,$$

$$[\text{So}] = 2^{7/12}\nu, \quad [\text{La}] = 2^{9/12}\nu, \quad [\text{Ti}] = 2^{11/12}\nu, \quad [\text{Do}] = 2^{12/12}\nu = 2\nu.$$

The frequency increases in a geometric sequence by the ratio $2^{2/12}$ from [Do] to [Mi]. But, it only increases by the ratio $2^{1/12}$ from

[Mi] to [Fa]. Then it increases by the ratio $2^{2/12}$ for [Fa] to [Ti], before it increases by the ratio $2^{1/12}$ again from [Ti] to [Do]. The increase in the ratio of frequencies by $2^{2/12} (= 1.122 \cdots)$ is called a "whole tone", while the increase by $2^{1/12} (= 1.059 \cdots)$ is called a "half tone". In this way, there are places where the scale increases by a whole tone and places where it increases by a half tone. However, in spite of this, when expressing the scale note in a score, it is written as if it increases uniformly. When I studied music for the first time, I wondered about this irrational and inconvenient notation; I thought there may be some historical reason for it.

In the case of wind instruments, the most troublesome thing is that the pitch of the same [Do] sounds are different depending on the instruments, and they do not match the pitch of a piano. Therefore, when playing the saxophone with piano accompaniment, either the saxophonist or the piano player must transpose the score to match the sound pitch. There is another reason to transpose the score. The piano has a range of seven octaves. By comparison, the saxophone typically has a range of only two and a half octaves. There are some good professional saxophonists who can use the *fradio* playing method to extend the range to four octaves. But, this is very difficult for an amateur. Therefore, depending on the score, it will be impossible to play as it is, so the whole or a part of the score must be transposed. This transposition is not just a parallel translation, because there is a half tone in the middle of the scale. A sharp ♯ or a flat ♭ is used to raise or lower by a half tone. Professional musicians are used to this and will not find it inconvenient. But, it can be a source of confusion for amateurs, wondering where and how they will be attached. At the beginning of my learning the saxophone, I thought about composing a completely new score by myself which represents the 12 scales of all half tones. But later, I learned that there is already a "chromatic notation" based on this idea which does not use any sharp ♯ nor flat ♭. However, even I, who is stubborn, is gradually getting used to transposing scores after all these years. Although this story has veered away from mathematics, I felt it would be interesting to look at music from a physicist's viewpoint.

5.3 String Vibration with Rigidity

In the previous section, we dealt with how the string deforms with time when the string is vibrating. The string discussed there was thought to be like a chain with no rigidity at all. In fact, the strings treated in mathematical physics usually do not have rigidity. However, since the string of stringed instruments such as the piano, harp, and violin has some non-zero rigidity, so we look at whether actual string vibration would have the vibration mode as described in the previous section. Here, we will consider how the string vibrates when it has rigidity.

5.3.1 *Introduction to the equation*

As is well known, when we put the displacement in the direction perpendicular to the string as $U(x,t)$ at coordinate x and time t, the equation for string vibration is

$$S\rho\frac{\partial^2 U(x,t)}{\partial t^2} = T\frac{\partial^2 U(x,t)}{\partial x^2}. \tag{5.3.1}$$

Here, S, ρ and T are the cross-sectional area, volume density and tension of the string, respectively. In particular, the product $S\rho$ represents the linear density of the string. On the other hand, there is the **Bernoulli–Euler model** as an equation that expresses the perpendicular vibration of a rod that has rigidity without tension; it is written as

$$S\rho\frac{\partial^2 U(x,t)}{\partial t^2} = -EI\frac{\partial^4 U(x,t)}{\partial x^4}. \tag{5.3.2}$$

Here, E and I are **Young's modulus** and the **moment of inertia of area**, respectively. From these two equations, the equation when both tension and rigidity are contained is supposed to be

$$S\rho\frac{\partial^2 U(x,t)}{\partial t^2} = T\frac{\partial^2 U(x,t)}{\partial x^2} - EI\frac{\partial^4 U(x,t)}{\partial x^4}. \tag{5.3.3}$$

However, this equation is not in great demand, and does not appear in ordinary textbooks, nor on the Internet. Although the method of deriving this equation is not described here, it is recognized as a correct one within the linear approximation.

5.3.2 *Eigenvalues and eigenfunctions*

To solve equation (5.3.3), the displacement $U(x,t)$ is assumed to be a separation type with the variables x and t,

$$U(x,t) = X(x)P(t). \tag{5.3.4}$$

Substituting this into the equation and dividing it appropriately,

$$\frac{1}{S\rho X}\left[T\frac{d^2X}{dx^2} - EI\frac{d^4X}{dx^4}\right] = \frac{1}{P}\frac{d^2P}{dt^2}. \tag{5.3.5}$$

Since this left-hand side is a function of only x and the right-hand side is a function of only t, this equality means the value of this expression must be a constant. If we set this value as $-\omega^2$, it becomes two separate equations,

$$EI\frac{d^4X}{dx^4} - T\frac{d^2X}{dx^2} - S\rho\omega^2X = 0, \qquad \frac{d^2P}{dt^2} + \omega^2P = 0. \tag{5.3.6}$$

To begin with, we solve this first equation. Here, setting $X(x) = e^{ikx}$ so as to satisfy the equation, we have

$$EIk^4 + Tk^2 - S\rho\omega^2 = 0. \tag{5.3.7}$$

This is conjectured to be the fourth-degree equation of k, but in reality it is a quadratic one of k^2 and can be easily solved. Here we define two positive constants μ, ν depending on ω as

$$\mu = \sqrt{\frac{\sqrt{T^2 + 4EIS\rho\omega^2} - T}{2EI}}, \qquad \nu = \sqrt{\frac{\sqrt{T^2 + 4EIS\rho\omega^2} + T}{2EI}}. \tag{5.3.8}$$

Using these constants, the solutions of equation (5.3.7) become

$$k = \pm\mu, \quad k = \pm i\nu. \tag{5.3.9}$$

From the linearity of the equation, the linear combinations of e^{ikx} with these values k substituted become the solutions. If they are expressed in the form of real functions, we have the solutions $\cos(\mu x)$, $\sin(\mu x)$, $\cosh(\nu x)$ and $\sinh(\nu x)$. Similarly, the solutions of the time-dependent part $P(t)$ are $\cos(\omega t)$ and $\sin(\omega t)$ from the second equation of (5.3.6).

Here, as the initial condition, the string is pulled up symmetrically around the center of the string and released quietly without the initial velocity. Under this initial condition, the displacement $U(x,t)$ should obviously remain symmetrical when viewed from the center of the string, even when it is vibrating. This means that if the middle point of the string is chosen as the origin of the x-axis, the x-dependent part $X(x)$ will be an even function. Therefore, $X(x)$ is written with A and B as arbitrary constants,

$$X(x) = A\cos(\mu x) + B\cosh(\nu x). \tag{5.3.10}$$

Furthermore, since the initial velocity is set to zero, the time-dependent part $P(t)$ is suitable for $\cos(\omega t)$. Now the displacement $U(x, t)$ is represented as

$$U(x, t) = X(x)P(t) = \big[A\cos(\mu x) + B\cosh(\nu x)\big]\cos(\omega t). \qquad (5.3.11)$$

Next, let the string length be 2ℓ, and therefore the range of x is $[-\ell, \ell]$. Assuming that the end $x = \ell$ is an embedded fixed end, we set

$$X(\ell) = 0, \qquad \frac{dX(x)}{dx}\bigg|_{x=\ell} = 0. \qquad (5.3.12)$$

These are the boundary conditions and applying these two conditions to (5.3.10), we have

$$A\cos(\mu\ell) + B\cosh(\nu\ell) = 0, \qquad -A\mu\sin(\mu\ell) + B\nu\sinh(\nu\ell) = 0. \qquad (5.3.13)$$

In order for both A and/or B not to be zero, the coefficient determinant of these equations must be zero, that is, we have an equation

$$\begin{vmatrix} \cos(\mu\ell) & \cosh(\nu\ell) \\ -\mu\sin(\mu\ell) & \nu\sinh(\nu\ell) \end{vmatrix} = 0,$$

therefore,

$$\cos(\mu\ell)\tanh(\nu\ell) + \frac{\mu}{\nu}\sin(\mu\ell) = 0. \qquad (5.3.14)$$

From now on, the value of ω contained in μ and ν will be determined as the **eigenvalue**. In this sense, this equation is the **eigenvalue equation**. Then, ω will be decided in a discrete manner. Let it be ω_k $(k = 1, 2, 3, \ldots)$ from the smallest positive value, and the corresponding μ and ν are designated by μ_k and ν_k, respectively.

In the following, the values of A and B are set so that the first expression of (5.3.13) is satisfied,

$$A = 1, \quad B = -\frac{\cos(\mu\ell)}{\cosh(\nu\ell)}. \qquad (5.3.15)$$

We define the function $X(x)$ of (5.3.10) with this selection as the **eigenfunction**,

$$F_k(x) = \cos(\mu_k x) - \frac{\cos(\mu_k \ell)}{\cosh(\nu_k \ell)}\cosh(\nu_k x). \qquad (5.3.16)$$

However, this is not yet normalized. The general solution of equation (5.3.3) is the superposition of these eigenfunctions using the arbitrary constant C_k,

$$U(x,t) = \sum_{k=1}^{\infty} C_k F_k(x) \cos(\omega_k t). \tag{5.3.17}$$

5.3.3 *Orthogonality and normalization of eigenfunctions*

The eigenfunction $F_k(x)$ defined in (5.3.16) satisfies the first equation of (5.3.6), therefore, we have

$$-S\rho\,\omega_k{}^2 F_k = T\frac{d^2 F_k}{dx^2} - EI\frac{d^4 F_k}{dx^4}. \tag{5.3.18}$$

We consider another equation where the eigenvalue number is changed from k to k'

$$-S\rho\,\omega_{k'}{}^2 F_{k'} = T\frac{d^2 F_{k'}}{dx^2} - EI\frac{d^4 F_{k'}}{dx^4}. \tag{5.3.19}$$

Multiplying equation (5.3.18) by $F_{k'}$, equation (5.3.19) by F_k, and subtracting both sides, we have

$$
\begin{aligned}
-S\rho(\omega_k{}^2 - \omega_{k'}{}^2)F_k F_{k'} = \frac{d}{dx}\Big[& T\big(F_{k'} F_k{}' - F_k F_{k'}{}'\big) \\
& - EI\big(F_{k'} F_k{}''' - F_k F_{k'}{}'''\big) \\
& + EI\big(F_{k'}{}' F_k{}'' - F_k{}' F_{k'}{}''\big)\Big].
\end{aligned} \tag{5.3.20}
$$

Here, the prime represents the derivative. Integrating both sides of this expression from $-\ell$ to ℓ with x

$$
\begin{aligned}
-S\rho(\omega_k{}^2 - \omega_{k'}{}^2) & \int_{-\ell}^{\ell} F_k F_{k'}\, dx \\
& = \Big[T\big(F_{k'} F_k{}' - F_k F_{k'}{}'\big) - EI\big(F_{k'} F_k{}''' - F_k F_{k'}{}'''\big) \\
& \quad + EI\big(F_{k'}{}' F_k{}'' - F_k{}' F_{k'}{}''\big)\Big]_{-\ell}^{\ell}.
\end{aligned} \tag{5.3.21}
$$

Since the eigenfunctions F_k, $F_{k'}$ and their first-order derivatives become zero at $x = \pm\ell$, the right-hand side of this expression is zero. Therefore, when $\omega_k \neq \omega_{k'}$ on the left-hand side, we obtain,

$$\int_{-\ell}^{\ell} F_k(x) F_{k'}(x)\, dx = 0. \tag{5.3.22}$$

This gives the orthogonality of the eigenfunctions. When $\omega_k = \omega_{k'}$, we can use directly the integration formulas

$$\int \cos^2(ax)dx = \frac{x}{2} + \frac{1}{2a}\sin(ax)\cos(ax),$$

$$\int \cosh^2(bx)dx = \frac{x}{2} + \frac{1}{2b}\sinh(bx)\cosh(bx),$$

(5.3.23)

$$\int \cos(ax)\cosh(bx)dx = \frac{1}{a^2+b^2}\left[b\cos(ax)\sinh(bx) + a\sin(ax)\cosh(bx)\right].$$

(5.3.24)

Summarizing these results, the **orthogonal relation** of the eigenfunctions is obtained,

$$\int_{-\ell}^{\ell} F_k(x)F_{k'}(x)dx = N_k^2 \delta_{k,k'},$$

(5.3.25)

where the **normalization constant** N_k^2 is defined as

$$N_k^2 = \ell\left[1 + \frac{\cos^2(\mu_k\ell)}{\cosh^2(\nu_k\ell)}\right] + \frac{1}{\mu_k}\sin(\mu_k\ell)\cos(\mu_k\ell) + \frac{1}{\nu_k}\tanh(\nu_k\ell)\cos^2(\mu_k\ell).$$

(5.3.26)

From now on, $F_k(x)/N_k$ becomes a **normalized eigenfunction**.

5.3.4 *Initial value problems*

As the simplest method, we assume that the string is pulled up at the center by U_0 and released without velocity. Let us find out what happens to the string shape $U(x,0)$. The equation is time-independent, therefore what needs to be solved is the equation where the time derivative is set to zero in (5.3.3),

$$T\frac{\partial^2 U(x,0)}{\partial x^2} - EI\frac{\partial^4 U(x,0)}{\partial x^4} = 0.$$

(5.3.27)

Suppose that $U(x,0) = e^{ikx}$, then we have

$$Tk^2 + EIk^4 = 0.$$

(5.3.28)

We define here the constant λ having the dimension of the reciprocal of length

$$\lambda = \sqrt{\frac{T}{EI}},$$

(5.3.29)

and the solution of (5.3.28) is

$$k = 0 \text{ (multiple root)}, \quad k = \pm i\lambda. \tag{5.3.30}$$

From this, $U(x,0)$ is written with arbitrary constants a, b, c, d,

$$U(x,0) = a + b|x| + c\cosh(\lambda x) + d\sinh(\lambda|x|). \tag{5.3.31}$$

The absolute value of x is given to make it an even function. Furthermore, we add the boundary conditions,

$$U(0,0) = U_0, \quad \left.\frac{dU(x,0)}{dx}\right|_{x=0} = 0, \quad U(\ell,0) = 0, \quad \left.\frac{dU(x,0)}{dx}\right|_{x=\ell} = 0. \tag{5.3.32}$$

Notice that the derivative at $x = 0$ is set to zero. This is because the equation with rigidity does not assume that the derivative will be discontinuous and it will be in a sharply bent shape. Therefore, in the case of an even function, the derivative must inevitably be zero at the point $x = 0$. At the point $x = \ell$, the embedded fixed end is used as before. From these conditions, we have

$$a + c = U_0, \quad b + d\lambda = 0,$$
$$a + b\ell + c\cosh(\lambda\ell) + d\sinh(\lambda\ell) = 0, \tag{5.3.33}$$
$$b + c\lambda\sinh(\lambda\ell) + d\lambda\cosh(\lambda\ell) = 0.$$

Solving these equations for the coefficients,

$$a = \frac{\lambda\ell\sinh(\lambda\ell) - \cosh(\lambda\ell) + 1}{\lambda\ell\sinh(\lambda\ell) - 2\cosh(\lambda\ell) + 2}U_0,$$

$$b = \frac{-\lambda\sinh(\lambda\ell)}{\lambda\ell\sinh(\lambda\ell) - 2\cosh(\lambda\ell) + 2}U_0,$$

$$c = \frac{1 - \cosh(\lambda\ell)}{\lambda\ell\sinh(\lambda\ell) - 2\cosh(\lambda\ell) + 2}U_0, \tag{5.3.34}$$

$$d = \frac{\sinh(\lambda\ell)}{\lambda\ell\sinh(\lambda\ell) - 2\cosh(\lambda\ell) + 2}U_0.$$

Substituting these solutions into (5.3.31) and arranging it using the addition theorem, we obtain the initial value

$$U(x,0) = \frac{U_0}{\lambda\ell\sinh(\lambda\ell) - 2\cosh(\lambda\ell) + 2}\Big[\lambda(\ell - |x|)\sinh(\lambda\ell) + 1 - \cosh(\lambda\ell)$$
$$+ \cosh(\lambda x) - \cosh\big(\lambda(\ell - |x|)\big)\Big]. \tag{5.3.35}$$

5.3.5 *Final solution*

To find the final solution, we need to find the undetermined constant C_k in (5.3.17). To do this, when we put $t = 0$ in this expression, it should match $U(x, 0)$ in (5.3.35), that is,

$$\sum_{k=1}^{\infty} C_k F_k(x) = U(x, 0). \tag{5.3.36}$$

Multiplying both sides by $F_{k'}(x)$, integrating with respect to x and using the orthogonal relation of the eigenfunction (5.3.25), we can find the coefficient C_k as

$$C_k = \frac{1}{N_k^2} \int_{-\ell}^{\ell} F_k(x) U(x, 0) dx. \tag{5.3.37}$$

On returning this expression to (5.3.17), the displacement $U(x, t)$ is ultimately solved:

$$U(x, t) = \sum_{k=1}^{\infty} \frac{1}{N_k^2} \left(\int_{-\ell}^{\ell} F_k(x') U(x', 0) dx' \right) F_k(x) \cos(\omega_k t). \tag{5.3.38}$$

Of course, the integral contained in this expression is feasible, but the result is too cumbersome to present here.

5.3.6 *Solution when EI is infinitesimally small*

When the limit of $EI \to 0$ is taken in the solution obtained here, it must be confirmed whether it matches the solution when the rigidity does not exist. The parameters μ, ν defined in (5.3.8) have the relationship

$$\mu\nu = \sqrt{\frac{S\rho}{EI}}\omega. \tag{5.3.39}$$

Here, the wave propagation velocity of the normal wave equation (5.3.1) is defined as

$$c = \sqrt{\frac{T}{S\rho}}. \tag{5.3.40}$$

By using λ of (5.3.29), the relation (5.3.39) is rewritten as

$$\mu\nu = \frac{\omega}{c}\lambda. \tag{5.3.41}$$

Up to this point, the expression is strictly valid. Here, we assume that EI is sufficiently small, and use the asymptotic forms of μ, ν of (5.3.8),[2]

$$\mu \simeq \frac{\omega}{c}, \quad \nu \simeq \lambda, \tag{5.3.42}$$

and the relation (5.3.41) is reproduced beautifully from these expressions. Also, in the limit of $EI \to 0$, it becomes $\nu \to \infty$. The eigenvalue equation (5.3.14) becomes $\cos(\mu\ell) = 0$, that is, $\cos(\omega\ell/c) = 0$. From now on, the eigenvalue ω_k will be taken with a non-negative integer k.

$$\omega_k = c(k + \tfrac{1}{2})\frac{\pi}{\ell}, \quad k = 0, 1, 2, \ldots. \tag{5.3.43}$$

This result is in agreement with the one obtained directly from equation (5.3.1) where there is no rigidity. Also, this ω_k divided by 2π agrees with the ν_k of (5.1.4).

5.3.7 *Graph display by numerical analysis*

Here, we will perform a numerical analysis on a harp string. Harp strings can be made from nylon or steel. Here, we will use nylon string for our analysis. However, we do not have the equipment to obtain the values of the various parameters. Let the cross-section of the string be a disc with a radius r, then the cross-sectional area S and the moment of inertia of area I are, respectively,

$$S = \pi r^2, \quad I = 4\int_0^r x^2\sqrt{r^2 - x^2}\,dx = \frac{\pi}{4}r^4. \tag{5.3.44}$$

These values are determined only by the radius r.

Here, we take the central [La] sound string as an example. Its length is 64 cm, therefore, half of that is $\ell = 0.32$ m, and the radius of the cross-section is $r = 0.75$ mm $= 0.00075$ m. From the Internet, we adopt the volume density ρ as 1.14×10^3 kg/m^3, and Young's modulus E as 7 GPa. The value of this E varies considerably depending on the type of nylon. Harp strings are fairly stiff, so the larger value is taken. The problem is how to take the tension T. We could not find any literature for its value; therefore, we had no choice but to work backwards from a known value. The frequency of the central [La] sound is typically 440 Hz. Therefore, setting ω

---

[2]Strictly speaking, it should be said that $EIS\rho\omega^2$ is sufficiently smaller than T^2.

contained in μ, ν of expression (5.3.8) to $\omega = 2\pi \times 440$, the eigenvalue equation (5.3.14) is solved for the tension T contained as an unknown number. Once T is determined, the rest of the eigenvalues can be determined:

$$S\rho = 2.0145 \times 10^{-3}\,\text{kg/m}, \quad EI = 1.7395 \times 10^{-3}\,\text{N} \cdot \text{m}^2,$$

$$T = 632.3536\,\text{N}. \tag{5.3.45}$$

What is surprising about this number is that the tension T of one string is about the weight of an adult. It is no wonder that the harp nylon strings sometimes break.

Based on these data, the eigenvalue equation (5.3.14) is numerically solved to obtain 10 ω's, which are divided by 2π to get the frequencies

$$\frac{\omega_k}{2\pi} = 440.000,\ 1320.350,\ 2201.752,\ 3084.903,\ 3970.502,\ 4859.240,$$

$$5751.808,\ 6648.890,\ 7551.164,\ 8459.301. \tag{5.3.46}$$

The frequency of the fundamental wave $\omega_1/2\pi$ becomes just 440 Hz, because the tension T was decided so that it would be this value. Subsequent frequencies are approximately three times, five times, and odd number times the fundamental frequency. When the rigidity is added, it is no longer exactly an odd number multiple. Try to find the sequence of differences between these values,

$$\frac{\omega_{k+1} - \omega_k}{2\pi} = 880.3504,\ 881.4016,\ 883.1516,\ 885.5984,\ 888.7387,$$

$$892.5682,\ 897.0818,\ 902.2736,\ 908.1368. \tag{5.3.47}$$

These differences will increase for a while. This is a feature different from expression (5.3.43) when there is no rigidity.

When the eigenvalues are found, we can find the eigenfunctions $F_k(x)$ of (5.3.16), the normalization constants $N_k{}^2$ of (5.3.26), initial waveform $U(x,0)$ of (5.3.35) and the linear combination coefficients C_k of (5.3.37). Finally, the displacement $U(x,t)$ at an arbitrary time of expression (5.3.38) is obtained. In this case, the integral contained in C_k of (5.3.37) is carried out by numerical calculation, where the initial value $U(x,0)$ is decided from (5.3.35).

Figure 5.10 shows a graph of displacement $U(x,t)$. This figure shows the first four frames which are divided into 12 equal parts of one period of the fundamental wave. That is, let $1/440\,\text{sec} \equiv t_1$, the parts (1), (2), (3) and (4) in the figure correspond to the time $t = t_1 \times (0,\ 1,\ 2$ and $3)/12$, respectively.

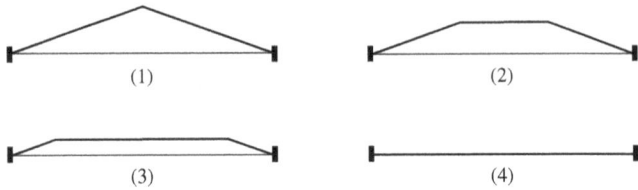

Fig. 5.10

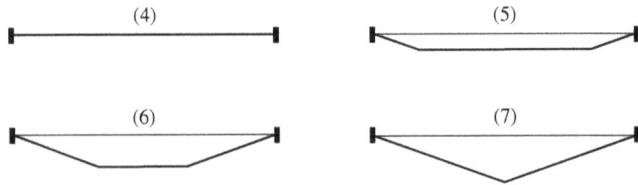

Fig. 5.11

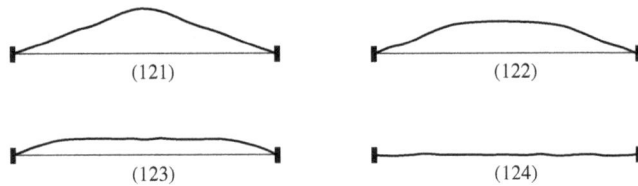

Fig. 5.12

As we can see, this is similar to the figure with no rigidity shown in §5.1, the shape of the string becomes flat from the top. Despite the analysis that takes rigidity into account, the effect is not noticeable. This is because the value of the coefficient EI, which represents the rigidity in equation (5.3.3), is too small. Figure 5.11 shows the graph when the time has passed.

In this figure, each of (4), (5), (6) and (7) corresponds to time $t = t_1 \times (3,\ 4,\ 5$ and $6)/12$, respectively. Looking at this figure, it seems that it is the same as when rigidity is not considered at all.

Let us skip the middle parts and look at the parts 10 cycles ahead: the parts (121), (122), (123) and (124) in Fig. 5.12 correspond to time $t = t_1 \times (120,\ 121,\ 122$ and $123)/12$, respectively, while the parts (124), (125), (126) and (127) in Fig. 5.13 correspond to time $t = t_1 \times (123,\ 124,\ 125$ and $126)/12$, respectively.

As we can see from these figures, as time goes by, the effect of adding rigidity begins to appear. It can be seen that the displacement $U(x,t)$

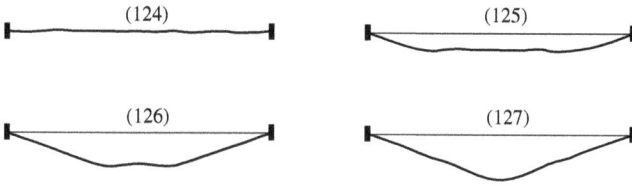

Fig. 5.13

turns into a rounded deformation. As we can see from the final solution of (5.3.38), each of the fundamental wave and the overtone waves are periodic functions. But when their superpositions are taken, it does not become a periodic function as a whole, because the frequencies of the overtone vibrations are not integral multiples of the frequency of the fundamental vibration. In particular, in (124) of Fig. 5.13, the displacement is not completely zero, and in (125) and (126) rippling occurs. These are phenomena caused by overtone vibration of non-integral multiples, which could not occur without rigidity. At this point, the effect of adding rigidity is clearly visible. However, at the point when the frequency of overtone does not become an integral multiple of the fundamental vibration, the word **"overtone vibration"** may have to be abandoned.

Here we have described harp nylon strings. Next we will also briefly describe piano strings. All piano strings are steel metal strings. The string length of the central [La] note is 39 cm, so half-length is $\ell = 19.5$ cm $= 0.195$ m, the radius of the cross-section of the string is $r = 0.6$ mm $= 0.0006$ m, the density ρ and Young's modulus E as taken from the Internet are $\rho = 7.87 \times 10^3$ kg/m^3, and $E = 200$ GPa, respectively. Using these values and taking the frequency of the [La] sound to be 440 Hz, we get

$$S\rho = 8.9007 \times 10^{-3}\,\text{kg/m}, \quad EI = 2.0357 \times 10^{-2}\,\text{N} \cdot \text{m}^2,$$

$$T = 999.0816\,\text{N}. \tag{5.3.48}$$

It is surprising that the string receives a tension force of about 100 kgw. The effect of rigidity is more remarkable, because the value of EI is larger than that of the harp nylon strings.

Finding the eigenvalues ω_k in this case, and dividing them by 2π to get the frequencies, we have

$$\frac{\omega_k}{2\pi} = 440.000, \ 1326.996, \ 2234.781, \ 3176.565, \ 4164.671, \ 5210.320,$$

$$6323.516, \ 7513.024, \ 8786.411, \ 10\,150.137. \tag{5.3.49}$$

This is larger than that in expression (5.3.46). Although not shown in a figure, the displacement here deforms flat from the top during the first cycle like when there is no rigidity, but from about the second cycle, it changes to a rounded deformation.

Whether it is a harp or a piano, the string that produces a low note is not a simple wire, but a wound string in which one core wire is wound with another wire. We did not initially understand why such a wound string was used, but this analysis has revealed the reason. Considering the non-rigidity expression (5.3.43), to make the value of ω as small as possible, we either increase the string length ℓ, or decrease the wave propagation velocity c. To decrease this c, we either decrease T from (5.3.40), or increase the linear density $S\rho$. However, if T is made too small, the sound itself will not be produced. Therefore, if the linear density is increased, that is, if a thicker string is used, the rigidity will also increase and the overtones will no longer be overtones. As a result, it seems that a wound string is convenient for increasing the linear density without increasing the rigidity.

Difficulties in Numerical Calculation (1)

In this analysis, I had unexpected difficulties in the numerical calculation. To define an eigenfunction at (5.3.15), I chose initially, $A = \cosh(\nu\ell)$, $B = -\cos(\mu\ell)$ and set the eigenfunction as $F_k(x) = \cosh(\nu_k\ell)\cos(\mu_k x) - \cos(\mu_k\ell)\cosh(\nu_k x)$ so that the first expression in (5.3.13) was satisfied. However, when I tried to program with this F_k, some errors occurred no matter how many times I repeated the program. I did not notice at first that the value of the hyperbolic function is quite large, for example, about 10^{64}. In the subtraction between such large numbers, I noticed that the number of digits had exceeded the significant digits and a ridiculous answer was returned.

The same mistake occurred for expression (5.3.35) which represents the initial waveform $U(x,0)$. At first, the calculation was done by decomposing the part $\cosh\big(\lambda(\ell - |x|)\big)$, but the same error occurred. When this is decomposed, it becomes the product of hyperbolic functions which is a very large number, and the subtraction between them causes a big mistake. This kind of error can only be understood by actually trying it. I learned a lot from these calculations.

5.4 Membrane Vibration

In the previous section, we described the string vibration in the case where the initial waveform of the string is given. In this section, we describe the membrane vibration that is a two-dimensional extension of the string vibration.

At first, we assume a drum membrane, and propound the problem of how the membrane vibrates when the center point of the circular membrane is pulled up and then released without initial velocity. However, we recognize that this is a mistake; in principle, if we try to pull up only one point of the membrane, it will be pulled up infinitely in the theoretical sense. In the following, we will analyze from the standpoint of the wave equation.

5.4.1 *Attempt to solve the wave equation*

As is well known, when the displacement in the direction perpendicular to the membrane at coordinates (x, y) and time t is denoted by $U(x, y, t)$, the wave equation is written as

$$\frac{\partial^2}{\partial t^2} U(x, y, t) = c^2 \left(\frac{\partial^2}{\partial x^2} + \frac{\partial^2}{\partial y^2} \right) U(x, y, t), \quad c = \sqrt{\frac{T}{\rho}}. \tag{5.4.1}$$

Here, c is the wave propagation velocity which is given by the second expression where ρ is the surface density of the membrane and T is the tension per unit length along the membrane.

The membrane handled here is a circular membrane with the radius ℓ at the periphery fixed, and the coordinate origin is set at the center of the circular membrane. Also, since only the phenomenon that is rotationally symmetric with respect to the coordinate origin is dealt with here, the two-dimensional Laplacian is written as

$$\frac{\partial^2}{\partial x^2} + \frac{\partial^2}{\partial y^2} = \frac{\partial^2}{\partial r^2} + \frac{1}{r} \frac{\partial}{\partial r}, \tag{5.4.2}$$

where r is the distance from the origin. The displacement is also written as $U(r, t)$ according to this symmetry. Then equation (5.4.1) becomes

$$\frac{\partial^2}{\partial t^2} U(r, t) = c^2 \left(\frac{\partial^2}{\partial r^2} + \frac{1}{r} \frac{\partial}{\partial r} \right) U(r, t). \tag{5.4.3}$$

First, let us find a static solution for this equation. Then, the left-hand side of the equation becomes zero, so the displacement which is denoted as

$S(r)$ satisfies the equation

$$\left(\frac{d^2}{dr^2} + \frac{1}{r}\frac{d}{dr}\right)S(r) = 0. \tag{5.4.4}$$

A solution which satisfies the boundary condition $S(\ell) = 0$ is

$$S(r) = C\log\left(\frac{r}{\ell}\right), \tag{5.4.5}$$

where C is a constant of integration. As we can see from this expression, if only the origin of $r = 0$ is pulled up, then the displacement will be infinite. This is not good because only one point is pulled up. If we pull up a domain with a certain area, it should not be infinite. Here, we assume that the initial displacement $S(r)$ is a constant value U_0 inside a suitable radius r_0 smaller than the membrane radius ℓ, and outside of radius r_0, expression (5.4.5) shall be followed,

$$S(r) = \begin{cases} U_0, & 0 \le r \le r_0 \\ \dfrac{U_0}{\log(r_0/\ell)}\log(r/\ell), & r_0 \le r \le \ell. \end{cases} \tag{5.4.6}$$

Setting the initial displacement in this way means forcibly pressing the membrane by hand at the point $r = r_0$. Therefore, equation (5.4.4) is not of course satisfied at this point. On the other hand, in the range other than $r = r_0$, no external force is applied to the membrane, so equation (5.4.4) is satisfied.

Under this initial condition, let us return to equation (5.4.3) and find the solution after releasing the membrane. We assume the displacement $U(r,t)$ is the separation of variables,

$$U(r,t) = R(r)P(t), \tag{5.4.7}$$

then the equation is deformed to

$$\frac{1}{c^2 P}\frac{d^2 P}{dt^2} = \frac{1}{R}\left(\frac{d^2 R}{dr^2} + \frac{1}{r}\frac{dR}{dr}\right). \tag{5.4.8}$$

Let the value of this expression, that is, the separation constant be $-k^2$, then the equation is separated into two equations,

$$\frac{d^2 P}{dt^2} + c^2 k^2 P = 0, \qquad \frac{d^2 R}{dr^2} + \frac{1}{r}\frac{dR}{dr} + k^2 R = 0. \tag{5.4.9}$$

From this first equation, excluding the constants attached to the whole, we have

$$P(t) = \cos(ckt). \tag{5.4.10}$$

Here, the sine function is excluded because the initial velocity is set to zero. We can solve the second equation of (5.4.9) by using the zeroth-order Bessel function,

$$R(r) = J_0(kr). \tag{5.4.11}$$

At this time, the second-kind Bessel function $N_0(kr)$ is also a solution. But it is excluded from the divergence at $r = 0$, because it is required to find a non-divergent solution for all r that satisfies equation (5.4.9) after releasing the membrane, unlike when determining the initial displacement $S(r)$.

Here, we impose a boundary condition that the displacement becomes zero at $r = \ell$, therefore, $k\ell$ must be the zero point of the Bessel function. That is, if we put the n-th zero point of the zeroth-order Bessel function as ξ_n,

$$k = \frac{\xi_n}{\ell}, \quad n = 1, 2, 3, \dots . \tag{5.4.12}$$

The solutions of (5.4.10) and (5.4.11) are rewritten as

$$P(t) = \cos\left(\frac{\xi_n ct}{\ell}\right), \quad R(r) = J_0\left(\frac{\xi_n r}{\ell}\right). \tag{5.4.13}$$

The general solution of equation (5.4.3) is represented by the superposition of the products of these functions

$$U(r,t) = \sum_{n=1}^{\infty} C_n J_0\left(\frac{\xi_n r}{\ell}\right) \cos\left(\frac{\xi_n ct}{\ell}\right). \tag{5.4.14}$$

Here, the coefficient C_n is determined from the initial conditions. Before doing this, the period t_n in the nth-term of this solution and its reciprocal frequency ν_n are

$$t_n = \frac{2\pi\ell}{\xi_n c}, \quad \nu_n = \frac{\xi_n c}{2\pi\ell}. \tag{5.4.15}$$

We have numerically calculated the values of the factor ξ_n/π that appears in the frequency ν_n, listed up to 10 pieces from $n = 1$ to 10 with a precision of 4 digits after the decimal point,

$$\frac{\xi_n}{\pi} = 0.7654, \ 1.7571, \ 2.7545, \ 3.7533, \ 4.7526, \ 5.7522, \ 6.7518,$$

$$7.7516, \ 8.7514, \ 9.7513. \tag{5.4.16}$$

The Bessel function is approximated asymptotically by the trigonometric functions as the value of the variable increases. Therefore, it can be seen from this expression that the sequence of the difference of ξ_n/π listed here asymptotically becomes 1.

The period and frequency are, when $n = 1$,

$$t_1 = \frac{1}{0.7654} \times \frac{2\ell}{c}, \quad \nu_1 = 0.7654 \times \frac{c}{2\ell}, \tag{5.4.17}$$

respectively. However, for $n \geq 2$ in the vibration mode, since the frequencies are not integral multiples of the frequency when $n = 1$, ν_1 given by (5.4.17) does not become the entire frequency of the displacement $U(r,t)$. That is, it merely means the frequency of one term. In this sense, in the case of membrane vibration, the meanings of the terms "**fundamental frequency**" and "**fundamental period**" are considered to be lost. The profound sound of hitting the drums may be due to this phenomenon.

Next, let us determine the coefficient C_n which appears in expression (5.4.14). Since the displacement when $t = 0$ is given by the initial condition (5.4.6),

$$S(r) = \sum_{n=1}^{\infty} C_n J_0\left(\frac{\xi_n r}{\ell}\right) \tag{5.4.18}$$

must be satisfied. We multiply both sides of this expression by $J_0(\xi_{n'} r/\ell)r$ and integrate with r, using the orthogonal relation with the zero point of the Bessel function,

$$\int_0^\ell J_0\left(\frac{\xi_n r}{\ell}\right) J_0\left(\frac{\xi_{n'} r}{\ell}\right) r\,dr = \frac{\ell^2}{2}\left[J_1(\xi_n)\right]^2 \delta_{n,n'}. \tag{5.4.19}$$

Then we have

$$C_n = \frac{2}{\ell^2\left[J_1(\xi_n)\right]^2} \int_0^\ell S(r) J_0\left(\frac{\xi_n r}{\ell}\right) r\,dr$$

$$= \frac{2U_0}{\ell^2\left[J_1(\xi_n)\right]^2} \Big[\int_0^{r_0} J_0\left(\frac{\xi_n r}{\ell}\right) r\,dr$$

$$+ \frac{1}{\log(r_0/\ell)} \int_{r_0}^\ell \log\left(\frac{r}{\ell}\right) J_0\left(\frac{\xi_n r}{\ell}\right) r\,dr\Big]. \tag{5.4.20}$$

Both integrations on the right-hand side can be performed with the aid of

$$\int z J_0(az)dz = \left(\frac{z}{a}\right) J_1(az), \quad \int J_1(az)dz = -\left(\frac{1}{a}\right) J_0(az), \tag{5.4.21}$$

where integration by parts is necessary for the last term. The result is an unexpectedly simple form because the cancellation occurs. Indeed, we have

$$C_n = -\frac{2U_0}{\log(r_0/\ell)} \frac{J_0(\xi_n r_0/\ell)}{\left[\xi_n J_1(\xi_n)\right]^2}. \tag{5.4.22}$$

Returning this to expression (5.4.14), we obtain the final solution for $U(r,t)$:

$$U(r,t) = -\frac{2U_0}{\log(r_0/\ell)} \sum_{n=1}^{\infty} \frac{J_0(\xi_n r_0/\ell)}{\left[\xi_n J_1(\xi_n)\right]^2} J_0\left(\frac{\xi_n r}{\ell}\right) \cos\left(\frac{\xi_n ct}{\ell}\right). \tag{5.4.23}$$

In the case of string vibration, the product of the trigonometric functions was included, and it was treated well with the product-to-sum formula. But, in the present case, it is the product of the Bessel function and the trigonometric functions. There is no formula to handle this well, so it would be impossible to simplify the expression further analytically.

5.4.2 *Numerical calculation*

In the following, we will perform numerical calculations based on expression (5.4.23). Here, the sum of n is taken up to $n = 20$, and we use the next formulas for numerical calculation for ξ_n, $J_1(\xi_n)$,

$$\xi_n = \frac{\pi a}{4}\left[1 + \frac{2}{(\pi a)^2} - \frac{62}{3(\pi a)^4} + \frac{15\,116}{15(\pi a)^6} - \frac{12\,554\,474}{105(\pi a)^8}\right.$$
$$\left. + \frac{8\,368\,654\,292}{315(\pi a)^{10}} - \cdots\right], \quad (a = 4n - 1), \tag{5.4.24}$$

$$J_1(\xi_n) = (-1)^{n+1}\frac{2^{3/2}}{\pi\sqrt{a}}\left[1 - \frac{56}{3(\pi a)^4} + \frac{9664}{5(\pi a)^6} - \frac{7\,381\,280}{21(\pi a)^8} + \cdots\right],$$
$$(a = 4n - 1). \tag{5.4.25}$$

The resulting graphs are shown in Figs. 5.14 to 5.17. In these figures, they are drawn three-dimensionally, taking the r-axis horizontally to the right, the time t-axis diagonally upward and the U-axis in the upward direction. Furthermore, the hidden lines elimination are processed, so the shaded areas are not drawn.

Here, let r_0 be equal to 0.1ℓ. Figures 5.14 to 5.17 show t from 0 to t_1, t_1 to $2t_1$, $2t_1$ to $3t_1$ and $3t_1$ to $4t_1$, respectively, where t_1 is defined in (5.4.17). As can be seen from these figures, the displacement $U(r,t)$ appears to change cyclically with a period of t_1 approximately, but, strictly speaking, it is not a periodic function.

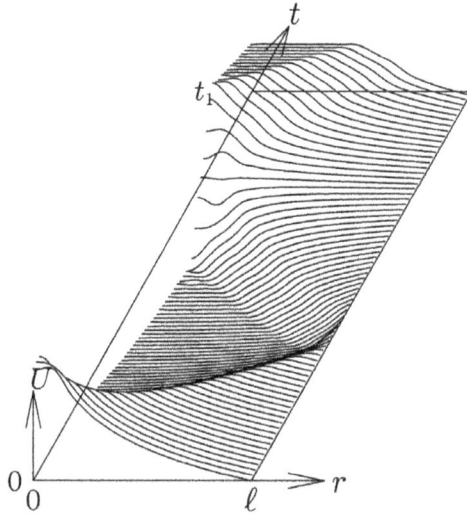

Fig. 5.14 $t = 0$ to t_1

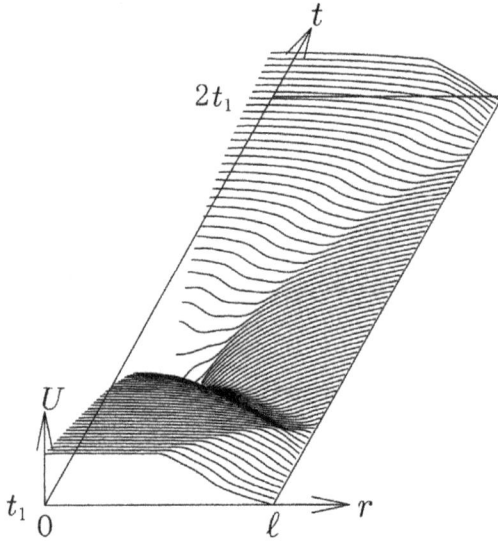

Fig. 5.15 $t = t_1$ to $2t_1$

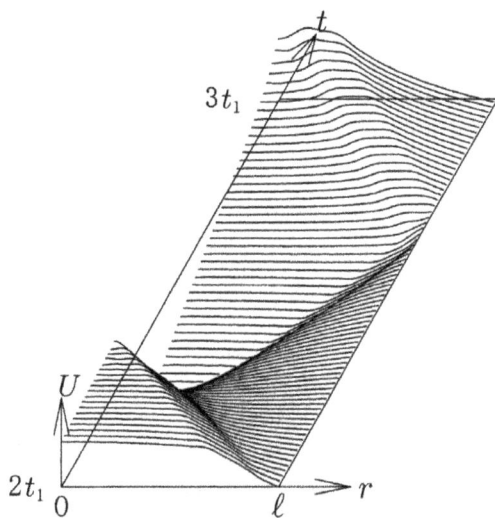

Fig. 5.16 $t = 2t_1$ to $3t_1$

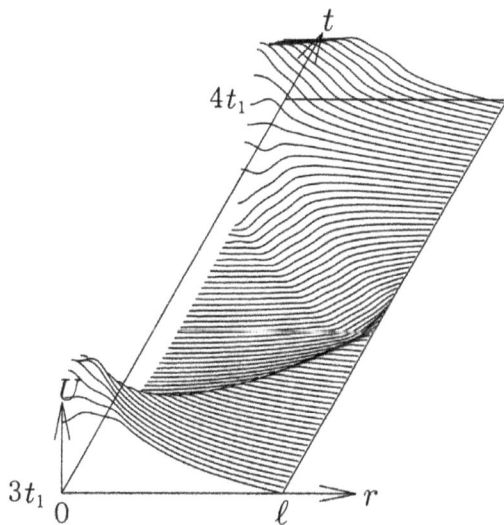

Fig. 5.17 $t = 3t_1$ to $4t_1$

Here, we dealt with the vibration of a simple single membrane. For the actual drums, it will be a much more complicated vibration than the one dealt with here, because the resonance vibration occurs in the air inside the cylinder and the membrane on the opposite side.

Membrane Vibration of Musical Instruments

We have dealt with here the initial value problem of membrane vibration. It was found that membranes do not have strict periodic vibrations, unlike strings. I describe here what I discovered about this. There are many types of musical instruments: roughly classified, there are string instruments such as the guitar, violin, and piano; percussion instruments such as the drum and cymbals; and wind instruments like the saxophone and clarinet. String instruments make use of the vibration of a one-dimensional object string. Wind instruments are physically similar as they make use of the one-dimensional vibration of the air inside a tube. On the other hand, drums and cymbals have two-dimensional vibrating bodies. I have wondered for a long time, why there is no percussion instrument that can produce a melody. Percussion instruments are used exclusively as instruments for playing rhythms. However, if we line up a lot of big and small drums and hit them, we should be able to play a melody. But, I have never seen or heard anything like that. If we tried to do something like this, it would take up a large space, so it may not be feasible. But, as far as I know, nobody in the world has tried this, so perhaps there might be some other reason. The reason may be that the melody will not be beautiful. As we have seen here, the membrane vibration cannot produce a perfect periodic vibration.

I recently learned that there is a percussion instrument called a steelpan drum. This instrument was invented in the Republic of Trinidad and Tobago, an island country in the Caribbean. It is shaped like a pan about one meter in diameter, with about 30 large and small protuberances at the bottom. The sounds produced depend on which of the protuberances is hit, and it seems that you can play songs on it. You can view them on the Internet, but I feel that the sounds are not as good as what we normally hear from other music instruments.

Though not considered a music instrument, there is also the glass-xylophone or glass-harp. This is a line-up of many glasses filled with water at various heights. In the case of a glass-xylophone, we can play a melody by hitting the glasses with a stick, while in the case of a glass-harp, we rub the glasses with our fingers. I have heard these instruments, and they sound really beautiful.

5.5 Analysis of Wave Phenomenon in Nature

We can see the wave phenomenon in nature, in the waves of the sea, and in the following examples:

(1) Wind ripples on sand as a result of the sand being blown by the wind. The principle of this phenomenon has already been clarified.

(2) On a slightly larger scale, sand dunes. These have a certain unique shape.

(3) Wave-shaped irregularities on snow. In Hokkaido, in the northern part of Japan where this author lives, winter snow is common. No matter how much the wind blows in a snow field, a snow pattern does not occur. Is this because the snow is so light that it can fly a considerable distance? Instead, on snowy roads with vehicular traffic, wave-shaped irregularities (bumps) with a wavelength of about 10 cm can be found about 50 m before and after traffic light junctions. This is thought to be because cars produce a large horizontal force to the snow surface as they stop or start at the traffic lights. But the details of this mechanism are unknown.

(4) Cirrocumulus clouds (mackerel sky) in the autumn sky. We cannot help but wonder why they have such beautiful shapes. The clouds are sometimes connected lengthwise in the horizontal direction and shaped like a wave. They are said to occur by Benard convection. Therefore, the wave phenomenon may not be involved here.

(5) The rolling illusion of waves as wind blows through a grass meadow. It is strange that the phase of the wave is neatly aligned perpendicular to the direction of the wind.

(6) Density of traffic on highways. This may not be a natural phenomenon, but there are places on a highway where the density of traffic is high or low. This sparseness and denseness of traffic can propagate as a wave.

In this section, we will explain the Werner model for the "wind pattern phenomenon" of wind ripples and sand dunes. Finally, we will explain the wave phenomenon on highways as simply as possible.

5.5.1 *Phenomenon of wind ripples*

First, we describe the **wind ripple** phenomenon that occurs on sandy ground. This wind ripple is also known as a **"wind pattern"**. The wind ripple phenomenon occurs when the wind blows on sandy ground and the sand is blown away. The flying sand is the first factor for the formation of a wind ripple, but a beautiful wind ripple with a uniform phase cannot be achieved by itself. To make a beautiful wind pattern, it is necessary to consider the effect that the sand collapses from high to low. Here, for ease of explanation, sand **"flying"** is introduced first and sand **"collapse"** will be introduced later.

Set the two-dimensional coordinates (x, y) on the sandy ground. The wind shall blow in the positive direction of the x-axis, and it is assumed that a wave is generated in the x-axial direction. However, at the beginning, it should be uniform in the y direction, so it is treated as a one-dimensional coordinate of only the x-axis.

Then, the height of the sand surface at coordinate x and time t is set as $h(x, t)$. However, this $h(x, t)$ is measured based on the average height when there is no wind pattern, and $h(x, t)$ is sufficiently small compared to the wavelength of the wind pattern. Also, let the distance at which the sand is blown at coordinate x and time t be $L(x, t)$, and assume that the sand which is blown from a high place will fly a longer distance. We set

$$L(x, t) = \ell_0 + bh(x, t), \tag{5.5.1}$$

where ℓ_0 is a quantity determined by the strength of the wind, and b is a positive constant. Furthermore, let the amount of sand flying per unit length and per unit time be $N(x, t)$ at coordinate x and time t, and also, let the landing amount be $M(x, t)$. Let ξ be the starting point for the sand that lands on point x, and because the sand flies a distance of L, we have

$$\xi = x - L(\xi, t). \tag{5.5.2}$$

Notice that L on the right-hand side is $L(\xi, t)$, and not $L(x, t)$. Hereafter, the time when the sand is flying is ignored as it is smaller than the change time of $h(x, t)$. The variable ξ becomes a function of x in the form $\xi = \xi(x)$ from this relation.

Now, let $d\xi$ be an infinitesimal length, the amount of sand that flies from the interval $[\xi, \xi + d\xi]$ in a unit time is $N(\xi, t)d\xi$, and if this amount lands on the interval $[x, x + dx]$, it must be equal to $M(x, t)dx$. Then we have the relation

$$M(x,t) = N(\xi,t)\frac{d\xi}{dx}. \tag{5.5.3}$$

Here, we formulate an equation for time evolution. With dt as an infinitesimal time, subtracting the amount of sand $N(x,t)dxdt$ that decreases in interval $[x, x + dx]$ within the time dt, from the amount of sand $M(x,t)dxdt$ that increases in the same interval, the result is equal to the amount of sand $[h(x, t + dt) - h(x, t)]dx$ which is caused by the change in height in this interval:

$$[h(x, t+dt) - h(x,t)]dx = [M(x,t) - N(x,t)]dxdt. \tag{5.5.4}$$

Dividing both sides by $dxdt$ and rewriting, we have

$$\frac{\partial h(x,t)}{\partial t} = M(x,t) - N(x,t). \tag{5.5.5}$$

This corresponds to the **equation of continuity** in hydrodynamics. Furthermore, by substituting (5.5.3), M is expressed by N,

$$\frac{\partial h(x,t)}{\partial t} = N(\xi,t)\frac{d\xi}{dx} - N(x,t). \tag{5.5.6}$$

Here, differentiating (5.5.2) with respect to x, and using (5.5.1), $d\xi/dx$ becomes

$$\frac{d\xi}{dx} = 1 - b\frac{\partial h(\xi,t)}{\partial x}. \tag{5.5.7}$$

Then, (5.5.6) is rewritten as

$$\frac{\partial h(x,t)}{\partial t} = N(\xi,t)\left[1 - b\frac{\partial h(\xi,t)}{\partial x}\right] - N(x,t). \tag{5.5.8}$$

Here, we set a big assumption, that is, the amount of sand N flying per unit time and per unit length is assumed to be constant regardless of the coordinate or time,

$$N(\xi,t) = N(x,t) \equiv N_0. \tag{5.5.9}$$

Then, equation (5.5.8) becomes

$$\frac{\partial h(x,t)}{\partial t} = -v_0 \frac{\partial h(\xi,t)}{\partial x}, \tag{5.5.10}$$

where the constant v_0 is defined as

$$v_0 = N_0 b. \tag{5.5.11}$$

Furthermore, this $h(\xi,t)$ on the right-hand side is approximated by using relations (5.5.2) and (5.5.1), because h is small enough,

$$h(\xi,t) = h\big(x - \ell_0 - bh(\xi,t), t\big) \cong h(x - \ell_0, t). \tag{5.5.12}$$

Equation (5.5.10) becomes a linear one,

$$\frac{\partial h(x,t)}{\partial t} = -v_0 \frac{\partial h(x - \ell_0, t)}{\partial x}. \tag{5.5.13}$$

If there is no ℓ_0 in h on the right-hand side, this is a wave equation that represents a one-dimensional right-propagating wave. What does the existence of this ℓ_0 mean? Here, let k be a positive wavenumber and ν be a frequency. We try to put

$$h(x,t) = e^{i(kx - \nu t)}. \tag{5.5.14}$$

Substituting this expression into equation (5.5.13), we have

$$\nu = v_0 k \big[\cos(k\ell_0) - i \sin(k\ell_0) \big], \tag{5.5.15}$$

where ν is solved as a complex number. Returning this expression to (5.5.14), we then find

$$h(x,t) = e^{-v_0 k \sin(k\ell_0)t} \cdot e^{ik[x - v_0 \cos(k\ell_0)t]}. \tag{5.5.16}$$

Actually, the linear combinations of the real and imaginary parts of this expression are the solutions. In any case, when $\sin(k\ell_0) > 0$, it converges to zero over time and no wind pattern is formed. On the contrary, when $\sin(k\ell_0) < 0$, the solution will diverge over time. Of course, $\sin(k\ell_0) < 0$ should be enough to create a wind pattern, but once the wind pattern begins to form, it will progress steadily and become unstable. To stop this, it is necessary to add the effect that the sand "collapses" on the surface and softens the change in height. Here, it is assumed that the amount of sand that collapses in a unit time is proportional to the height gradient. That is, with a positive constant λ, we put it as $\lambda \partial h / \partial x$. Consider the interval $[x, x + dx]$. Subtracting the amount of sand $\lambda [\partial h/\partial x]_x dt$ that goes out by the collapse at the point x within an infinitesimal time dt, from the amount of sand $\lambda[\partial h/\partial x]_{x+dx}dt$ that enters this interval at point $x + dx$, the result represents the amount of change by the collapse in this interval within the time dt.

Here, for the sake of convenience later, we will return to equation (5.5.8), which was derived only by the flying sand, not the collapse. Adding the amount of change due to this collapse to the right-hand side of (5.5.8), the equation becomes

$$\frac{\partial h(x,t)}{\partial t} = N(\xi,t)\left[1 - b\frac{\partial h(\xi,t)}{\partial x}\right] - N(x,t) + \lambda\frac{\partial^2 h(x,t)}{\partial x^2}. \qquad (5.5.17)$$

This is the equation for preserving the total amount of sand when both "flying" and "collapse" are incorporated. This newly-added term of the second derivative is called the **dissipative term**. If we compare it to the equation of heat conduction, this effect can be thought of as the same as the heat moving from high to low temperature through the dissipation effect.

Equation (5.5.17) is used for the **sand dunes** described in §5.5.3. In the case of the wind pattern discussed here, as mentioned above, the first and second terms on the right-hand side is approximated as the right-hand side of (5.5.13), and the equation to be solved is

$$\frac{\partial h(x,t)}{\partial t} = -v_0\frac{\partial h(x - \ell_0, t)}{\partial x} + \lambda\frac{\partial^2 h(x,t)}{\partial x^2}. \qquad (5.5.18)$$

Substituting (5.5.14) for this equation, we have

$$\nu = v_0 k \cos(k\ell_0) - ik\left[v_0 \sin(k\ell_0) + \lambda k\right]. \qquad (5.5.19)$$

Returning this to (5.5.14), we have

$$h(x,t) = e^{-k[v_0 \sin(k\ell_0)+\lambda k]t} \cdot e^{ik[x - v_0 \cos(k\ell_0)t]}. \qquad (5.5.20)$$

This has the effect of preventing the wind pattern from growing too much at $v_0 \sin(k\ell_0) < 0$ by using the term $\lambda k > 0$. In the actual wind pattern, it grows with $v_0 \sin(k\ell_0) + \lambda k < 0$, and when it grows to some extent, the value of λ also increases. As a result, this value approaches zero and growth stops.

It should be noted that the moving speed V of the wind pattern is

$$V = v_0 \cos(k\ell_0). \qquad (5.5.21)$$

When $\sin(k\ell_0) < 0$, the velocity V of this wind pattern can take both positive and negative values or a value of zero, therefore, the wave does not always move in the direction of the wind.

Here, let us find out in more detail the conditions for forming a wind pattern. The factor appearing in the first exponential function on the right-hand side of (5.5.20) is set as

$$f = v_0 \sin(k\ell_0) + \lambda k. \tag{5.5.22}$$

If $f > 0$, the wind pattern is not formed. In order for the wind pattern to form, it must be $f < 0$. Here we set $f = 0$, which is the boundary value, and consider the conditions for the parameters to have a positive root for k. There are 3 parameters, v_0, λ, ℓ_0 in expression (5.5.22). However if we let the unknown variable be $k\ell_0$, and deform $f = v_0\left[\sin(k\ell_0) + \lambda/(v_0\ell_0) \cdot (k\ell_0)\right]$, whether the positive root exists or not is determined by only one parameter $\lambda/(v_0\ell_0)$. When this value is large enough, there is no positive root. Also, when this value is small enough, there are multiple positive roots. Therefore, when we move the value of $\lambda/(v_0\ell_0)$ from the larger one to the smaller one, and when it has roots for the first time, it becomes a multiple root. Therefore, we set two expressions

$$f = 0, \quad \frac{df}{dk} = 0. \tag{5.5.23}$$

By solving this, the conditions for the parameters can be obtained. Eliminating the trigonometric functions from these two expressions, we find

$$k\ell_0 = \sqrt{\left(\frac{v_0\ell_0}{\lambda}\right)^2 - 1} \equiv \alpha. \tag{5.5.24}$$

Here, the value on the right-hand side is set as α. Substituting this into the expression $df/dk = 0$, we have

$$\cos(\alpha) = -\frac{1}{\sqrt{1 + \alpha^2}}. \tag{5.5.25}$$

This equation cannot be solved analytically, so we try to find the minimum positive root numerically. We obtain $\alpha = 2.3962\cdots$, and therefore

$$\frac{v_0\ell_0}{\lambda} = \sqrt{1 + \alpha^2} = 2.5965\cdots. \tag{5.5.26}$$

This is the minimum condition for a wind pattern to form, and is generally

$$\frac{v_0\ell_0}{\lambda} > 2.5965\cdots. \tag{5.5.27}$$

From now on, the larger v_0, ℓ_0 are and the smaller λ is, the easier it is to create a wind pattern. But if v_0, ℓ_0 become too large, the number of allowed wavenumbers k will increase, and it will be difficult to form a wind pattern. Also, if λ is set too small, we will not get a clean wave with the same phase.

So far, we have described the coordinate in one-dimension. Here, we extend to two-dimensional coordinates with (x, y). In equation (5.5.18), the second derivative of the last dissipative term on the right-hand side is extended to the two-dimensional Laplacian,

$$\frac{\partial h(x, y, t)}{\partial t} = -v_0 \frac{\partial h(x - \ell_0, y, t)}{\partial x} + \lambda \left(\frac{\partial^2}{\partial x^2} + \frac{\partial^2}{\partial y^2} \right) h(x, y, t). \quad (5.5.28)$$

This last term represents the effect of sand collapsing in all directions. It is due to this term that the phase of the wave is neatly aligned horizontally (y direction). An example of a numerical solution based on this analysis is shown in the following subsection.

5.5.2 *Numerical analysis of wind ripples*

In order to convert the differential equation (5.5.28) into a difference equation, we introduce small quantities Δt and $\Delta x, (\equiv) \Delta y$ as the units of time and length. Moreover, the time $t/\Delta t$ and lengths $x/\Delta x$, $y/\Delta y$ measured in these units are redefined as dimensionless integers t and x, y, respectively. Correspondingly, we also introduce dimensionless parameters; the distance ℓ_0, the coefficients v_0 and λ contained in this equation are redefined as

$$\frac{\ell_0}{\Delta x} \mapsto \ell_0, \quad v_0 \frac{\Delta t}{\Delta x} \mapsto v_0, \quad \lambda \frac{\Delta t}{\Delta x^2} \mapsto \lambda. \quad (5.5.29)$$

With these replacements, the differential equation (5.5.28) is converted to a difference equation:

$$h(x, y, t+1) - h(x, y, t) = -\frac{v_0}{2} \left[h(x - \ell_0 + 1, y, t) - h(x - \ell_0 - 1, y, t) \right]$$
$$+ \lambda \left[h(x+1, y, t) - 2h(x, y, t) + h(x-1, y, t) \right]$$
$$+ \lambda \left[h(x, y+1, t) - 2h(x, y, t) + h(x, y-1, t) \right]. \quad (5.5.30)$$

Hereafter, we will solve this equation numerically.

Here, the point (x, y) is placed on the grid point of 200×200, and at that boundary, we will impose periodic boundary conditions in both the x and y directions. Also, for the parameters ℓ_0, v_0, λ, the wind pattern can be the most beautiful if we set

$$\ell_0 = 15, \quad v_0 = 0.01, \quad \lambda = 0.01. \quad (5.5.31)$$

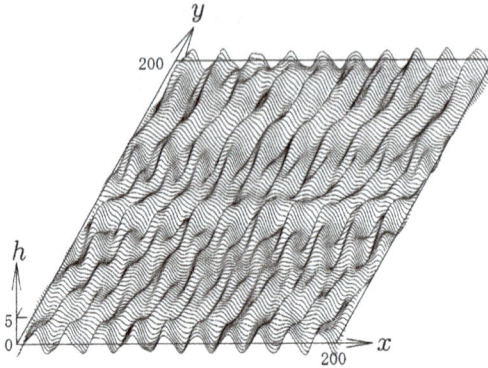

Fig. 5.18   Wind ripples based on (5.5.30)

Fig. 5.19   Picture of wind ripples (Source: Noboru Nakanishi)

The height h of the wind pattern at $t = 0$ is taken by a random number for each point. With these initial values, it is calculated numerically according to equation (5.5.30) until time $t = 2300$. The height h at that time is shown in Fig. 5.18.

Figure 5.18 is drawn three-dimensionally, with the x-axis being taken horizontally to the right, y-axis diagonally upward and the height $h(x, y)$ in the upward direction. From this figure, it can be seen that the phases of the waves are aligned in the y direction. Also, it seems natural that these phases are switching in some places, so we can see the natural phenomenon. For comparison, actual wind ripples are shown in Fig. 5.19 which is a photograph taken in the Merzouga desert of Morocco.

As time passes, the height will become higher and higher and eventually diverge. In the actual wind pattern, as mentioned above, as the height increases, the value of the coefficient λ also increases. Therefore, it is thought that the growth of the wind pattern stops when it grows to some extent.

5.5.3 *Mathematics of sand dunes*

We can find many treatises about wind pattern phenomenon on the Internet. Many of them also mention the **sand dune** phenomenon. First, we thought that the shape of sand dunes would be irregular, but apparently they are not. A typical example is the "**Barchan dune**". The schematic diagram is shown in Fig. 5.20.

In this figure, the wind is blowing from left to right, and a crescent-shaped dune is formed. The area facing the windward side of this dune has a gentle slope and its angle is about 15 degrees. The area facing the leeward side has a steep slope and the angle is about 30 to 35 degrees. The size of this Barchan dune is about 30 m in height and the width of the base reaches about 370 m.

This Barchan dune was discovered by Russian naturalist Alexander von Middendorf in 1881 in a desert in Tajikistan. The name Barchan is derived

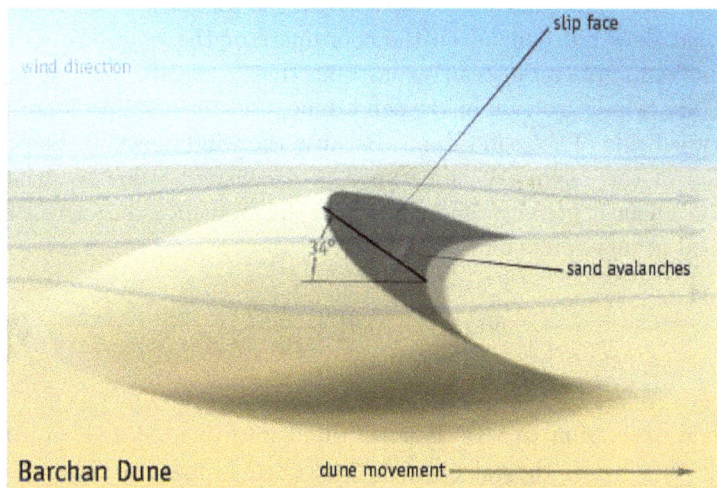

Fig. 5.20 Barchan dune
(Source: https://commons.wikimedia.org/wiki/File.Barchan.jpg)

from Russian, and means a **crescent-shaped dune**. When I first saw the picture of this Barchan dune, I wondered why it had such a beautiful shape. Moreover, I was surprised that this shape can be reproduced with a mathematical model.

5.5.4 *Introduction to the equation*

The equation for the sand dune is a slight modification of the one for the wind pattern in the previous section. The wind ripple or the wind dune is a phenomenon that occurs when the wind blows sand from a sandy ground. Although this "flying" sand is the first factor for the formation of the wind pattern and the dune, it cannot be beautiful by itself. As discussed in the previous section, in order to create the beautiful wind pattern and dune, it is also necessary to consider the effect of the sand collapsing from high to low.

The basic equation is the same as equation (5.5.17). We will quote it again here:

$$\frac{\partial h(x,t)}{\partial t} = N(\xi,t)\left[1 - b\frac{\partial h(\xi,t)}{\partial x}\right] - N(x,t) + \lambda\frac{\partial^2 h(x,t)}{\partial x^2}. \qquad (5.5.32)$$

In the case of the wind pattern in the previous section, the amount of sand flying per unit time and unit length $N(x,t)$ is assumed to be a constant value that does not depend on the coordinate or the time. We withdraw here this assumption; instead we assume that a certain amount of sand flies on the windward side of the sand dune, and the sand does not fly on the leeward side of the sand dune. Because the wind blows in the positive direction of the x-axis, N is assumed as a constant value N_0 when the height gradient is positive, and zero when the gradient is negative. This is expressed as an expression,

$$N(x,t) = N_0\theta\left(\frac{\partial h(x,t)}{\partial x}\right), \qquad (5.5.33)$$

where θ is a unit step function.

When expression (5.5.33) is substituted into (5.5.32) and an attempt is made to solve it, the existence of ξ becomes a problem. The only way to resolve this problem would be to use an approximation with h small enough. First, for $h(\xi,t)$ included in equation (5.5.32), using expressions

(5.5.2) and (5.5.1), we approximate it as

$$h(\xi, t) = h\big(x - L(\xi, t), t\big) \cong h(x - \ell_0, t). \tag{5.5.34}$$

Also at first, we approximated the $N(\xi, t)$ part to $N(\xi, t) \cong N(x - \ell_0, t)$ and tried to solve it numerically. But we found that the approximation was too stringent and the total amount of sand decreased over time. This is probably because the amount of sand that lands is underestimated. So, we take the approximation one step further, that is, we set

$$N(\xi, t) \cong N\big(x - \ell_0 - bh(x - \ell_0, t), t\big). \tag{5.5.35}$$

From the above consideration, the equation becomes

$$\frac{\partial h(x, t)}{\partial t} = N\big(x - \ell_0 - bh(x - \ell_0, t), t\big)\left(1 - b\frac{\partial h(x - \ell_0, t)}{\partial x}\right)$$

$$-N(x, t) + \lambda \frac{\partial^2 h(x, t)}{\partial x^2}. \tag{5.5.36}$$

So far, we have described the coordinate in one-dimension, but we extend here the coordinates to two-dimensions (x, y). In equation (5.5.36), the second derivative of the last dissipative term is extended to the two-dimensional Laplacian. Therefore the equation to be solved is

$$\frac{\partial h(x, y, t)}{\partial t} = N\big(x - \ell_0 - bh(x - \ell_0, y, t), y, t\big)\left(1 - b\frac{\partial h(x - \ell_0, y, t)}{\partial x}\right)$$

$$-N(x, y, t) + \lambda\left(\frac{\partial^2}{\partial x^2} + \frac{\partial^2}{\partial y^2}\right)h(x, y, t). \tag{5.5.37}$$

This last term describes the effect of sand collapsing in all directions. It plays an important role in forming the shape of the entire Barchan dune. When the term $N\big(x - \ell_0 - bh(x - \ell_0, y, t), y, t\big)$ contained in this equation is substituted into expression (5.5.33), it becomes a complicated equation that includes an unknown function among other unknown functions, and it will no longer be possible to solve analytically.

5.5.5 *Numerical calculation*

We convert here the differential equation (5.5.37) into a difference equation. As derived in the previous subsection, introducing a small unit of time Δt and a small unit of length $\Delta x, (\equiv)\Delta y$, we make variables t and

x, y dimensionless. In addition, we make the parameters h, ℓ_0, N_0 and λ dimensionless as

$$\frac{h}{\Delta x} \mapsto h, \quad \frac{\ell_0}{\Delta x} \mapsto \ell_0, \quad N_0 \frac{\Delta t}{\Delta x} \mapsto N_0, \quad \lambda \frac{\Delta t}{\Delta x^2} \mapsto \lambda. \tag{5.5.38}$$

With this replacement, the differential equation (5.5.37) is transformed into a forward type difference equation,

$$\begin{aligned}
h(x, & y, t + 1) - h(x, y, t) \\
= & N\big(x - \ell_0 - bh(x - \ell_0, y, t), y, t\big) \\
& \times \left[1 - \frac{b}{2}\big[h(x - \ell_0 + 1, y, t) - h(x - \ell_0 - 1, y, t)\big]\right] - N(x, y, t) \\
& + \lambda\big[h(x + 1, y, t) - 2h(x, y, t) + h(x - 1, y, t)\big] \\
& + \lambda\big[h(x, y + 1, t) - 2h(x, y, t) + h(x, y - 1, t)\big].
\end{aligned} \tag{5.5.39}$$

Hereafter, we will solve this equation numerically. We set here the value of the parameters

$$\ell_0 = 15, \quad N_0 = 0.01, \quad b = 0.5, \quad \lambda = 0.02. \tag{5.5.40}$$

The time evolution of $h(x, y, t)$ by the numerical solution method is shown in Fig. 5.20(a)–(f). In these figures, (x, y) is placed on the grid points of 200×200, and the figures are drawn three-dimensionally with the x-axis in the horizontal direction, the y-axis in the diagonally upward direction, and the h-axis in the upward direction. It should be noted that the boundaries of x and y are periodic boundaries.

In Fig. 5.21(a) at $t = 0$, the initial value $h(x, y, 0)$ is given as a uniform random number from -0.5 to $+0.5$. This evolved over time, and in Fig. 5.21(b) at $t = 5000$, the same wind pattern as the previous one is formed.

In Fig. 5.21(c) at $t = 10\,000$, we can see what is called eggs of three dunes in the center of the graph. Here, these three dunes are named A, B and C from left to right. Later we will look at how these three dunes evolve over time. In Fig. 5.21(d) at $t = 15\,000$, it can be seen that all of these A, B and C dunes move in the positive direction of the x-axis. C is on the far right in Fig. 5.21(c), but it is on the left in Fig. 5.21(d). This is because the periodic boundary condition was adopted. Also, A and C are growing, but B is not so large.

In Fig. 5.21(e) at $t = 20\,000$, B and C are united into one Barchan dune. Actually, since B did not grow as much as C, C has taken over B.

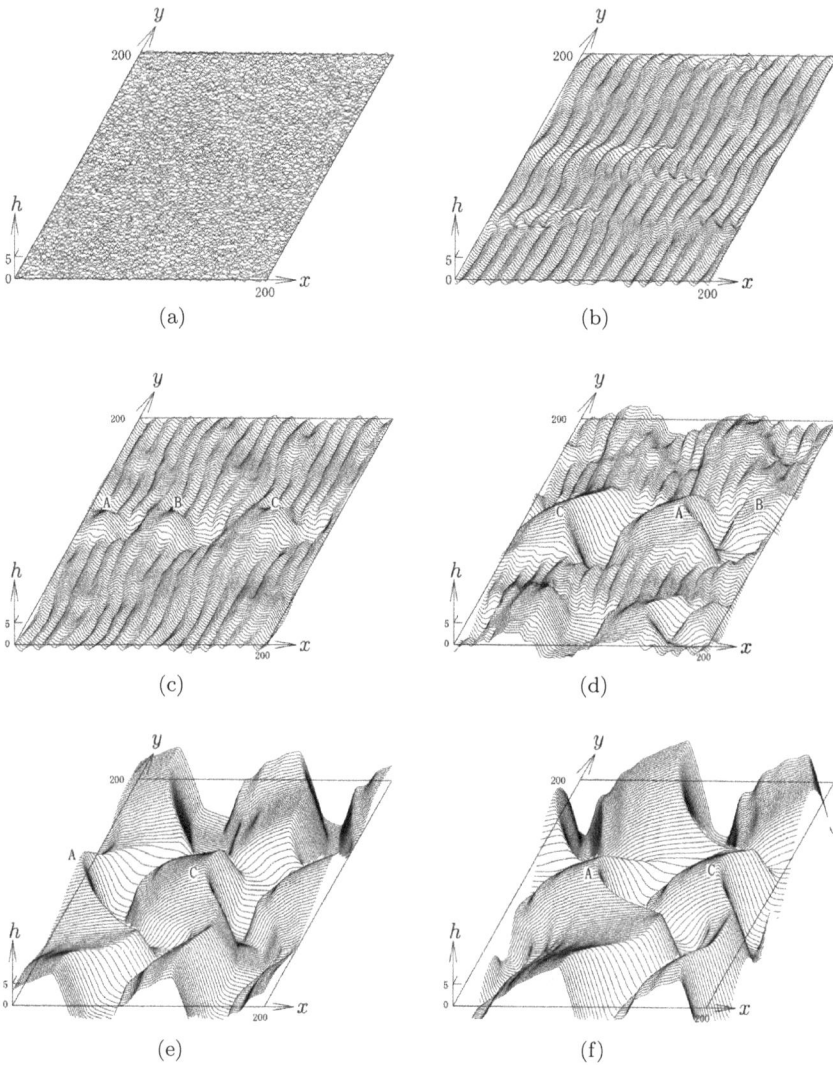

Fig. 5.21 (a) $t = 0$, (b) $t = 5000$, (c) $t = 10\,000$, (d) $t = 15\,000$, (e) $t = 20\,000$, (f) $t = 25\,000$

This C has the closest shape to the Barchan dunes shown in Fig. 5.20. At $t = 25\,000$ in Fig. 5.21(f), in addition to these A, B and C dunes, there are many other Barchan dunes. They are connected in the horizontal direction (y direction) and are shaped like a winding mountain range. This kind of dunes is called Mega-Barhan dunes, and the sand dunes we see in general are probably of this shape.

In these figures, the (x, y)-axis scale and the h-axis scale are changed for readability. Note that the h-axis scale are changed for $10/3$ times larger than the (x, y)-axis scale.

Difficulties in Numerical Calculation (2)

My motto is to solve analytically as much as possible. But the numerical calculation has become the main work at this time. When I am calculating numerically with the same parameters, I noticed that the patterns that came out were different each time. At first, I thought that my computer was broken and I wondered what to do. I thought about it carefully; the calculation uses random numbers for the initial data, so if I execute it repeatedly, the seeds for making random numbers will be different. I noticed that the pattern was different each time. I decided to use the one that is closest to the schematic diagram of the Barchan dune. If we use random numbers for the initial value, multiple Barchan dunes will be created and they will be concatenated eventually. If we take the initial values well, we should be able to create only one Barchan dune as shown in Fig. 5.20. However, this is a fairly difficult problem, and I tried various things, but everything failed.

In this numerical calculation, it took about 40 minutes to calculate up to $t = 25\,000$. It was quite painful to stare at the computer during that time. I regretted I did not have a faster computer.

5.6 Wave Phenomenon on Highways

On a highway, vehicles tend to crowd together, creating high and low density parts. Moreover, this density propagates as a wave. This is a **wave phenomenon on the highway**. Why does this phenomenon occur? We consider here the flow of cars like hydrodynamics. Let the car density at location x and time t be $\rho(x, t)$, and the velocity of the cars be $v(x, t)$. By considering the interval $[x, x + dx]$, where dx is a small distance, at point x, the number of cars entering this interval within a small time dt is $\rho(x, t)v(x, t)dt$, and at point $x + dx$, the number of cars leaving this interval is $\rho(x + dx, t)v(x + dx, t)dt$. The difference becomes the change in the number of cars $\left[\rho(x, t + dt) - \rho(x, t)\right]dx$ after the time dt has passed in this interval, so we have

$$\left[\rho(x,t+dt)-\rho(x,t)\right]dx = \rho(x,t)v(x,t)dt-\rho(x+dx,t)v(x+dx,t)dt. \quad (5.6.1)$$

Dividing both sides by $dxdt$ and rewriting, we obtain

$$\frac{\partial\rho(x,t)}{\partial t} = -\frac{\partial\left[\rho(x,t)v(x,t)\right]}{\partial x}. \quad (5.6.2)$$

This is the **equation of continuity** itself in hydrodynamics.

As the first step, let us assume here that the car speed $v(x,t)$ is a function of the density $\rho(x,t)$. When the density is low, the speed is increased, and when the density is high, the speed is decreased. However, there is a speed limit on the highway, so even if the density is low, it is not possible to increase the speed indefinitely. We assume here that v is a quadratic function of ρ as the simplest model,

$$v(x,t) = v_0\left[1 - \left(\frac{\rho(x,t)}{\rho_0}\right)^2\right], \quad (5.6.3)$$

where v_0 is the speed limit, and ρ_0 is the density at its maximum and the car cannot move. From this expression, when $\rho \to 0$, the velocity v becomes v_0, and when $\rho \to \rho_0$, v becomes 0; therefore the required conditions are somewhat satisfied.

However, what happens on an actual highway is not so simple. On a highway, if traffic is heavy, the car must slow down, and if traffic is light, the car will speed up. This means that the speed of the car also depends on the density gradient $\partial\rho/\partial x$. Suppose there are two locations where the density ρ is low and high, respectively, and suppose that the density gradients at these two locations are the same. Where the density is low at first, the speed is high. Then, when the traffic ahead is heavy, the car has to slow down a lot. On the other hand, where the density is originally high, the speed is low. Then, when the traffic ahead is heavy, the amount of deceleration should be small. Under these circumstances, as the second step, we add a term that depends on the density gradient, to expression (5.6.3),

$$v(x,t) = v_0\left[1 - \left(\frac{\rho(x,t)}{\rho_0}\right)^2\right] - \frac{\lambda}{\rho(x,t)}\frac{\partial\rho(x,t)}{\partial x} \quad (5.6.4)$$

where λ is a positive constant. In this expression, we have inserted a factor $\rho(x,t)$ to the denominator of the last term. This means that the smaller the density is, the greater the amount of speed increase or decrease depending on the sign of the density gradient.

Here, we introduce further expansion as the third step. The density gradient is determined by the position of the car in front. In actual situations, the driver should pay attention to the positions of several cars ahead of his own. This means that it is necessary to consider even the higher differential coefficients of the density. Here, the second derivative of density $\partial^2 \rho / \partial x^2$ is also incorporated. When this second derivative is positive, it is predicted that traffic will be heavy ahead, so the speed must decrease. When the second derivative is negative, it is predicted that traffic will be light ahead, so the speed may increase. That is, expression (5.6.4) is further extended to the one with an additional term,

$$v(x,t) = v_0 \left[1 - \left(\frac{\rho(x,t)}{\rho_0} \right)^2 \right] - \frac{\lambda}{\rho(x,t)} \frac{\partial \rho(x,t)}{\partial x} - \frac{\mu}{\rho(x,t)} \frac{\partial^2 \rho(x,t)}{\partial x^2},$$

$$(5.6.5)$$

where μ is a positive constant. Substituting expression (5.6.5) into (5.6.2), we have

$$\frac{\partial \rho(x,t)}{\partial t} = -v_0 \frac{\partial}{\partial x} \left[\rho(x,t) \left(1 - \frac{\rho^2(x,t)}{\rho_0^2} \right) \right] + \lambda \frac{\partial^2 \rho(x,t)}{\partial x^2} + \mu \frac{\partial^3 \rho(x,t)}{\partial x^3}.$$

$$(5.6.6)$$

This is the equation to be solved. The first term on the right-hand side is a nonlinear term, and the second term is the **dissipation term** introduced in the wind pattern and sand dune in the previous section; it has the effect of softening the extreme changes in density. In addition, the third term is called the **dispersion term** and has the effect of dispersing the density.

This equation is generically difficult to solve, but, the "stationary wave" solution when either λ or μ is zero can be easily obtained.

5.6.1 *Case of $\lambda \neq 0$, $\mu = 0$*

The equation becomes

$$\frac{\partial \rho(x,t)}{\partial t} = -v_0 \frac{\partial}{\partial x} \left[\rho(x,t) \left(1 - \frac{\rho^2(x,t)}{\rho_0^2} \right) \right] + \lambda \frac{\partial^2 \rho(x,t)}{\partial x^2}. \qquad (5.6.7)$$

This equation is known as **Burgers' equation**[3] in which the order of the equation is raised by one degree. This **stationary wave** solution can be

---

[3] When the wave function is denoted by u, Burgers' equation is written as $\partial u / \partial t + u \partial u / \partial x - \nu \partial^2 u / \partial x^2 = 0$.

easily obtained. Let c be a constant, and we assume

$$\rho = \rho(\xi), \quad \xi = x - ct, \tag{5.6.8}$$

then the equation becomes

$$-c\frac{d\rho}{d\xi} = -v_0\frac{d}{d\xi}\left[\rho\left(1 - \frac{\rho^2}{\rho_0^2}\right)\right] + \lambda\frac{d^2\rho}{d\xi^2}. \tag{5.6.9}$$

Integrating this once with respect to ξ, we have

$$\lambda\frac{d\rho}{d\xi} = \left(v_0 - c - \frac{v_0}{\rho_0^2}\rho^2\right)\rho, \tag{5.6.10}$$

where we assume that $\rho = 0$, $d\rho/d\xi = 0$ at $\xi \to -\infty$; accordingly the constant of integration becomes zero. In addition, setting $0 < c < v_0$, separating the variables and integrating, we obtain

$$\rho = \frac{\rho_1 e^{\alpha\xi}}{\sqrt{1 + e^{2\alpha\xi}}}, \quad \text{where} \quad \rho_1 = \rho_0\sqrt{\frac{v_0 - c}{v_0}}, \quad \alpha = \frac{v_0 - c}{\lambda}. \tag{5.6.11}$$

This solution is a **kink-solution** which satisfies $\rho = 0$ for $\xi \to -\infty$, and $\rho = \rho_1$ for $\xi \to \infty$. The car speed $v(x, t)$ at this time, on substituting this solution into (5.6.4), becomes a constant value $v = c$ that does not depend on location or time. However, the fact that this v becomes a constant value c is seen without substituting this solution into (5.6.4); it is obtained easily by integrating (5.6.2) under the assumption (5.6.8). In fact, (5.6.10) becomes expression (5.6.4) with $v = c$. This kink will proceed at a speed of c without changing its shape as the car moves (see the following subsection).

5.6.2 *Case of* $\lambda = 0$, $\mu \neq 0$

The equation is

$$\frac{\partial\rho(x, t)}{\partial t} = -v_0\frac{\partial}{\partial x}\left[\rho(x, t)\left(1 - \frac{\rho^2(x, t)}{\rho_0^2}\right)\right] + \mu\frac{\partial^3\rho(x, t)}{\partial x^3}. \tag{5.6.12}$$

This equation is essentially the same as the **modified Korteweg–deVries (mKdV) equation**,[4] though they are a little different. To find the stationary wave solution of this equation, we assume the relation (5.6.8), and integrating with respect to ξ, we have

$$\mu\frac{d^2\rho}{d\xi^2} = (v_0 - c)\rho - \frac{v_0}{\rho_0^2}\rho^3. \tag{5.6.13}$$

---

[4]When the wave function is denoted by u, the mKdV equation is written as $\partial u/\partial t + 6u^2\partial u/\partial x + \partial^3 u/\partial x^3 = 0$.

Here, we assume that $\rho = 0$, $d\rho/d\xi = 0$, $d^2\rho/d\xi^2 = 0$ as $\xi \to \pm\infty$. In addition, multiplying both sides by $2(d\rho/d\xi)$ and integrating with respect to ξ, we have

$$\left(\frac{d\rho}{d\xi}\right)^2 = \frac{\beta^2}{\rho_2^2}(\rho_2^2 - \rho^2)\rho^2,$$

$$\text{where} \quad \rho_2 = \sqrt{2}\rho_1 = \rho_0\sqrt{\frac{2(v_0 - c)}{v_0}}, \quad \beta = \sqrt{\frac{v_0 - c}{\mu}}. \tag{5.6.14}$$

From this, $d\rho/d\xi$ is obtained. Then, since its variables are separated, it can be integrated; we obtain the solution

$$\rho(\xi) = \frac{\rho_2}{\cosh(\beta\xi)}. \tag{5.6.15}$$

This is a **soliton-type** solution. The car speed $v(x,t)$ at this time is a constant value $v = c$, same as for the kink-solution. This can also be confirmed by substituting this solution into (5.6.5) with $\lambda = 0$. Therefore, this soliton moves at a speed of c along with the flow of cars. However, this soliton is different from the usual one. That is, depending on the value of c, the higher the height of the soliton, the slower the speed is, and the lower the height of the soliton, the faster the speed is.

5.6.3 *Numerical analysis of highways*

Next, we show some graphs drawn by numerical calculation of equation (5.6.6) which shows the wave phenomenon on highways. As done in the wind pattern in the previous section, let Δt, Δx be the small time and the small distance, respectively, we will measure time and distance using these as units. Also, the constant v_0, λ and μ are redefined in the dimensionless form:

$$v_0\frac{\Delta t}{\Delta x} \mapsto v_0, \quad \lambda\frac{\Delta t}{\Delta x^2} \mapsto \lambda, \quad \mu\frac{\Delta t}{\Delta x^3} \mapsto \mu. \tag{5.6.16}$$

With this transformation, equation (5.6.6) becomes a difference equation

$$\rho(x, t+1) - \rho(x, t)$$

$$= -\frac{v_0}{2}\left[\rho(x+1,t)\left(1 - \frac{\rho^2(x+1,t)}{\rho_0^2}\right) - \rho(x-1,t)\left(1 - \frac{\rho^2(x-1,t)}{\rho_0^2}\right)\right]$$

$$+ \lambda\left[\rho(x+1,t) - 2\rho(x,t) + \rho(x-1,t)\right]$$

$$+ \frac{\mu}{2}\left[\rho(x+2,t) - 2\rho(x+1,t) + 2\rho(x-1,t) - \rho(x-2,t)\right]. \tag{5.6.17}$$

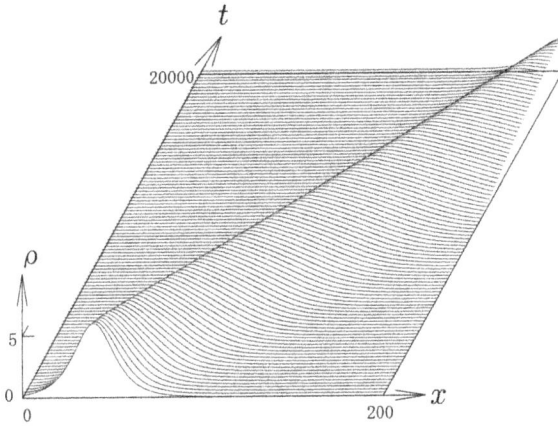

Fig. 5.22 $\lambda = 0.01$, $\mu = 0.001$

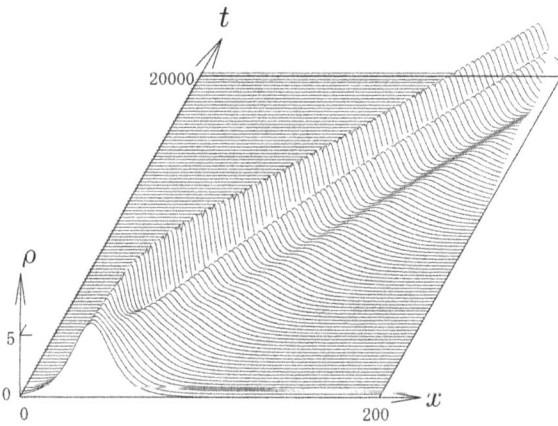

Fig. 5.23 $\lambda = 0.001$, $\mu = 0.01$

Numerical calculations based on this equation are shown in Fig. 5.22 and Fig. 5.23.

In these figures, the values of the parameters v_0, ρ_0, λ, μ are set, respectively, as

$$v_0 = 0.01, \quad \rho_0 = 10, \quad \lambda = 0.01, \quad \mu = 0.001 \quad \text{in Fig. 5.22}$$
$$v_0 = 0.01, \quad \rho_0 = 10, \quad \lambda = 0.001, \quad \mu = 0.01 \quad \text{in Fig. 5.23.} \tag{5.6.18}$$

In Figs. 5.22 and 5.23, the values of v_0, ρ_0 are the same, but the values of λ and μ are 10 times different, respectively. These figures are the same as before; they are shown three-dimensionally with the x-axis in the right horizontal direction, the t-axis in the diagonally upward direction, and the ρ-axis in the upward direction. The x-axis is set from 0 to 200, and both ends are free boundaries. In addition, the time is calculated up to 20 000 steps, and during that time, a total of 101 curves are drawn every 200 steps. We set the initial values, for both Fig. 5.22 and Fig. 5.23, as

$$\rho(x,0) = \frac{6}{\cosh\left((x - 40)/10\right)}. \tag{5.6.19}$$

From Fig. 5.22, it can be seen that the shape of the density ρ approaches the kink-solution over time. Also, in Fig. 5.23, the soliton which is set in the initial state is not a perfect soliton in this system; in the middle of the time evolution, it can be seen to split into three solitons. As mentioned before, in this system, as the lower soliton is faster, we can see that it goes ahead on the x axis.

Problems Involving Bessel Functions

The Bessel function plays the leading role in special functions. In this chapter, we deal with problems where the Bessel function works well, even in unexpected places. First, we explain Kepler's equation which was first solved by Friedrich Wilhelm Bessel using the Bessel function. Next we deal with the buckling problem of an elastic rod in which a special form of the Bessel function, called the Airy function, plays an important role. In addition, we discuss a string vibration problem in the case in which the thickness changes linearly, and the eigenvalue problems of Schrödinger equations with linear potential. Furthermore, we explain some problems in which the Bessel function works well, for example, the vibration of a one-dimensional lattice and the vibration of a keyboard percussion instrument.

6.1 Bessel Function and Kepler's Equation

When we think of the Bessel function, we are probably reminded of a function that appears in the radial component when solving the wave equation in terms of polar coordinates. An integer-order Bessel function appears in the case of a two-dimensional wave equation, and a half-integer-order Bessel function appears in the case of a three-dimensional wave equation. The Bessel function might often be thought of as a function created to solve the wave equation with polar coordinates, but that is not the case.

Friedrich Wilhelm Bessel (1784–1846) came up with the Bessel function to solve **Kepler's equation** which describes the motion of the planets. This equation was a great important problem at that time. The Bessel function, however, existed even before Bessel. It is said that Daniel Bernoulli (1700–1782) first used it to solve Riccati-type differential equations in 1724. When Bernoulli discussed the vibration of a chain suspended under gravity, the zeroth-order Bessel function was also used (see Chapter 2). After that, it was studied by Euler, Lagrange, Fourier, Poisson *et al.*, and was expanded to the integer-order Bessel function. However, this function still did not have a name. In 1824, when Bessel published a treatise that solved the Kepler problem, he summarized the results of this function up to that point. Therefore, this function came to be named the Bessel function after him. This function was subsequently studied by Hankel, Lommel, Neumann, Schläfli *et al.*, and extended to a non-integer-order Bessel function and it had become the most important function among special functions, in the first half of the 20th century.

As ordinary textbooks neither discuss the derivation of Kepler's equation nor its solution by Bessel, little is known about this method in general. Here we establish the Kepler problem and then present the process leading to its solution in an as-easy-to-follow manner as possible without any prior knowledge.

First, as preparation for solving planetary motion, we describe how velocity and acceleration, and elliptical orbits are expressed in polar coordinates. Next, we derive the planetary motion as an elliptical orbit, and derive Kepler's equation to find the position of the planet as a function of time. Furthermore, after preparing the Bessel function, we describe the exact solution of the equation obtained by using the Bessel function. We also describe some formulas for the Bessel function as by-products of this process.

6.1.1 *Preparing to solve planetary motion*

Velocity and acceleration in a polar coordinate system

We take a stationary Cartesian coordinate system (x, y) on a two-dimensional plane, and consider the movement of a point mass. Let the position vector of this point mass be r, and we examine the relationship with Cartesian coordinates when polar coordinates (r, θ) are introduced. Let the unit vector in the r direction and the θ direction be e_r and e_θ, respectively, when the point mass is at position r. Of course, these are vectors that change over time as the mass moves. Let these two vectors e_r, e_θ be expressed by unit vectors i, j of the direction x, y in Cartesian coordinates, and we have

$$e_r = \cos\theta \; i + \sin\theta \; j, \quad e_\theta = -\sin\theta \; i + \cos\theta \; j. \tag{6.1.1}$$

Differentiating these relations with respect to time t, we can easily obtain

$$\frac{de_r}{dt} = \frac{d\theta}{dt}e_\theta, \quad \frac{de_\theta}{dt} = -\frac{d\theta}{dt}e_r. \tag{6.1.2}$$

Next, we express the velocity dr/dt of the point mass in terms of the polar coordinate system. By noting that the position vector is written as $r = re_r$, and by using relation (6.1.2), we have

$$\frac{dr}{dt} = \frac{dr}{dt}e_r + r\frac{d\theta}{dt}e_\theta. \tag{6.1.3}$$

Further we differentiate with respect to time to obtain acceleration

$$\frac{d^2 r}{dt^2} = \frac{d^2 r}{dt^2}e_r + 2\frac{dr}{dt}\frac{d\theta}{dt}e_\theta + r\frac{d^2\theta}{dt^2}e_\theta - r\left(\frac{d\theta}{dt}\right)^2 e_r, \tag{6.1.4}$$

and summarizing, we have

$$\frac{d^2 r}{dt^2} = \left[\frac{d^2 r}{dt^2} - r\left(\frac{d\theta}{dt}\right)^2\right]e_r + \frac{1}{r}\frac{d}{dt}\left(r^2\frac{d\theta}{dt}\right)e_\theta. \tag{6.1.5}$$

Display of an ellipse in a polar coordinate system

An ellipse is the locus of point P such that the sum of the distances from the two focus points F_1, F_2 is constant. Here, we assume that the sum of the distances is $2a$, and the distance between the two focus points is $2ae$ ($0 \le e < 1$).

When the point P is taken as shown in Fig. 6.1, we assume the distance of $F_1 P$ is r, and the angle between $F_1 P$ and the base line of polar coordinates

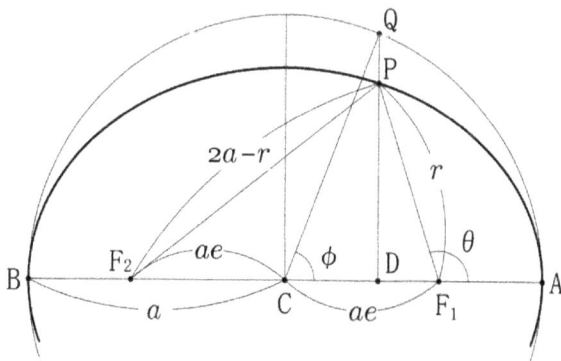

Fig. 6.1 Ellipse

is θ. Then, the distance between F_2P is $2a - r$, and we let the foot of the perpendicular line drawn from P to the base line be the point D. Applying the Pythagorean theorem to a triangle F_2DP, we have

$$(2ae + r\cos\theta)^2 + (r\sin\theta)^2 = (2a - r)^2. \tag{6.1.6}$$

By rearranging this, we obtain

$$r = \frac{\ell}{1 + e\cos\theta}, \quad \ell = a(1 - e^2). \tag{6.1.7}$$

This is the expression of the ellipse in the polar coordinate system. The e is called **eccentricity**, and the ℓ is called **semi-latus rectum** which is equal to the value of r when $\theta = \pi/2$. It is obvious that a is the **semi-major axis** of this ellipse. And since the **semi-minor axis** b is $b = a\sqrt{1 - e^2}$, the semi-latus rectum ℓ is also rewritten as $\ell = b^2/a$. The closer eccentricity e is to 0, the closer the ellipse becomes a circle, and the closer e is to 1, the thinner the ellipse becomes.

Let the center of the ellipse be the point C. By extending the line segment DP, let the intersection with a circle with radius a be the point Q. Then, let the angle between CQ and the baseline be ϕ. This angle ϕ is called **eccentric anomaly**, while θ is called **true anomaly**. These terms, used by astronomers, seem to have some origins. As we can see from Fig. 6.1, the cosine of the eccentric anomaly ϕ is given by

$$\cos\phi = \frac{ae + r\cos\theta}{a}. \tag{6.1.8}$$

With the aid of (6.1.7), this becomes

$$\cos\phi = \frac{e + \cos\theta}{1 + e\cos\theta}. \tag{6.1.9}$$

Solving this in reverse, we have

$$\cos\theta = \frac{\cos\phi - e}{1 - e\cos\phi}. \tag{6.1.10}$$

Substituting this equation back to relation (6.1.7), we obtain

$$r = a(1 - e\cos\phi). \tag{6.1.11}$$

This relation is represented without using fractions. For later use, from (6.1.10), we find

$$\sin\theta = \sqrt{1 - \cos^2\theta} = \frac{\sqrt{1 - e^2}\,\sin\phi}{1 - e\cos\phi}. \tag{6.1.12}$$

This relation shows that there is no sign problem when θ and ϕ are extended to arbitrary values, because when the sign of $\sin\theta$ changes, the sign of $\sin\phi$ also changes at the same time.

6.1.2 *Motion of a planet*

Derivation of an elliptical orbit

The equation of motion of a planet with mass m which is rotating around the sun with mass M, is written as

$$m\frac{d^2\boldsymbol{r}}{dt^2} = -G\frac{Mm}{r^2}\boldsymbol{e}_r, \tag{6.1.13}$$

where G is the gravitational constant. If we decompose this equation into the r direction and the θ direction using relation (6.1.5), we have

$$\frac{d^2r}{dt^2} - r\left(\frac{d\theta}{dt}\right)^2 = -\frac{GM}{r^2}, \tag{6.1.14}$$

and

$$\frac{d}{dt}\left(r^2\frac{d\theta}{dt}\right) = 0. \tag{6.1.15}$$

First, equation (6.1.15) yields a constant,

$$r^2\frac{d\theta}{dt} = \text{const.} \equiv h. \tag{6.1.16}$$

Areal velocity is the rate at which an area is swept by a planet around the sun per unit time. This is equal to $h/2$, and the relation is called the **law of constant areal velocity**.

From these equations, r and θ should be calculated as functions of time, but this is cumbersome, so we will not do it at the moment. Let us first find the relationship between r and θ. We replace the time derivative with the angle θ derivative by using the law of constant areal velocity. Specifically, we put

$$\frac{dr}{dt} = \frac{d\theta}{dt}\frac{dr}{d\theta} = \frac{h}{r^2}\frac{dr}{d\theta}, \tag{6.1.17}$$

and the second derivative with respect to time

$$\frac{d^2r}{dt^2} = \frac{d}{dt}\left(\frac{h}{r^2}\frac{dr}{d\theta}\right) = \frac{d\theta}{dt}\frac{d}{d\theta}\left(\frac{h}{r^2}\frac{dr}{d\theta}\right) = \frac{h}{r^2}\frac{d}{d\theta}\left(\frac{h}{r^2}\frac{dr}{d\theta}\right); \tag{6.1.18}$$

furthermore, substituting this expression into equation (6.1.14), we obtain

$$\frac{h}{r^2}\frac{d}{d\theta}\left(\frac{h}{r^2}\frac{dr}{d\theta}\right) - r\left(\frac{h}{r^2}\right)^2 = -\frac{GM}{r^2}. \tag{6.1.19}$$

At first glance, this equation looks difficult to solve. Here we define h^2/GM with the length ℓ as

$$\ell \equiv \frac{h^2}{GM}, \tag{6.1.20}$$

and rearranging a little, we have

$$\frac{d}{d\theta}\left(\frac{\ell}{r^2}\frac{dr}{d\theta}\right) - \left(\frac{\ell}{r} - 1\right) = 0. \tag{6.1.21}$$

Here, we transform the second term on the left-hand side as

$$\frac{\ell}{r} - 1 = \zeta, \tag{6.1.22}$$

whose derivative is

$$-\frac{\ell}{r^2}dr = d\zeta. \tag{6.1.23}$$

Then equation (6.1.21) is expressed in terms of ζ as

$$\frac{d^2\zeta}{d\theta^2} + \zeta = 0. \tag{6.1.24}$$

As is well known, the solution of this equation is represented by trigonometric functions, and it becomes, with e and δ_0 as constants of integration,

$$\zeta = e\cos(\theta - \delta_0). \tag{6.1.25}$$

Returning to the original r, we obtain the solution

$$r = \frac{\ell}{1 + e\cos(\theta - \delta_0)}. \tag{6.1.26}$$

This is the ellipse in polar coordinates of (6.1.7), if it is rotated around the focus point F_1 by angle δ_0. If the reference point for measuring the angle θ is taken to the **perihelion** (the point closest to the sun, point A in Fig. 6.1), this δ_0 can always be excluded. Also, when the constant e corresponding to eccentricity is $0 \le e < 1$, the movement of the planet becomes an elliptical orbit that periodically rotates around the sun without being sucked into or separated from the sun infinitely. It should be noted that when this e is equal to 1 or larger than 1, its orbit becomes a parabola or a hyperbola which corresponds to the orbit of a comet. However, this situation will not be dealt with here.

Derivation of Kepler's equation

Next, let us find the relationship with time t. By integrating the constant areal velocity (6.1.16) over time t, we have

$$ht = \int_0^\theta r^2 d\theta', \tag{6.1.27}$$

where the time t is measured from the perihelion in the same manner as θ. By substituting r of (6.1.26) with $\delta_0 = 0$ in this relation, we have

$$ht = \int_0^\theta \frac{\ell^2}{\left(1 + e\cos\theta'\right)^2} d\theta'. \tag{6.1.28}$$

This integration can be carried out as it is, but it becomes a very cumbersome expression. In order to work this in a smart way, we convert from the true anomaly θ to the eccentric anomaly ϕ in relation (6.1.28). To do so, we take the derivative of (6.1.10),

$$\sin\theta \, d\theta = \frac{(1 - e^2)\sin\phi}{(1 - e\cos\phi)^2} d\phi, \tag{6.1.29}$$

and furthermore, substituting $\sin\theta$ of (6.1.12), we have

$$d\theta = \frac{\sqrt{1 - e^2}}{1 - e\cos\phi} d\phi. \tag{6.1.30}$$

We substitute this relation and relation (6.1.11), which represents r as a function of ϕ, into relation (6.1.27), and we obtain the form that does not include fractions,

$$ht = a^2\sqrt{1 - e^2} \int_0^\phi (1 - e\cos\phi')d\phi'. \tag{6.1.31}$$

This can be easily integrated to give

$$ht = a^2\sqrt{1 - e^2}(\phi - e\sin\phi). \tag{6.1.32}$$

Since a is the semi-major axis of the ellipse and the semi-minor axis b is $b = a\sqrt{1 - e^2}$, the area of the ellipse S is equal to $S = \pi ab = \pi a^2\sqrt{1 - e^2}$.

Further, the area S divided by the areal velocity $h/2$ becomes the period T, and we have

$$T = \frac{2\pi a^2 \sqrt{1 - e^2}}{h}. \tag{6.1.33}$$

Furthermore, rewriting relation (6.1.32) in terms of this period T, we obtain

$$\phi - e\sin\phi = \frac{2\pi t}{T}. \tag{6.1.34}$$

This is called Kepler's equation. However, Kepler himself did not derive this equation. Johannes Kepler (1571–1630) lived before Isaac Newton (1642–1727). Kepler may have found this equation empirically by arranging the vast amount of observational data left by Tycho Brahe (1546–1601). The man who had derived Kepler's equation according to Newtonian mechanics is Joseph-Louis Lagrange (1736–1813). This was in 1770, 140 years after Kepler's death.

6.1.3 *Exact solution of Kepler's equation*

Equation (6.1.34) connects the time t and the eccentric anomaly ϕ. However, even if we try to find ϕ from this equation, it is not easy because it is a transcendental equation. Bessel solved this problem in 1817.

Preparation of the Bessel function

Before solving this equation, we must have the basic knowledge of the Bessel function. The function $\exp[(z/2)(t - 1/t)]$ which is called the generating function of the Bessel function is divided into two products and is carried out by the Taylor expansion,

$$\exp\left[\frac{z}{2}\left(t - \frac{1}{t}\right)\right] = \left[\sum_{k=0}^{\infty} \frac{1}{k!}\left(\frac{zt}{2}\right)^k\right]\left[\sum_{\ell=0}^{\infty} \frac{1}{\ell!}\left(-\frac{z}{2t}\right)^\ell\right]. \tag{6.1.35}$$

Removing these parentheses and aligning with the power of t, that is, making it into the form of a Laurent expansion, the right-hand side becomes

$$= \sum_{n=0}^{\infty}\left[\sum_{m=0}^{\infty} \frac{(-1)^m}{m!(n+m)!}\left(\frac{z}{2}\right)^{n+2m}\right]t^n + \sum_{n=1}^{\infty}\left[\sum_{m=0}^{\infty} \frac{(-1)^{n+m}}{m!(n+m)!}\left(\frac{z}{2}\right)^{n+2m}\right]t^{-n}. \tag{6.1.36}$$

The coefficient attached to t^n of the first term of this expression is just the definition of the Bessel function of n-th order for the non-negative integer n,

$$J_n(z) = \sum_{m=0}^{\infty} \frac{(-1)^m}{m!(n+m)!} \left(\frac{z}{2}\right)^{n+2m}. \tag{6.1.37}$$

Furthermore, if we extend this definition to a negative integer n, as the factorial of negative integer is infinite, we have the relation

$$J_{-n}(z) = \sum_{m=0}^{\infty} \frac{(-1)^m}{m!(-n+m)!} \left(\frac{z}{2}\right)^{-n+2m} = \sum_{m=n}^{\infty} \frac{(-1)^m}{m!(-n+m)!} \left(\frac{z}{2}\right)^{-n+2m}$$

$$= \sum_{m=0}^{\infty} \frac{(-1)^{n+m}}{(n+m)!m!} \left(\frac{z}{2}\right)^{n+2m} = (-1)^n J_n(z). \tag{6.1.38}$$

As a result, this expansion formula becomes

$$\exp\left[\frac{z}{2}\left(t - \frac{1}{t}\right)\right] = \sum_{n=-\infty}^{\infty} J_n(z)t^n. \tag{6.1.39}$$

This is called the generating function display of the Bessel function.

Next, we divide both sides of expression (6.1.39) by t^{n+1}, and consider an integral that goes around the origin of the complex t plane. From Cauchy's integral theorem, the right-hand side is $2\pi i J_n(z)$, and the result is

$$2\pi i J_n(z) = \oint \frac{1}{t^{n+1}} \exp\left[\frac{z}{2}\left(t - \frac{1}{t}\right)\right] dt. \tag{6.1.40}$$

Here we convert the integral variable from t to ϕ as $t = e^{i\phi}$,[1] and we have

$$J_n(z) = \frac{1}{2\pi} \int_{-\pi}^{\pi} e^{-i(n\phi - z\sin\phi)} d\phi. \tag{6.1.41}$$

Dividing this integration interval into $[-\pi, 0]$ and $[0, \pi]$, and changing the sign of ϕ in the integration of the interval $[-\pi, 0]$, we obtain

$$J_n(z) = \frac{1}{\pi} \int_0^{\pi} \cos(n\phi - z\sin\phi)d\phi. \tag{6.1.42}$$

This expression was found by Bessel and is called **Bessel's integral representation formula**. This formula contains a linear combination of ϕ and $\sin\phi$. It already has an atmosphere that seems to be related to Kepler's equation.

---

[1] In order to distinguish it from the eccentricity e, we denote Napier's constant by the non-italic e.

Solution of Kepler's equation

In the following, the right-hand side of Kepler's equation (6.1.34) is set as dimensionless time τ

$$\tau \equiv \frac{2\pi t}{T}. \tag{6.1.43}$$

Then the equation is rewritten as

$$\phi - e \sin \phi = \tau. \tag{6.1.44}$$

From this equation, we see that when $\tau = n\pi$, ϕ is also $n\pi$ for any integer n. Even if the signs of ϕ and τ are changed at once, the equation is invariant. Therefore, ϕ must be an odd-function of τ. Furthermore, because the equation is invariant for the translation with $\phi \mapsto \phi + 2\pi$, $\tau \mapsto \tau + 2\pi$, it can be seen that $\phi - \tau$ is a periodic function with a period of 2π. Therefore, we can expand $\phi - \tau$ in the Fourier-sine series:

$$\phi - \tau = \sum_{n=1}^{\infty} A_n \sin(n\tau). \tag{6.1.45}$$

Multiplying both sides by $\sin(m\tau)$ and integrating with respect to τ with the aid of the orthogonality of sine function

$$\int_0^\pi \sin(n\tau) \sin(m\tau) d\tau = \frac{\pi}{2} \delta_{n,m}, \tag{6.1.46}$$

we find that the coefficients A_n are represented by integral

$$A_n = \frac{2}{\pi} \int_0^\pi (\phi - \tau) \sin(n\tau) d\tau. \tag{6.1.47}$$

Integrating by parts, and noting that $\phi - \tau$ becomes zero at $\tau = 0$, π, we have

$$A_n = \frac{2}{\pi} \int_0^\pi \frac{\cos(n\tau)}{n} \left(\frac{d\phi}{d\tau} - 1\right) d\tau = \frac{2}{\pi} \int_0^\pi \frac{\cos(n\tau)}{n} \frac{d\phi}{d\tau} d\tau. \tag{6.1.48}$$

In this final integral, Kepler's equation (6.1.44) can be used to change the integral variable from τ to ϕ. Then it becomes

$$A_n = \frac{2}{\pi n} \int_0^\pi \cos\left[n(\phi - e \sin \phi)\right] d\phi. \tag{6.1.49}$$

This is the same type as the Bessel integral representation formula presented in expression (6.1.42), and as a result, we have

$$A_n = \frac{2}{n} J_n(ne). \tag{6.1.50}$$

From the above results, ϕ is solved in the form of a trigonometric series with the Bessel function as the coefficient, that is, we have the solution

$$\phi = \tau + \sum_{n=1}^{\infty} \frac{2}{n} J_n(ne) \sin(n\tau). \tag{6.1.51}$$

Since ϕ was found as a function of time t, if we want to know $\cos \phi$ as a function of t, we have only to take the cosine of $\phi(t)$. However, there is a more intelligent way using Fourier-cosine series,

$$\cos \phi = B_0 + \sum_{n=1}^{\infty} B_n \cos(n\tau). \tag{6.1.52}$$

To find B_0 first, integrating both sides with respect to τ from 0 to π, and carrying out integration by parts, we have

$$\pi B_0 = \int_0^\pi \cos \phi \, d\tau = \tau \cos \phi \Big|_0^\pi + \int_0^\pi \tau \sin \phi \frac{d\phi}{d\tau} d\tau. \tag{6.1.53}$$

When $\tau = \pi$, ϕ is also π, and for the second term, using Kepler's equation to convert the τ integral to the ϕ integral, we have

$$\pi B_0 = -\pi + \int_0^\pi (\phi - e \sin \phi) \sin \phi \, d\phi, \tag{6.1.54}$$

and carrying out this integral, we obtain

$$B_0 = -\frac{e}{2}. \tag{6.1.55}$$

To determine B_n for which n is 1 or greater, we multiply both sides of (6.1.52) by $\cos(m\tau)$ and use the orthogonality relation of the cosine function,

$$\int_0^\pi \cos(n\tau) d\tau = 0, \quad \int_0^\pi \cos(n\tau) \cos(m\tau) d\tau = \frac{\pi}{2} \delta_{n,m}, \quad n, m \geq 1. \tag{6.1.56}$$

Then we have

$$\frac{\pi}{2} B_n = \int_0^\pi \cos(n\tau) \cos \phi \, d\tau. \tag{6.1.57}$$

We integrate by parts in a similar way as before, eliminate τ by using Kepler's equation and change to ϕ integral from τ integral. Furthermore, it is divided into two terms using the product-to-sum formula of trigonometric functions. Finally, using the Bessel integral representation formula, we obtain

$$B_n = \frac{1}{n}\left[J_{n-1}(ne) - J_{n+1}(ne)\right]. \tag{6.1.58}$$

From the above results, $\cos\phi$ is expressed as a function of τ:

$$\cos\phi = -\frac{e}{2} + \sum_{n=1}^{\infty} \frac{1}{n}\left[J_{n-1}(ne) - J_{n+1}(ne)\right]\cos(n\tau). \tag{6.1.59}$$

Substituting this result into relation (6.1.11), the radial component r is presented as a function of time:

$$r = a\left(1 + \frac{e^2}{2}\right) - ae\sum_{n=1}^{\infty} \frac{1}{n}\left[J_{n-1}(ne) - J_{n+1}(ne)\right]\cos(n\tau). \tag{6.1.60}$$

There is a question whether the cosine of the true anomaly θ can be found in the same way. Deforming expression (6.1.10), we have

$$\cos\theta = -\frac{1}{e} + \left(\frac{1-e^2}{e}\right)\frac{1}{1 - e\cos\phi}. \tag{6.1.61}$$

To find the second term, we differentiate Kepler's equation (6.1.44) with respect to τ,

$$\frac{1}{1 - e\cos\phi} = \frac{d\phi}{d\tau}. \tag{6.1.62}$$

This right-hand side can be obtained from solution (6.1.51) by differentiating with respect to τ, so we have

$$\frac{1}{1 - e\cos\phi} = 1 + 2\sum_{n=1}^{\infty} J_n(ne)\cos(n\tau). \tag{6.1.63}$$

Substituting this into relation (6.1.61), we obtain the result,

$$\cos\theta = -e + \frac{2(1-e^2)}{e}\sum_{n=1}^{\infty} J_n(ne)\cos(n\tau). \tag{6.1.64}$$

However, this equation has poorer convergence than expressions (6.1.51) and (6.1.59).

Extra formula for the Bessel function

Since $\phi = 0$ for $\tau = 0$, we have a formula for the Bessel function from (6.1.63):

$$1 + 2 \sum_{n=1}^{\infty} J_n(ne) = \frac{1}{1-e}. \tag{6.1.65}$$

Similarly, from (6.1.59) with $\tau = 0$, we obtain

$$\sum_{n=1}^{\infty} \frac{1}{n} [J_{n-1}(ne) - J_{n+1}(ne)] = 1 + \frac{e}{2}. \tag{6.1.66}$$

With the help of the differential formula of the Bessel function

$$J_{n-1}(z) - J_{n+1}(z) = 2 \frac{d}{dz} J_n(z), \tag{6.1.67}$$

we can rewrite (6.1.66) by integrating with respect to e to get

$$\sum_{n=1}^{\infty} \frac{2}{n^2} J_n(ne) = e + \frac{e^2}{4}, \tag{6.1.68}$$

where the constant of integration has been chosen so that both sides vanish for $e = 0$.

Expressions (6.1.65) and (6.1.66) appear in a mathematics formula collection as the relations for Kapteyn expansion. These three formulas, including formula (6.1.68), seem a little suspicious, so they were numerically examined by taking the sum over n up to 30. Although the convergence becomes worse as the value of e approaches 1, the values of both sides are in good agreement within the range of $e < 1$. From this result, these formulas were definitely genuine and unquestionable. However, they do not match at all in $e > 1$. It would be natural, because this situation assumes an elliptical orbit from the beginning. That is, in these expressions, the analytic continuation is not possible from the area of $e < 1$ to the area of $e > 1$. However, what are these formulas useful for?

Significance of Finding an Exact Solution

Kepler's equation that is dealt with here is the world's first transcendental equation for which an exact solution was sought. The process of arriving at this result, that is, switching from the true anomaly θ to the eccentric anomaly ϕ is said to be brilliantly technical. This is probably due to the historical background that began with Kepler.

However, when we think about it, the difficulty of the transcendental equation may just be in the Bessel function. Even if this is said to be an exact solution, when trying to find this value numerically, no one would seek ϕ according to the formula obtained by this Bessel function. Just finding the value of the Bessel function is difficult. Moreover, we have to sum the infinite series. If we want to find the solution numerically, it is much quicker to find it directly from Kepler's equation by using the method of successive approximation.

What is the significance of finding an exact solution? Nowadays, computer simulation using the finite element method is well developed, and we have entered an era in which this method is numerically sufficient without any knowledge of the mathematical logic. The era of discussions using special functions seem to be already over, and it may be that only nostalgists still indulge in such discussions. However, the beauty of the formula connected by "=" and the beauty of the formula connected by "≃" are completely different. As a simple example, this is just like approximating an irrational number with a rational number. Certainly, an irrational number can be approximated to any extent by a rational number. However, an irrational number and a rational number are essentially different.

Nowadays, perhaps special functions are like classical music masterpieces worth listening to or classical painting masterpieces worth looking at.

6.2 Bessel Function and Airy Function

The title of this section is "Bessel Function and Airy Function"; however the Airy function is a kind of Bessel function. It may be correct to say "Airy function as Bessel function". In particular, a Bessel function of order $\pm 1/3$ is sometimes called the **Airy function**. The Airy function is named after George Biddell Airy (1801–1892) who was the Chief of the Royal Observatory, Greenwich in England. He discovered the "**Airy integral**" in developing the theory of the rainbow (1838). It was also this Airy who defined the meridian that passes through the Greenwich Observatory as the Prime Meridian, and the starting point for Coordinated Universal Time (UTC).

A second-order linear ordinary differential equation which contains a first-order term of independent variable can be solved by using a Bessel function of order $\pm 1/3$. Here we look at three such examples.

6.2.1 *Buckling of an elastic rod by gravity*

What is buckling by gravity?

How high can a steel column stand when it is erected straight upright? In order not to be crushed by its own gravity, there must be enough compressive strength at the bottom of the column to exceed its gravity. Let the compressive strength of steel be $\sigma_y = 200\,\text{MPa}$, the density be $\rho = 8000\,\text{kg/m}^3$, and the gravitational acceleration be $g \cong 10\,\text{m/s}^2$. If the height is h m, the pressure applied to the bottom part is $\rho g h$ which must be less than the compressive strength. The maximum height is therefore $h = \sigma_y/\rho g = 2500\,\text{m}$. This is a theoretical value, under certain conditions, such as if there is no wind. However, even if there is no wind, the column is unstable and will bend due to its own gravity. It is better to lower the potential energy due to gravity by bending than by reducing its height. This is called **buckling by gravity**. Let us investigate this phenomenon in more detail.

Introduction to the equation, and the solution

Assume an elastic rod with height h stands vertically from the x axis upward and its bottom end as the origin. The bottom end is embedded, and its lateral displacement and tilt angle are zero.

Let the **volume density**, **Young's modulus**, **cross-sectional area** and **moment of inertia of area** of this rod be ρ, E, S and I, respectively. Now assume that this rod is inclined and let the angle of inclination of the principal axis and the lateral displacement at the point x be $\theta(x)$ and $V(x)$, respectively. The relationship between these two quantities is

$$V(x) = \int_0^x \sin\theta(x')dx'. \tag{6.2.1}$$

Since shear deformation is not considered here, the angle of inclination of the cross-section is also θ. At this time, at point x, the **strain energy caused by the bending** U_1 and the **potential energy of gravity** U_2 per unit volume are given by

$$U_1 = \frac{EI}{2S}\theta_x{}^2, \quad U_2 = \rho g \int_0^x \cos\theta(x')dx', \tag{6.2.2}$$

respectively,[2] where the subscript x of the θ_x represents the derivative of that variable. Hereafter, this notation is followed. Therefore, the energy of

---

[2]Please refer to an appropriate textbook for the derivation method of U_1.

the entire elastic rod U is written as

$$U = \int_0^h (U_1 + U_2)\,S\,dx = \frac{EI}{2}\int_0^h \theta_x{}^2\,dx + \rho g S \int_0^h \left[\int_0^x \cos\theta(x')\,dx'\right]dx.$$

$$(6.2.3)$$

In the second term of the right-hand side, by exchanging the order of integration of x' and x, the equation is rewritten as

$$U = \frac{EI}{2}\int_0^h \theta_x{}^2\,dx + \rho g S \int_0^h (h - x)\cos\theta(x)\,dx. \qquad (6.2.4)$$

Now we take the variation of this energy U with respect to θ, and we obtain the equation as the Euler–Lagrange equation

$$EI\theta_{xx} + \rho g S(h - x)\sin\theta = 0. \qquad (6.2.5)$$

Furthermore, the linear approximation $\sin\theta \cong \theta$ is adopted as the angle of inclination θ is small, so the equation is rewritten as

$$EI\theta_{xx} + \rho g S(h - x)\theta = 0. \qquad (6.2.6)$$

Since this is a linear homogeneous differential equation, it can be solved.

Before that, let us make the variables dimensionless. First, the height h of the elastic rod is used as the unit of length, and the variable x is changed to

$$\frac{x}{h} \mapsto x. \qquad (6.2.7)$$

In addition, we introduce a dimensionless quantity μ which is a measure of the ease of bending

$$\mu = \sqrt{\frac{\rho g S h^3}{EI}}. \qquad (6.2.8)$$

Then equation (6.2.6) can be further rewritten as

$$\theta_{xx} + \mu^2(1 - x)\theta = 0. \qquad (6.2.9)$$

To solve this equation, we transform the independent variable x and the dependent variable θ as follows:

$$z = \frac{2}{3}\mu(1 - x)^{3/2}, \quad \theta(x) = z^{1/3}f(z). \qquad (6.2.10)$$

This transformation is something we can do because we know the answer. As a result, we have

$$\frac{d^2 f}{dz^2} + \frac{1}{z}\frac{df}{dz} + \left(1 - \frac{1}{9z^2}\right)f = 0. \qquad (6.2.11)$$

The solution of this equation can be found in terms of the **Bessel function of order** $\pm 1/3$, with C and D as arbitrary constants:

$$f(z) = C J_{-1/3}(z) + D J_{1/3}(z). \tag{6.2.12}$$

Transforming this back to the original variable, we obtain the final solution

$$\theta(x) = \sqrt{1-x}\left[C J_{-1/3}\left(\frac{2\mu}{3}(1-x)^{3/2}\right) + D J_{1/3}\left(\frac{2\mu}{3}(1-x)^{3/2}\right)\right], \tag{6.2.13}$$

where the overall coefficient factor $(2\mu/3)^{1/3}$ has been absorbed into the arbitrary constants C and D. Although equation (6.2.9) is simple, the solution is complicated. The Bessel function of order $\pm 1/3$ was first studied by Airy, and is called the **Airy function**. The exact definition of the Airy function is given in the following section.

By using the differential formula of the Bessel function

$$\frac{d}{dz}J_\nu(z) = \frac{\nu}{z}J_\nu(z) - J_{\nu+1}(z) = J_{\nu-1}(z) - \frac{\nu}{z}J_\nu(z), \tag{6.2.14}$$

we differentiate solution (6.2.13) with respect to x,

$$\theta_x(x) = C\mu(1-x)J_{2/3}\left(\frac{2\mu}{3}(1-x)^{3/2}\right) - D\mu(1-x)J_{-2/3}\left(\frac{2\mu}{3}(1-x)^{3/2}\right). \tag{6.2.15}$$

Now, we impose the boundary condition on this solution. Since the bottom end $x = 0$ of this elastic rod is embedded, the angle of inclination is zero, and in addition, at the upper end $x = 1$, the **bending moment** is zero, or in other words, the **curvature** is zero. These conditions are expressed as

$$\theta(0) = 0, \quad \theta_x(1) = 0. \tag{6.2.16}$$

First, we apply the condition $\theta_x(1) = 0$ to expression (6.2.15). The first term of the right-hand side becomes zero at $x = 1$. As for the second term, a non-vanishing constant remains at $x = 1$, as is seen from the Taylor expansion of the Bessel function.[3] Therefore, it must be $D = 0$. After that, we apply $\theta(0) = 0$ to expression (6.2.13), and we have

$$J_{-1/3}\left(\frac{2\mu}{3}\right) = 0, \tag{6.2.17}$$

that is, $2\mu/3$ must be the zero point of the Bessel function of order $-1/3$. Let these zero points be $2\mu_i/3$, $(i = 1, 2, 3, \ldots)$ in increasing order. We list

---

[3] See (6.2.68) given considerably later.

the first six values:

$$\frac{2\mu_i}{3} = 1.866, \quad 4.987, \quad 8.166, \quad 11.493, \quad 14.554, \quad 17.700. \quad (6.2.18)$$

Rewriting $\theta(x)$ as $\theta(x, \mu_i)$ in order to clarify its dependence on μ_i, we have the final solution,

$$\theta(x, \mu_i) = C\sqrt{1 - x}\, J_{-1/3}\left(\frac{2\mu_i}{3}(1 - x)^{3/2}\right). \quad (6.2.19)$$

This represents the solution of buckling by gravity.

Specific example of buckling by gravity

Let us consider a specific example of this buckling by gravity. In the variables included in the definition of μ (6.2.8), ρ and E are constants determined by the choice of material. Here, it is assumed to be steel, and we put

$$\rho = 8000 \, \mathrm{kg/m}^3, \quad E = 200 \, \mathrm{GPa}. \quad (6.2.20)$$

Also, S and I are constants determined by the shape of the cross-section. Here, it is a column whose cross-section is a disc with a radius r, so we put

$$S = \pi r^2, \quad I = 4\int_0^r y^2\sqrt{r^2 - y^2}\, dy = \frac{\pi}{4}r^4. \quad (6.2.21)$$

At this time, S/I becomes, for a column,

$$\frac{S}{I} = \frac{4}{r^2}. \quad (6.2.22)$$

In addition to this, we consider a cylinder with an outer radius r and an inner radius $r - d$, that is, its thickness being d. However, it is assumed that this thickness is very thin, and we can calculate S and I with the condition $d \ll r$. The result is

$$S = \pi[r^2 - (r - d)^2] \cong 2\pi r d, \quad I = \frac{\pi}{4}[r^4 - (r - d)^4] \cong \pi r^3 d. \quad (6.2.23)$$

In this case, S/I becomes

$$\frac{S}{I} = \frac{2}{r^2}. \quad (6.2.24)$$

This result does not depend on the thickness of the cylinder d. Whether it buckles or not is determined only by the outer radius, regardless of the thickness of the material that makes the cylinder. This may seem strange, but this is probably because the thinner the material, the less it is susceptible to

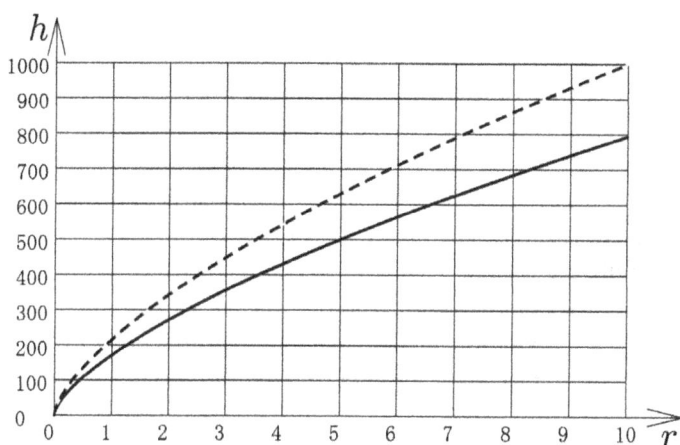

Fig. 6.2 Buckling by gravity

gravity. These values (6.2.20), (6.2.22), (6.2.24) and $g = 9.8\,\text{m/s}^2$ are substituted in μ of (6.2.8), and let this value μ be equal to μ_1 which initially causes buckling. As a result, we can calculate the relationship between the height h and the radius r when buckling occurs. This is shown in Fig. 6.2. Here, the units of r and h are meter, and the case of the column is shown by a solid line, and the case of a cylinder is shown by a dotted line. From this graph, even if the radius is $r = 10\,\text{m}$ which is quite large for a building, the buckling height h for a column is about 800 m, and for a cylinder it is about 1000 m. This value is considerably smaller than the height of 2500 m, where the bottom end of the elastic rod breaks due to its own gravity. In other words, it means that a vertically standing elastic rod buckles by gravity before it collapses by gravity.

By the way, the tallest building in the world is a building in Dubai, United Arab Emirates with a height of 828 m. This building gets thinner as it goes up. If the thickness is uniform, the height is about 400 m at best.

Dynamic analysis is required to determine how buckling by gravity occurs in time. However, when dynamic analysis is performed on this problem, it becomes a fourth-order differential equation which becomes a very cumbersome problem. Here, we will only give an overview of the numerical analysis of this problem. Let θ be $\theta(x, t)$ at coordinates x and time t, and let this time-dependent part be $\cos(\omega t)$. Substituting into the equation and adding boundary conditions, the value of frequency ω is obtained as the eigenvalue. Let this be ω_i, $(i = 1, 2, 3, \ldots)$. This value depends on μ: the larger the value of μ, the smaller the value of each w_i. Then, when $\mu = \mu_1$, it becomes $w_1 = 0$. This corresponds to the static solution described here. If μ is made larger than that, w_1 which was previously obtained as a positive real number, becomes an imaginary number. Furthermore, in general, when $\mu = \mu_i, (i = 2, 3, \ldots)$, then $w_i = 0$. If the value of μ is increased further, the value of w_i becomes an imaginary number. That the value of w is imaginary means that the trigonometric function $\cos(wt)$ turns into the hyperbolic function $\cosh(|w|t)$. This means that the amplitude increases exponentially and collapses over time.

In the next section, we will introduce a completely different model in which the calculation formula derived here can be used as it is.

6.2.2 *Vibration of string whose density changes linearly*

As shown in the previous section, the second-order differential equations containing the first-order term of an independent variable can be solved by using the Bessel function of order $\pm 1/3$. If we look for such an example in classical mechanics, there is the vibration of the string whose density changes linearly depending on the location. This model is easiest to understand and dynamic analysis can be performed on it.

Introduction to the equation, and the solution

Now, a string with a density ρ is stretched with a tension T for a finite length ℓ. Here, this density ρ changes from place to place. The x-axis is taken along the string, and one end point is taken as the origin. Let the lateral displacement at coordinate x, time t be $V(x, t)$, and the density be $\rho(x)$. Within the range of linear approximation, we have the wave equation

$$\rho(x)V_{tt} = TV_{xx}. \tag{6.2.25}$$

Here, the density of the string is assumed to be a linear expression of x,

$$\rho(x) = \rho_0\left(1 - \frac{x}{\ell}\right). \tag{6.2.26}$$

That is, at $x = 0$, $\rho = \rho_0$. This value decreases linearly and is set to zero at the other end $x = \ell$. Here we define a constant c having the dimension of velocity, and a constant τ having the dimension of time, as

$$c = \sqrt{\frac{T}{\rho_0}}, \quad \tau = \frac{\ell}{c}. \tag{6.2.27}$$

By using ℓ and τ as the units of length and time, respectively, we transpose the variables x, t and V to dimensionless ones,

$$\frac{x}{\ell} \mapsto x, \quad \frac{t}{\tau} \mapsto t, \quad \frac{V}{\ell} \mapsto V, \tag{6.2.28}$$

then the equation becomes

$$(1 - x)V_{tt} = V_{xx}. \tag{6.2.29}$$

To solve this equation, we separate the displacement $V(x, t)$ for x and t, and assuming that the time-dependent part has a single frequency ω, we put

$$V(x, t) = X(x)\sin(\omega t). \tag{6.2.30}$$

The equation of this displacement becomes

$$X_{xx} + \omega^2(1 - x)X = 0. \tag{6.2.31}$$

If we replace ω with μ, and X with θ, this equation is exactly the same as equation (6.2.9). Therefore, if we set the boundary conditions here the same as in the previous section, that is, $x = 0$ is the fixed end, and $x = 1$ is the free end,

$$X(0) = 0, \quad X_x(1) = 0, \tag{6.2.32}$$

we can use (6.2.9) to (6.2.19) as they are, by replacing μ with ω, and θ with X. We summarize the result here again. The frequency ω is determined by the **eigenvalue** equation

$$J_{-1/3}\left(\frac{2\omega}{3}\right) = 0. \tag{6.2.33}$$

Let the solutions of this equation be the eigenvalues ω_i, $(i = 1, 2, 3, \ldots)$. Their first 6 values are the same as (6.2.18)

$$\frac{2\omega_i}{3} = 1.866, \quad 4.987, \quad 8.166, \quad 11.493, \quad 14.554, \quad 17.700. \tag{6.2.34}$$

Also, we rewrite the **eigenfunction** as $X(x, \omega_i)$ to clarify the ω_i dependency of $X(x)$,

$$X(x, \omega_i) = \sqrt{1 - x}\, J_{-1/3}\left(\frac{2\omega_i}{3}(1 - x)^{3/2}\right). \tag{6.2.35}$$

Here, the constant C in expression (6.2.19) is set to 1, however, this is not normalized.

Normalization of eigenfunction
To normalize the eigenfunction, we prepare two ω, ω' that are not necessarily eigenvalues. Corresponding to these, we express the function $X(x)$

as $X(x, \omega)$ or $X(x, \omega')$. We prepare the equations of the form (6.2.31) that these functions satisfy,

$$X_{xx}(x, \omega) + \omega^2(1 - x)X(x, \omega) = 0, \quad X_{xx}(x, \omega') + \omega'^2(1 - x)X(x, \omega') = 0.$$
$$(6.2.36)$$

Multiplying this first equation by $X(x, \omega')$, and the second equation by $X(x, \omega)$, and taking the difference of both, we have

$$\frac{d}{dx}\big[X(x, \omega')X_x(x, \omega) - X(x, \omega)X_x(x, \omega')\big]$$
$$+ (\omega^2 - \omega'^2)(1 - x)X(x, \omega)X(x, \omega') = 0. \qquad (6.2.37)$$

Integrating this equation, we obtain

$$\int_0^1 (1 - x)X(x, \omega)X(x, \omega')dx$$

$$= \frac{1}{\omega^2 - \omega'^2}\big[X(x, \omega)X_x(x, \omega') - X(x, \omega')X_x(x, \omega)\big]\Big|_0^1. \quad (6.2.38)$$

Here, when ω and ω' are distinct eigenvalues ω_i, ω_j, since the boundary condition (6.2.32) is satisfied, the right-hand side of this equation becomes zero. That is, the orthogonality of the eigenfunctions that belong to different eigenvalues is shown. If the two eigenvalues are the same, first we set $\omega' = \omega_i$, and assuming $X_x(1, \omega) = 0$, this right-hand side becomes

$$\text{RHS} = \frac{-1}{\omega^2 - \omega_i{}^2}X(0, \omega)X_x(0, \omega_i). \qquad (6.2.39)$$

Here, if we take the limit $\omega \to \omega_i$, l'Hôpital's rule implies

$$\text{RHS} = \frac{-1}{2\omega_i}[\partial_\omega X(0, \omega)]\big|_{\omega=\omega_i}X_x(0, \omega_i) = \frac{1}{3}\Big[J_{2/3}\Big(\frac{2\omega_i}{3}\Big)\Big]^2. \qquad (6.2.40)$$

To summarize the above results, the orthogonality of eigenfunctions is obtained as

$$\int_0^1 (1 - x)X(x, \omega_i)X(x, \omega_j)dx = N_i{}^2\delta_{i,j}. \qquad (6.2.41)$$

Here, the normalization constant N_i is defined by

$$N_i = \frac{1}{\sqrt{3}}J_{2/3}\Big(\frac{2\omega_i}{3}\Big). \qquad (6.2.42)$$

Therefore, $X(x, \omega_i)/N_i$ is a normalized eigenfunction. The first 6 normalized eigenfunctions are shown in Fig. 6.3. The number of each curve is a

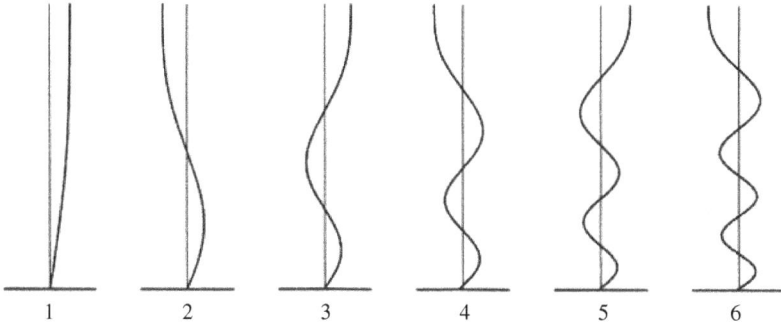

Fig. 6.3 Eigenfunctions

mode number. The number of zero points of each eigenfunction is equal to the mode number.

Solution when initial conditions are given

Here is an example of the simplest initial conditions. We set the initial displacement as zero at all points, and the initial velocity shall be given by $v_0(x)$ for each point,

$$V(x,0) = 0, \quad V_t(x,0) = v_0(x). \tag{6.2.43}$$

First, the displacement $V(x,t)$ can be expanded in terms of the eigenfunctions $X(x, \omega_i)$. Writing the expansion coefficients as K_i, we have

$$V(x,t) = \sum_{i=1}^{\infty} K_i X(x, \omega_i) \sin(\omega_i t). \tag{6.2.44}$$

This already satisfies the condition that the initial displacement is zero. Regarding the initial velocity, we differentiate this expression with respect to time t and set $t = 0$,

$$\sum_{i=1}^{\infty} \omega_i K_i X(x, \omega_i) = v_0(x). \tag{6.2.45}$$

Multiplying both sides by $(1-x)X(x, \omega_j)$, and integrating with respect to x, we can obtain the coefficients K_j, by using the orthogonality relation of (6.2.41)

$$K_j = \frac{1}{\omega_j N_j^2} \int_0^1 (1-x)v_0(x)X(x, \omega_j)dx. \tag{6.2.46}$$

Returning this to expression (6.2.44), we obtain the final solution

$$V(x,t) = \sum_{i=1}^{\infty} \frac{1}{\omega_i N_i^2} \left[\int_0^1 (1 - x')v_0(x')X(x',\omega_i)dx' \right] X(x,\omega_i) \sin(\omega_i t).$$

(6.2.47)

6.2.3 *Example in quantum mechanics*

In a quantum harmonic oscillator, the potential in the Schrödinger equation is in the form of a quadratic expression. This solution is represented by the product of the exponential function and the Hermite polynomial. This is familiar, because it can be found in ordinary quantum mechanics textbooks. However, when the potential becomes a linear expression, not many books have the solution. Here, we describe the solution of the Schrödinger equation with first-order potential which is the basis of the WKB method used in the theory of nuclear α decay.

Solution of the Schrödinger equation with first-order potential

In the Schrödinger equation of a one-dimensional electron wave,

$$\left[-\frac{\hbar^2}{2m}\frac{d^2}{dx^2} + V(x) \right]\psi = E\psi,$$

(6.2.48)

let the potential be expressed in the linear form

$$V(x) = V(0) + V'(0)x.$$

(6.2.49)

Here, we replace the independent variable x as

$$x + \frac{V(0) - E}{V'(0)} \quad \mapsto \quad x,$$

(6.2.50)

then the equation becomes

$$\left[-\frac{\hbar^2}{2m}\frac{d^2}{dx^2} + V'(0)x \right]\psi = 0.$$

(6.2.51)

The problem here is to find a solution that satisfies the condition $\psi \to 0$ as $|x| \to \infty$.

Here, it is assumed that $V'(0)$ is positive. Furthermore, since $\hbar^2/(2mV'(0))$ has a dimension of the cube of length, a quantity with a dimension of length is defined as

$$\ell \equiv \left(\frac{\hbar^2}{2mV'(0)} \right)^{1/3}.$$

(6.2.52)

Using this ℓ as the unit of length, coordinate x is transformed into a dimensionless one,

$$\frac{x}{\ell} \mapsto x. \qquad (6.2.53)$$

With this operation equation (6.2.51) becomes

$$\psi_{xx} - x\psi = 0. \qquad (6.2.54)$$

Since this equation contains the second derivative with respect to x and the linear term of x, it is the same as equation (6.2.9). Therefore, for $x > 0$, the same kind of transformation as in relation (6.2.10),

$$z = \frac{2}{3}x^{3/2}, \quad \psi = z^{1/3}g(z), \qquad (6.2.55)$$

transforms equation (6.2.54) into

$$\left[\frac{d^2}{dz^2} + \frac{1}{z}\frac{d}{dz} - \left(1 + \frac{1}{9z^2}\right)\right]g = 0. \qquad (6.2.56)$$

This equation can be solved by using the **modified Bessel function** of order $\pm 1/3$. Before solving this, let us mention a little about the modified Bessel function. First, as the solution of this equation, there exists

$$I_{\pm 1/3}(z) \equiv e^{\mp \pi i/6} J_{\pm 1/3}(iz) = \left(\frac{z}{2}\right)^{\pm 1/3} \sum_{n=0}^{\infty} \frac{(z/2)^{2n}}{n!\,\Gamma(\pm\frac{1}{3} + n + 1)}. \qquad (6.2.57)$$

However, both of these functions of order $\pm 1/3$ diverge as $z \to \infty$, so they cannot be adopted as the solution in the present case. Therefore, we consider the linear combined form of these two solutions,

$$K_{1/3}(z) \equiv \frac{\pi}{2}\frac{I_{-1/3}(z) - I_{1/3}(z)}{\sin(\pi/3)}. \qquad (6.2.58)$$

Since this $K_{1/3}(z)$ is a function that approaches zero exponentially as $z \to \infty$, it can be adopted as a solution in this case. In general, the functions $I_\alpha(z)$ and $K_\alpha(z)$ are called the first and second kind modified Bessel functions, respectively.

For our convenience later, adding an appropriate coefficient to the solution of equation (6.2.56),

$$g(z) = \frac{1}{\sqrt{3}\sqrt[3]{2/3}\,\pi}K_{1/3}(z), \qquad (6.2.59)$$

and returning to the original variable using relation (6.2.55), we obtain the solution

$$\psi(x) = \text{Ai}(x). \tag{6.2.60}$$

Here, the function $\text{Ai}(x)$ on the right-hand side is called the **Airy function for $x > 0$**, and is defined by

$$\text{Ai}(x) \equiv \frac{1}{\pi}\sqrt{\frac{x}{3}}K_{1/3}\left(\frac{2}{3}x^{3/2}\right), \quad x > 0. \tag{6.2.61}$$

The expression of this function is not good, but, this is nevertheless an entire function of x. In fact, expanding this function using relations (6.2.57) and (6.2.58) for the first two terms, we have

$$\text{Ai}(x) = \frac{1}{3^{2/3}\Gamma(\frac{2}{3})} - \frac{1}{3^{1/3}\Gamma(\frac{1}{3})}x + \cdots . \tag{6.2.62}$$

This constant term arises from $I_{-1/3}$, while the first-order term comes from $I_{1/3}$. The asymptotic form at $x \to \infty$ comes from the asymptotic expression for the modified Bessel function $I_\nu(z)$,

$$\text{Ai}(x) \approx \frac{e^{-\frac{2}{3}x^{3/2}}}{2\sqrt{\pi}x^{1/4}}. \tag{6.2.63}$$

This is a function that decreases exponentially as $x \to \infty$. If we return to the original variable using equations (6.2.49) and (6.2.50), the area of $x > 0$ is the one in which the energy E of the electron is smaller than the potential $E < V(x)$. Therefore, the electron cannot go there in classical mechanics, but in quantum mechanics, the electron waves seep into this area.

We have just found the solution for $x > 0$. Next, let us find the solution for $x < 0$. We transform the variables; instead of the relation (6.2.55), we have

$$z = \frac{2}{3}(-x)^{3/2}, \quad \psi = z^{1/3}g(z). \tag{6.2.64}$$

Then equation (6.2.54) becomes

$$\frac{d^2g}{dz^2} + \frac{1}{z}\frac{dg}{dz} + \left(1 - \frac{1}{9z^2}\right)g = 0. \tag{6.2.65}$$

This is a normal Bessel differential equation of order $\pm 1/3$. The solution is given by a linear combination of $J_{\pm 1/3}(z)$ with the coefficients C and D

$$g(z) = CJ_{-1/3}(z) + DJ_{1/3}(z). \tag{6.2.66}$$

Returning to the original ψ in relation (6.2.64), we have the solution

$$\psi(x) = \left(\frac{2}{3}\right)^{1/3}\sqrt{-x}\left[CJ_{-1/3}\left(\frac{2}{3}(-x)^{3/2}\right) + DJ_{1/3}\left(\frac{2}{3}(-x)^{3/2}\right)\right]. \quad (6.2.67)$$

In order that this ψ is a wave function of quantum mechanics, the function itself and its derivative must be continuous at $x = 0$. By using the series expansion of the Bessel function,

$$J_\nu(z) = \sum_{n=0}^{\infty}\frac{(-1)^n}{n!\Gamma(\nu+n+1)}\left(\frac{z}{2}\right)^{\nu+2n}, \quad (6.2.68)$$

if we expand the first and second terms of equation (6.2.67) near $x = 0$,

$$\sqrt{-x}J_{-1/3}\left(\frac{2}{3}(-x)^{3/2}\right) = \frac{1}{3^{-1/3}\Gamma(\frac{2}{3})} + \cdots,$$

$$\sqrt{-x}J_{1/3}\left(\frac{2}{3}(-x)^{3/2}\right) = -\frac{1}{3^{-2/3}\Gamma(\frac{1}{3})}x + \cdots. \quad (6.2.69)$$

When these are substituted into equation (6.2.67), to match the expansion formula (6.2.62) of Ai(x) at $x > 0$ up to the first-order term, it must be

$$C = D = \frac{1}{3}\left(\frac{2}{3}\right)^{-1/3}. \quad (6.2.70)$$

At this time, equation (6.2.67) becomes

$$\psi(x) = \text{Ai}(x), \quad (6.2.71)$$

and here, the Airy function for $x < 0$ is defined as

$$\text{Ai}(x) \equiv \frac{\sqrt{-x}}{3}\left[J_{-1/3}\left(\frac{2}{3}(-x)^{3/2}\right) + J_{1/3}\left(\frac{2}{3}(-x)^{3/2}\right)\right], \quad x < 0. \quad (6.2.72)$$

There are two Airy functions: (6.2.61) for $x > 0$, and (6.2.72) for $x < 0$. They look completely different, but they are just different representations of the same function, because both of these are the solutions of the same second-order differential equation, and when they are Taylor expanded at one point, if the zeroth and first-order terms coincide, all the terms should coincide. In fact, if we expand the Airy function of (6.2.61) by using the definition of $K_{1/3}(z)$ (6.2.58) and $I_{\pm1/3}(z)$ (6.2.57), we have

$$\text{Ai}(x) = \frac{1}{3^{2/3}}\sum_{n=0}^{\infty}\frac{x^{3n}}{n!\,\Gamma(\frac{2}{3}+n)3^{2n}} - \frac{1}{3^{4/3}}\sum_{n=0}^{\infty}\frac{x^{3n+1}}{n!\,\Gamma(\frac{4}{3}+n)3^{2n}}. \quad (6.2.73)$$

This is consistent with the Airy function of expression (6.2.72) which is expanded by using expression (6.2.68). In this sense, the Airy function is an entire function of x. Alternatively, this expansion formula is obtained easily if we expand $\psi(x)$ in equation (6.2.54) as

$$\psi(x) = \sum_{n=0}^{\infty} C_n x^{3n} + \sum_{n=0}^{\infty} D_n x^{3n+1}. \tag{6.2.74}$$

After substituting, align the powers of x and set all coefficients to zero. Since we have the recurrence formula for the coefficients C_n and D_n, and we set the initial values

$$C_0 = \frac{1}{3^{2/3}\Gamma(\frac{2}{3})}, \quad D_0 = -\frac{1}{3^{4/3}\Gamma(\frac{4}{3})}\left(= -\frac{1}{3^{1/3}\Gamma(\frac{1}{3})}\right), \tag{6.2.75}$$

and thus we reach the required expression.

It should be noted that the asymptotic form of this function (6.2.72) as $x \to -\infty$ is derived from that of the Bessel function $J_\nu(z)$, that is

$$\text{Ai}(x) \approx \frac{\sin\left(\frac{2}{3}(-x)^{3/2} + \frac{1}{4}\pi\right)}{\sqrt{\pi}(-x)^{1/4}}. \tag{6.2.76}$$

As x goes to $-\infty$, the amplitude slowly goes to zero while the vibration period gradually decreases. In addition, there is another Airy function of the second kind $\text{Bi}(x)$ which diverges as $x \to \infty$. However, it is not necessary here, so we will not explain it. Figure 6.4 shows a graph of the Airy function $\text{Ai}(x)$.

First, we try to draw all the parts by using the expansion formula (6.2.73), but, as the absolute value of x becomes larger, it will diverge

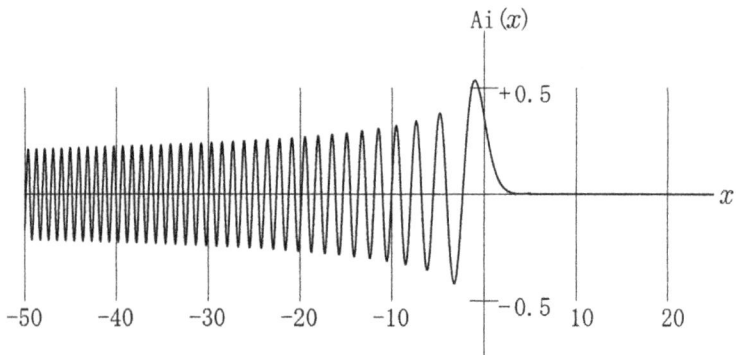

Fig. 6.4 Airy function $\text{Ai}(x)$

numerically. Here, we divide the range into 3 parts: $-50 \le x \le -14$, $-14 \le x \le 5$ and $5 \le x \le 25$ by using the asymptotic expression (6.2.76), the expansion formula (6.2.73) and the asymptotic expression (6.2.63), respectively.

Integral representation of the Airy function

The Airy function is easier to express by using the **Airy integral** than using the Bessel function. This function is expressed by the following integral:

$$\mathrm{Ai}(x) = \frac{1}{2\pi i} \int_C e^{\frac{1}{3}t^3 - xt} dt. \tag{6.2.77}$$

Here, the integration path C is from the infinity where the argument of t is $-\pi/3$ to the infinity where the argument is $+\pi/3$ on the complex plane, that is,

$$\int_C = \lim_{M \to \infty} \int_{Me^{-(\pi/3)i}}^{Me^{(\pi/3)i}} = \lim_{M \to \infty} \left(\int_0^{Me^{(\pi/3)i}} - \int_0^{Me^{-(\pi/3)i}} \right). \tag{6.2.78}$$

Since the main factor of the integrand becomes $e^{\frac{1}{3}t^3} \to e^{-\frac{1}{3}|t|^3}$ at both ends of the integration path, the convergence is very good. Therefore, the variable x can also be extended to complex numbers z. In what follows, we show that this function represented by the integral (6.2.77) reproduces the Airy function which we have dealt with so far. First, we can show that this integral representation satisfies equation (6.2.54), as follows:

$$\left(\frac{d^2}{dx^2} - x \right) \mathrm{Ai}(x) = \frac{1}{2\pi i} \int_C (t^2 - x) e^{\frac{1}{3}t^3 - xt} dt$$

$$= \lim_{M \to \infty} \frac{1}{2\pi i} \left[e^{\frac{1}{3}t^3 - xt} \right]_{Me^{-(\pi/3)i}}^{Me^{(\pi/3)i}} = 0. \tag{6.2.79}$$

Next, let us show that when this integral representation is expanded into powers of x, it becomes equation (6.2.73). First, the part e^{-xt} is expanded into a Taylor series and the order of summation and integration is exchanged, then we have

$$\mathrm{Ai}(x) = \frac{1}{2\pi i} \int_C e^{\frac{1}{3}t^3} \left(\sum_{k=0}^{\infty} \frac{(-xt)^k}{k!} \right) dt = \frac{1}{2\pi i} \sum_{k=0}^{\infty} \frac{(-1)^k}{k!} \left(\int_C e^{\frac{1}{3}t^3} t^k dt \right) x^k.$$

$$\tag{6.2.80}$$

By using this integration path as the final form of expression (6.2.78), and replacing $t \to e^{(\pi/3)i}t$, or $t \to e^{-(\pi/3)i}t$, expression (6.2.80) becomes

$$\text{Ai}(x) = \frac{1}{\pi} \sum_{k=0}^{\infty} \frac{(-1)^k}{k!} \sin\left(\tfrac{1}{3}(k+1)\pi\right) \left(\int_0^{\infty} e^{-\frac{1}{3}t^3} t^k \, dt \right) x^k. \qquad (6.2.81)$$

Furthermore, when the integration variable t is changed to $\xi = \frac{1}{3}t^3$, it becomes the definition of the gamma function, that is,

$$\int_0^{\infty} e^{-\frac{1}{3}t^3} t^k \, dt = 3^{\frac{1}{3}(k-2)} \int_0^{\infty} e^{-\xi} \xi^{\frac{1}{3}(k-2)} \, d\xi = 3^{\frac{1}{3}(k-2)} \Gamma\left(\tfrac{1}{3}(k+1)\right).$$

$$(6.2.82)$$

In addition, the $k!$ in the denominator of expression (6.2.81) can be rewritten, by using the triple product formula of the gamma function, as

$$k! = \Gamma(k+1) = \frac{3^{k+1}}{2\pi\sqrt{3}} \Gamma\left(\tfrac{1}{3}(k+1)\right)\Gamma\left(\tfrac{1}{3}(k+2)\right)\Gamma\left(\tfrac{1}{3}(k+3)\right). \qquad (6.2.83)$$

Therefore, expression (6.2.81) becomes

$$\text{Ai}(x) = 2\sqrt{3} \sum_{k=0}^{\infty} \frac{(-1)^k 3^{-\frac{1}{3}(2k+5)} \sin\left(\tfrac{1}{3}(k+1)\pi\right)}{\Gamma\left(\tfrac{1}{3}(k+2)\right)\Gamma\left(\tfrac{1}{3}(k+3)\right)} x^k. \qquad (6.2.84)$$

We decompose the summation over k into three parts for $k = 3n, 3n+1$ and $3n + 2$, n being a non-negative integer. When $k = 3n + 2$, the sine part of the above expression becomes zero. Therefore, we can check the remaining two cases. When $k = 3n$, we have

$$= \frac{1}{3^{2/3}} \sum_{n=0}^{\infty} \frac{x^{3n}}{n!\,\Gamma(\tfrac{2}{3}+n)3^{2n}}.$$

This is nothing but the first term of equation (6.2.73). Also, when $k = 3n + 1$, we have

$$= -\frac{1}{3^{4/3}} \sum_{n=0}^{\infty} \frac{x^{3n+1}}{n!\,\Gamma(\tfrac{4}{3}+n)3^{2n}}.$$

This is the second term of equation (6.2.73). Thus, equation (6.2.73) is reproduced as the sum of these expressions.

In the integral representation of the Airy function (6.2.77), we adopted the integration path from the infinity where the argument is $-\pi/3$ to the infinity where the argument is $+\pi/3$ on the complex plane. However, in reality, only the real part of t^3 should be negative. Strictly speaking, it can be taken from the infinity where the argument is between $-\pi/2$ and

$-\pi/6$ to the infinity where the argument is between $\pi/6$ and $\pi/2$. Here we try to adopt just the argument from $-\pi/2$ to $\pi/2$. In other words, we adopt the path along the imaginary axis. We replace t by it, the latter t being real in expression (6.2.77), and it becomes

$$\mathrm{Ai}(x) = \frac{1}{2\pi} \int_{-\infty}^{\infty} e^{-i(\frac{1}{3}t^3 + xt)} \, dt$$

$$= \frac{1}{2\pi} \int_{-\infty}^{\infty} \left[\cos\left(\frac{1}{3}t^3 + xt\right) - i \sin\left(\frac{1}{3}t^3 + xt\right) \right] dt$$

$$= \frac{1}{\pi} \int_{0}^{\infty} \cos\left(\frac{t^3}{3} + xt\right) dt. \tag{6.2.85}$$

The result is represented by the real integral. However, in the case of this integral representation, we must take care of the fact that it does not manifestly converge. Since the integrand of this integral does not have a definite limit as $t \to \infty$, so we use the identity of cosine function represented by derivative,

$$\cos\left(\frac{1}{3}t^3 + xt\right) = \left[\frac{d}{dt} \sin\left(\frac{1}{3}t^3 + xt\right)\right] \frac{1}{t^2 + x},$$

and by using the integration by parts, then the integral of (6.2.85) is rewritten as

$$\int^{\infty} \cos\left(\frac{t^3}{3} + xt\right) dt$$

$$= \frac{1}{t^2 + x} \sin\left(\frac{t^3}{3} + xt\right)\Big|^{\infty} + \int^{\infty} \frac{2t}{(t^2 + x)^2} \sin\left(\frac{t^3}{3} + xt\right) dt. \tag{6.2.86}$$

Thus the convergence is shown. Also, we can show that this function (6.2.85) satisfies equation (6.2.54),

$$\left(\frac{d^2}{dx^2} - x\right) \mathrm{Ai}(x) = -\frac{1}{\pi} \int_{0}^{\infty} (t^2 + x) \cos\left(\frac{t^3}{3} + xt\right) dt = -\frac{1}{\pi} \sin\left(\frac{t^3}{3} + xt\right)\Big|_{0}^{\infty}. \tag{6.2.87}$$

This last term is clearly zero at $t = 0$. Also, in the case of the limit as $t \to \infty$, it can be regarded as zero in the sense of a generalized function. Thus equation (6.2.54) is crucially satisfied.

Furthermore, if we make an expansion formula for the variable x using formula (6.2.85), we can calculate well up to the first-order of x, and as a result, equation (6.2.62) is reproduced. Therefore, it is guaranteed that this integral representation formula (6.2.85) coincides with the

Airy function Ai(x). If we want to know the higher-order expansion formula, we can integrate by parts repeatedly, however, we will not do this here.

In addition, if we use an imaginary part in the variable x in formula (6.2.85), the e^{-ixt} part diverges depending on the value of t. We must be careful to use a real number x, and be sure to pay attention to the convergence. Only then can we say that this integral representation formula of the Airy function is easy to use.

Orthogonality and completeness of the Airy function

The orthogonality of the Airy function is a little different. In problems where the electron is localized in a finite region (for example, a quantum harmonic oscillator, a square well potential problem, a hydrogen atom, etc.), the energy of the electron is determined in discrete values as an eigenvalue. And the wave function as an eigenfunction belonging to it has orthogonality.

Where did the energy disappear to in the problem dealt with here? If we go back to relations (6.2.53) and (6.2.50), the energy is in the form of the sum at coordinate x. Here, we do not translate x as in relation (6.2.50); instead we introduce the dimensionless energy as

$$\xi = \frac{V(0) - E}{\ell V'(0)}. \tag{6.2.88}$$

By using this ξ and the dimensionless x, the wave function as an eigenfunction is written as $\psi = $ Ai$(x + \xi)$. In this case, since the range in which the electron moves is infinite, the energy eigenvalue spectrum becomes continuous. Therefore, the orthogonality relation should not be the Kronecker delta, but Dirac's delta. That is, by using the integral representation formula (6.2.85), we easily obtain the orthogonality relation of the Airy functions belonging to two energy eigenvalues ξ, ξ'

$$\int_{-\infty}^{\infty} \text{Ai}(x + \xi)\text{Ai}(x + \xi')dx = \delta(\xi - \xi'). \tag{6.2.89}$$

In fact, if we substitute the one using the exponential function of expression (6.2.85) into the left-hand side of (6.2.89)

$$\int_{-\infty}^{\infty} \text{Ai}(x + \xi)\text{Ai}(x + \xi')dx$$

$$= \frac{1}{4\pi^2} \iiint_{-\infty}^{\infty} e^{-i[\frac{1}{3}(t^3 + t'^3) + (x + \xi)t + (x + \xi')t']} dt\,dt'\,dx, \tag{6.2.90}$$

taking out only the x integral of this, and using the generalized function formula, we have

$$\int_{-\infty}^{\infty} e^{-i(t+t')x} dx = -\lim_{M\to\infty} \frac{e^{-i(t+t')x}}{i(t+t')}\Bigg|_{-M}^{M} = 2\pi \lim_{M\to\infty} \frac{\sin\big((t+t')M\big)}{\pi(t+t')}$$

$$= 2\pi\delta(t+t'). \qquad (6.2.91)$$

Therefore, integral (6.2.90) becomes

$$\int_{-\infty}^{\infty} \mathrm{Ai}(x+\xi)\mathrm{Ai}(x+\xi')dx = \frac{1}{2\pi}\int_{-\infty}^{\infty} e^{-i(\xi-\xi')t} dt. \qquad (6.2.92)$$

This integral is similar to expression (6.2.91),

$$\frac{1}{2\pi}\int_{-\infty}^{\infty} e^{-i(\xi-\xi')t} dt = \lim_{M\to\infty} \frac{\sin\big((\xi-\xi')M\big)}{\pi(\xi-\xi')} = \delta(\xi-\xi'), \qquad (6.2.93)$$

and as a result, the orthogonality relation (6.2.89) is proved.

Also, if the roles of the variables are swapped,

$$\int_{-\infty}^{\infty} \mathrm{Ai}(x+\xi)\mathrm{Ai}(x'+\xi)d\xi = \delta(x-x'). \qquad (6.2.94)$$

Because this formula is considered an expression of completeness, if we make the transformation for an arbitrary continuous function $f(x)$ which is integrable in the range $(-\infty, \infty)$,

$$g(\xi) = \int_{-\infty}^{\infty} \mathrm{Ai}(\xi+x)f(x)dx, \qquad (6.2.95)$$

this inverse transformation is possible by using (6.2.94)

$$f(x) = \int_{-\infty}^{\infty} \mathrm{Ai}(x+\xi)g(\xi)d\xi. \qquad (6.2.96)$$

In these relations, the core part of the transformation is in the form of the sum $x+\xi$, and this is unprecedentedly strange.

For example, the Fourier-transform defined in $(-\infty, \infty)$ is written as

$$g(\xi) = \frac{1}{\sqrt{2\pi}}\int_{-\infty}^{\infty} e^{i\xi x} f(x)dx, \quad f(x) = \frac{1}{\sqrt{2\pi}}\int_{-\infty}^{\infty} e^{-ix\xi} g(\xi)d\xi, \qquad (6.2.97)$$

or the Fourier–Bessel-transform defined in $[0, \infty)$ is written as

$$g(\xi) = \int_{0}^{\infty} J_\nu(\xi x)f(x)x dx, \quad f(x) = \int_{0}^{\infty} J_\nu(x\xi)g(\xi)\xi d\xi, \quad \nu > -1.$$
$$(6.2.98)$$

Since these core parts are in the form of a product $x\xi$, it seems that the transformation in the form of the sum $x + \xi$ obtained here is completely different and rare.

Airy Function as Bessel Function

Mathematical Formula III[4] contains formulas for over 1000 Bessel functions. However, there are only two formulas related to the Airy function (integral). There is almost no literature dealing with the Airy function. Even if there is, it is only an official and perfunctory list without proof. There should be many more situations where this function works. Here, we have tried to explain and give as much proof as possible.

   The Airy integral representation (6.2.77) as a solution of the differential equation (6.2.54) is impressive and really well thought out. The x derivative is changed to a polynomial in the integral variable t by using an exponential function. If we expand this way of thinking, we can make an example of the fourth-order differential equation and its solution,

$$\psi_{xxxx} + \alpha\psi_{xx} - x\psi = 0, \quad \psi(x) = \int_D e^{\frac{1}{5}t^5 + \alpha\frac{1}{3}t^3 - xt} dt.$$

Here, the integration path D is from the infinity where the argument of t is $-\pi/5$ to the infinity where the argument for t is $\pi/5$. There is no merit in expanding further unless there is an application for it. After all, meaningful expansion cannot be done overnight.

6.3 Bessel Function and Semi-infinite One-dimensional Lattice

6.3.1 *Case of a fixed end: introduction to the equation*

In this section, we demonstrate how the Bessel function works when solving one-dimensional lattice vibration problems. We consider a system in which weights with mass m and springs with spring-constant k are connected one-dimensionally and alternately, as shown in Fig. 6.5.

---

[4]S. Moriguchi, K. Udagawa and S. Hitotsumatsu (Eds), *Mathematical Formula III*, (1956) Iwanami shoten (in Japanese).

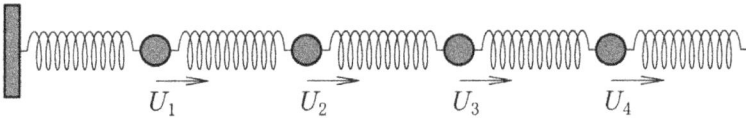

Fig. 6.5 One-dimensional lattice: case of a fixed end

Initially we set the number of weights to N. However, this number N will be expanded to infinite later when we consider the semi-infinite vibration system. These weights are numbered $1, 2, \ldots, N$ from left to right. When they are vibrating, let the displacement of the n-th weight parallel to the spring at time t be $U_n(t)$, and let us consider the equation of motion of these weights. Because the n-th weight receives a force of $k(U_{n+1} - U_n)$ from the right and a force of $-k(U_n - U_{n-1})$ from the left, the equation of motion becomes

$$m\frac{d^2}{dt^2}U_n = -k(-U_{n-1} + 2U_n - U_{n+1}), \quad n = 1, 2, \ldots, N. \quad (6.3.1)$$

If we set $n = 1$, we get the term U_0, and if we set $n = N$, we get the term U_{N+1}. These are the left and right ends of the spring. If these ends are always fixed, we put

$$U_0(t) = 0, \quad U_{N+1}(t) = 0. \quad (6.3.2)$$

These two expressions are the boundary conditions. In addition, initial displacements and initial velocities

$$U_n(0), \quad \left.\frac{dU_n}{dt}\right|_{t=0}, \quad n = 1, 2, \ldots, N, \quad (6.3.3)$$

shall be given appropriately, as the initial conditions.

To represent equation (6.3.1) in a matrix format, we define a displacement vector $U(t)$ and a coefficient matrix P as

$$U(t) = \begin{pmatrix} U_1(t) \\ U_2(t) \\ \vdots \\ U_{N-1}(t) \\ U_N(t) \end{pmatrix}, \quad P = \begin{pmatrix} 2 & -1 & 0 & \cdots & \cdots & \cdots & \cdots \\ -1 & 2 & -1 & 0 & \cdots & \cdots & \cdots \\ 0 & -1 & 2 & -1 & 0 & \cdots & \cdots \\ \vdots & \vdots & \ddots & \ddots & \ddots & \vdots & \vdots \\ \cdots & \cdots & \cdots & 0 & -1 & 2 & -1 \\ \cdots & \cdots & \cdots & \cdots & 0 & -1 & 2 \end{pmatrix}. \quad (6.3.4)$$

Using these quantities, equation (6.3.1) is rewritten as

$$m\frac{d^2}{dt^2}U(t) = -kPU(t).\tag{6.3.5}$$

Solution to the equation

Now, we consider the eigenvalue equation of this matrix P, that is, we diagonalize it,

$$PV = \lambda V, \quad V = {}^t(V_1, V_2, \ldots, V_N),\tag{6.3.6}$$

where λ and V are the **eigenvalue** and **eigenvector**, respectively, and the superscript "t" represents transposition, This is expressed in terms of components,

$$-V_{n-1} + 2V_n - V_{n+1} = \lambda V_n.\tag{6.3.7}$$

This formula holds only for $n = 2, 3, \ldots, N-1$. To make this hold for $n = 1$ and $n = N$, we must introduce V_0, V_{N+1} and set them to zero, just as for U,

$$V_0 = V_{N+1} = 0.\tag{6.3.8}$$

If we set $V_n = e^{in\theta}$ in equation (6.3.7), the eigenvalue λ is represented by a function of θ,

$$\lambda = 2(1 - \cos\theta) = 4\sin^2\left(\frac{\theta}{2}\right).\tag{6.3.9}$$

Since the same λ can be obtained for the sign replacement of θ, we can, in general, set

$$V_n = \alpha e^{in\theta} + \beta e^{-in\theta}\tag{6.3.10}$$

by using the arbitrary constants α, β. By imposing the boundary conditions (6.3.8) on this expression, we have

$$\alpha + \beta = 0, \quad \alpha e^{i(N+1)\theta} + \beta e^{-i(N+1)\theta} = 0.\tag{6.3.11}$$

Since the values of α and β cannot become zero at the same time, the value of the coefficient determinant of these equations must be zero, that is, we have

$$\begin{vmatrix} 1 & 1 \\ e^{i(N+1)\theta} & e^{-i(N+1)\theta} \end{vmatrix} = 0 \quad \Rightarrow \quad \sin\big((N+1)\theta\big) = 0.\tag{6.3.12}$$

Then, θ is determined as

$$\theta^{(\ell)} = \frac{\ell\pi}{N+1}, \quad \ell = 1, 2, \ldots, N. \tag{6.3.13}$$

From (6.3.9), the eigenvalue λ is

$$\lambda^{(\ell)} = 4\sin^2\left(\frac{\ell\pi}{2(N+1)}\right), \tag{6.3.14}$$

and the component $V_n^{(\ell)}$ of the eigenvector $V^{(\ell)}$ is determined as

$$V_n^{(\ell)} = 2\alpha \sin\left(\frac{n\ell\pi}{N+1}\right). \tag{6.3.15}$$

However, the original α here is replaced with $-i\alpha$, so that the imaginary number is not attached to this formula. Also, when solving equation (6.3.12), there may be a solution of $\theta = 0$ ($\ell = 0$), but in this case, both the eigenvalue and the eigenvector become zero, so we will exclude this case.

To normalize this eigenvector, let us require the following expression to determine the value of α,

$$\sum_{n=1}^{N} \left(V_n^{(\ell)}\right)^2 = 1. \tag{6.3.16}$$

By using the summation formula of the sine function,[5]

$$\sum_{n=1}^{N} \sin^2(nx) = \frac{N}{2} - \frac{\cos((N+1)x)\sin(Nx)}{2\sin x}, \tag{6.3.17}$$

and by substituting (6.3.15) into (6.3.16), we have

$$1 = \sum_{n=1}^{N} \left(V_n^{(\ell)}\right)^2 = (2\alpha)^2 \sum_{n=1}^{N} \sin^2\left(\frac{n\ell\pi}{N+1}\right) = (2\alpha)^2 \sum_{n=1}^{N+1} \sin^2\left(\frac{n\ell\pi}{N+1}\right). \tag{6.3.18}$$

Using formula (6.3.17) with N replaced by $N+1$, the α is determined as

$$(2\alpha)^2 = \frac{2}{N+1}. \tag{6.3.19}$$

Since, as a matter of course, the eigenvectors belonging to different eigenvalues are orthogonal, we obtain the orthogonality relation,

$$\sum_{n=1}^{N} V_n^{(\ell)} V_n^{(\ell')} = \delta_{\ell, \ell'}. \tag{6.3.20}$$

---

[5] S. Moriguchi, K. Udagawa and S. Hitotsumatsu (Eds), *Mathematical Formula II*, (1956) Iwanami Zensyo (in Japanese), p. 19.

The eigenvector $V^{(\ell)}$ arranged horizontally with respect to ℓ is defined as the matrix Λ,

$$
\Lambda = \begin{pmatrix}
V_1^{(1)} & V_1^{(2)} & \cdots & V_1^{(N)} \\
V_2^{(1)} & V_2^{(2)} & \cdots & V_2^{(N)} \\
\vdots & \vdots & \vdots & \vdots \\
V_N^{(1)} & V_N^{(2)} & \cdots & V_N^{(N)}
\end{pmatrix}.
\tag{6.3.21}
$$

The orthogonality and the completeness of eigenvectors can be rewritten as a compact form $\Lambda^\dagger \Lambda = \Lambda \Lambda^\dagger = 1$ (unit matrix). The coefficient matrix P of (6.3.4) is diagonalized by the matrix Λ,

$$
\Lambda^\dagger P \Lambda = \begin{pmatrix}
\lambda^{(1)} & 0 & \cdots & \cdots & \cdots \\
0 & \lambda^{(2)} & 0 & \cdots & \cdots \\
\vdots & 0 & \lambda^{(3)} & 0 & \cdots \\
\vdots & \vdots & \vdots & \ddots & \vdots \\
\cdots & \cdots & \cdots & \cdots & \lambda^{(N)}
\end{pmatrix}.
\tag{6.3.22}
$$

Here, multiplying $\Lambda^\dagger$ from the left-hand side of the equation of motion (6.3.5), and replacing the vector $U(t)$ with $W(t)$,

$$
\Lambda^\dagger U(t) = W(t), \quad W = {}^t(W^{(1)}, W^{(2)}, \ldots, W^{(N)}),
\tag{6.3.23}
$$

we obtain the decoupled equations in the component representation,

$$
m \frac{d^2}{dt^2} W^{(\ell)}(t) = -k\lambda^{(\ell)} W^{(\ell)}(t), \quad \ell = 1, 2, \ldots, N.
\tag{6.3.24}
$$

This general solution takes the form

$$
W^{(\ell)}(t) = A \sin\left(\omega_0 \sqrt{\lambda^{(\ell)}}\, t\right) + B \cos\left(\omega_0 \sqrt{\lambda^{(\ell)}}\, t\right),
\tag{6.3.25}
$$

where A and B are arbitrary constants, and ω_0 is an angular frequency which is defined by the weight and spring as

$$
\omega_0 = \sqrt{\frac{k}{m}}.
\tag{6.3.26}
$$

Furthermore, the constants A and B are represented by the initial value $W^{(\ell)}(0)$ and $\dot{W}^{(\ell)}(0)$,[6]

$$
A = \frac{\dot{W}^{(\ell)}(0)}{\omega_0 \sqrt{\lambda^{(\ell)}}}, \quad B = W^{(\ell)}(0),
\tag{6.3.27}
$$

---

[6] In the following, we use the Newtonian symbol with dot for the time derivative.

therefore, the solution is written as

$$W^{(\ell)}(t) = \frac{\dot{W}^{(\ell)}(0)}{\omega_0 \sqrt{\lambda^{(\ell)}}} \sin\left(\omega_0 \sqrt{\lambda^{(\ell)}}\, t\right) + W^{(\ell)}(0) \cos\left(\omega_0 \sqrt{\lambda^{(\ell)}}\, t\right)$$

$$= \dot{W}^{(\ell)}(0) \int_0^t dt' \cos\left(\omega_0 \sqrt{\lambda^{(\ell)}}\, t'\right) + W^{(\ell)}(0) \cos\left(\omega_0 \sqrt{\lambda^{(\ell)}}\, t\right)$$

$$= \int_0^t dt' \left[\dot{W}^{(\ell)}(0) + W^{(\ell)}(0)\delta(t - t')\right] \cos\left(\omega_0 \sqrt{\lambda^{(\ell)}}\, t'\right). \quad (6.3.28)$$

This last expression is a summary to simplify the description below.

Now that we have a solution for W, we have to revert it to the original solution for U. First, $W^{(\ell)}(0)$, $\dot{W}^{(\ell)}(0)$ are represented by $U_r(0)$, $\dot{U}_r(0)$ respectively using equation (6.3.23),

$$W^{(\ell)}(0) = \sum_{r=1}^N V_r^{(\ell)} U_r(0), \quad \dot{W}^{(\ell)}(0) = \sum_{r=1}^N V_r^{(\ell)} \dot{U}_r(0). \quad (6.3.29)$$

Next, from (6.3.23), $U = \Lambda W$, so the form is changed from W to U, that is, we obtain

$$U_n(t) = \sum_{\ell=1}^N \sum_{r=1}^N V_n^{(\ell)} V_r^{(\ell)} \int_0^t dt' \left[\dot{U}_r(0) + U_r(0)\delta(t - t')\right] \cos\left(\omega_0 \sqrt{\lambda^{(\ell)}}\, t'\right).$$

$$(6.3.30)$$

Finally, using equations (6.3.14), (6.3.15) and (6.3.19), we substitute the specific form of $\lambda^{(\ell)}$, $V_n^{(\ell)}$ and α for expression (6.3.30)

$$U_n(t) = \frac{2}{N+1} \sum_{\ell=1}^N \sum_{r=1}^N \sin\left(\frac{n\ell\pi}{N+1}\right) \sin\left(\frac{r\ell\pi}{N+1}\right)$$

$$\times \int_0^t dt' \left[\dot{U}_r(0) + U_r(0)\delta(t - t')\right] \cos\left(2\omega_0 t' \sin\left(\frac{\ell\pi}{2(N+1)}\right)\right).$$

$$(6.3.31)$$

Now we have found the solution to equation (6.3.1) when the number of weights N is finite.

Case of $N \to \infty$

Here, we consider the case where the limit of $N \to \infty$ is taken in equation (6.3.31). In this case, we replace the summation over ℓ by integrating over a new continuous variable x,

$$\frac{\ell\pi}{2(N+1)} \mapsto x, \qquad \frac{\pi}{2(N+1)} \mapsto dx, \tag{6.3.32}$$

where the range of the variable x is $(0, \pi/2)$. With this operation, equation (6.3.31) becomes

$$U_n(t) = \frac{4}{\pi} \sum_{r=1}^{\infty} \int_0^{\pi/2} dx \, \sin(2nx) \sin(2rx)$$

$$\times \int_0^t dt' \left[\dot{U}_r(0) + U_r(0)\delta(t - t') \right] \cos(2\omega_0 t' \sin x). \tag{6.3.33}$$

Further, by using the product-to-sum formula of trigonometric functions, and also the integral representation formula of the Bessel function

$$J_{2n}(z) = \frac{2}{\pi} \int_0^{\pi/2} dx \, \cos(2nx) \cos(z \sin x), \quad n = 0, 1, 2, \ldots, \tag{6.3.34}$$

we can rewrite the displacement $U_n(t)$ in the form

$$U_n(t) = \sum_{r=1}^{\infty} \int_0^t dt' \left[\dot{U}_r(0) + U_r(0)\delta(t - t') \right] \left(J_{2(n-r)}(2\omega_0 t') - J_{2(n+r)}(2\omega_0 t') \right),$$
$$\tag{6.3.35}$$

using the Bessel function.

6.3.2 *Law of conservation of energy*

Here, as a simple application example, we consider the case where the initial velocity v_0 is given a non-vanishing value only to the first weight, that is,

$$U_n(0) = 0, \quad n = 1, 2, \ldots, \quad \dot{U}_1(0) = v_0, \quad \dot{U}_n(0) = 0, \quad n = 2, 3, \ldots. \tag{6.3.36}$$

The solution for this case is written from expression (6.3.35)

$$U_n(t) = v_0 \int_0^t dt' \left[J_{2(n-1)}(2\omega_0 t') - J_{2(n+1)}(2\omega_0 t') \right]. \tag{6.3.37}$$

The kinetic energy of the first weight at the beginning of $t = 0$ is

$$E = \frac{1}{2} m v_0^2. \tag{6.3.38}$$

In what follows, we check that this energy is really equal to the total energy at any time t. In the present case, the sum of the spring energies and the kinetic energies of the weights $E(t)$ becomes

$$E(t) = \frac{1}{2}k\sum_{n=1}^{\infty}(U_n - U_{n-1})^2 + \frac{1}{2}m\sum_{n=1}^{\infty}\dot{U}_n^2. \tag{6.3.39}$$

This calculation will get bogged down depending on the method, so we will explain it in detail. First, the part $U_n - U_{n-1}$ can be rewritten in reverse using the differential formula of the Bessel function,

$$\frac{d}{dz}J_\nu(z) = \frac{1}{2}[J_{\nu-1}(z) - J_{\nu+1}(z)], \tag{6.3.40}$$

and we get

$$U_n - U_{n-1}$$

$$= v_0\int_0^t dt'\left[J_{2n-2}(2\omega_0 t') - J_{2n+2}(2\omega_0 t') - J_{2n-4}(2\omega_0 t') + J_{2n}(2\omega_0 t')\right]$$

$$= v_0\int_0^t dt'\left[-\left(J_{2n-3-1}(2\omega_0 t') - J_{2n-3+1}(2\omega_0 t')\right)\right.$$

$$\left. + \left(J_{2n+1-1}(2\omega_0 t') - J_{2n+1+1}(2\omega_0 t')\right)\right]$$

$$= \frac{v_0}{\omega_0}\int_0^t dt'\left[-\frac{d}{dt'}J_{2n-3}(2\omega_0 t') + \frac{d}{dt'}J_{2n+1}(2\omega_0 t')\right]$$

$$= -\frac{v_0}{\omega_0}[J_{2n-3}(2\omega_0 t) - J_{2n+1}(2\omega_0 t)]. \tag{6.3.41}$$

By using this relation, $E(t)$ of (6.3.39) becomes

$$E(t) = \frac{1}{2}mv_0^2\sum_{n=1}^{\infty}\left[\left(J_{2n-3} - J_{2n+1}\right)^2 + \left(J_{2n-2} - J_{2n+2}\right)^2\right], \tag{6.3.42}$$

where we have omitted the arguments of the Bessel function $2\omega_0 t$. Hereafter, this is rewritten as follows:

$$E(t) = \frac{1}{2}mv_0^2\left[\left(J_1 + J_3\right)^2 + \left(J_0 - J_4\right)^2 + \sum_{r=1}^{\infty}\left(J_r - J_{r+4}\right)^2\right]$$

$$= \frac{1}{2}mv_0^2\left[J_0^2 + 2\sum_{r=1}^{\infty}J_r^2 - J_2^2 + 2J_1 J_3 - 2J_0 J_4 - 2\sum_{r=1}^{\infty}J_r J_{r+4}\right].$$

$$\tag{6.3.43}$$

Here, we use the summation formula of the Bessel function

$$\sum_{r=-\infty}^{\infty}J_r^2(z) = 1, \quad \sum_{r=-\infty}^{\infty}r^\ell J_r(z)J_{r+m}(z) = 0, \quad \ell = 0, 1, \ldots, m-1,$$

$$m \text{ is a positive integer.} \tag{6.3.44}$$

The summation of this first expression is divided into cases where r is negative, zero, and positive, and using $J_{-r} = (-1)^r J_r$, we have

$$J_0^2 + 2\sum_{r=1}^{\infty} J_r^2 = 1. \qquad (6.3.45)$$

Thus the sum of the first and second terms in the brackets of expression (6.3.43) is 1. We expect the rest term to be zero for the total energy to be conserved. This is done by setting $\ell = 0$, $m = 4$ in the second formula (6.3.44), and dividing the sum of r into negative, zero, and positive cases, we have

$$J_2^2 - 2J_1 J_3 + 2J_0 J_4 + 2\sum_{r=1}^{\infty} J_r J_{r+4} = 0. \qquad (6.3.46)$$

From this result, we obtain the expected result,

$$E(t) = \frac{1}{2} m v_0^2. \qquad (6.3.47)$$

Thus, the law of conservation of energy is satisfied. Similarly, we can also discuss the conservation of energy when only the first weight is shifted at the beginning. However, the calculation in this case is very cumbersone because the number of terms increases tremendously. This case will not be dealt with here.

6.3.3 *Relationship between a semi-infinite lattice and a full-infinite lattice*

We have dealt with the vibration of a semi-infinite one-dimensional lattice. If we analyze this method with a full-infinite lattice extending from $-\infty$ to $+\infty$, the result is

$$U_n(t) = \sum_{r=-\infty}^{\infty} \int_0^t dt' \left[\dot{U}_r(0) + U_r(0)\delta(t - t') \right] J_{2(n-r)}(2\omega_0 t'). \qquad (6.3.48)$$

But we will not prove it again here.

In this expression, the summation over r is divided into zero, the positive part, and the negative part; further, we impose an initial condition that is odd with respect to r,

$$U_0(0) = 0, \quad U_{-r}(0) = -U_r(0), \quad \dot{U}_0(0) = 0, \quad \dot{U}_{-r}(0) = -\dot{U}_r(0),$$

$$(r = 1, 2, 3, \dots). \qquad (6.3.49)$$

Then, this expression becomes the solution (6.3.35) where the left end of the system is the fixed end. This solution means that the reflected wave is phase-inverted and propagates in the positive direction.

6.3.4 *Case of a forced-vibration end: introduction to the equation*

So far, we have considered the one-dimensional lattice that connects weights and springs in a semi-infinite shape, and we have dealt with the vibration problem in which the initial displacement and the initial velocity of each weight are given. Here, we analyze the case where one end of the semi-infinite lattice is forcibly vibrated.

We consider a system in which weights with mass m and springs with spring-constant k are connected one-dimensionally and alternately, as shown in Fig. 6.6. The difference from the earlier case is that the weight of the left end is connected to the diaphragm that vibrates forcibly through the spring. The equation of motion is the same as that in the earlier case except for $n = 1$,

$$m\frac{d^2}{dt^2}U_n = -k(-U_{n-1} + 2U_n - U_{n+1}), \quad n = 2, 3, \ldots, N. \quad (6.3.50)$$

We set here $U_{N+1}(t) = 0$, just as in the earlier case.

The first weight receives a force of $k(U_2 - U_1)$ from the right. We denote the forced displacement of the diaphragm at time t by $f(t)$; then the weight receives a force of $-k(U_1 - f(t))$ from the left. Therefore the equation of motion of the first weight is

$$m\frac{d^2}{dt^2}U_1 = -k(2U_1 - U_2) + kf(t). \quad (6.3.51)$$

If we write equations (6.3.50) and (6.3.51) together

$$m\frac{d^2}{dt^2}U_n = -k(-U_{n-1} + 2U_n - U_{n+1}) + kf(t)\delta_{n,1}, \quad n = 1, 2, \ldots, N. \quad (6.3.52)$$

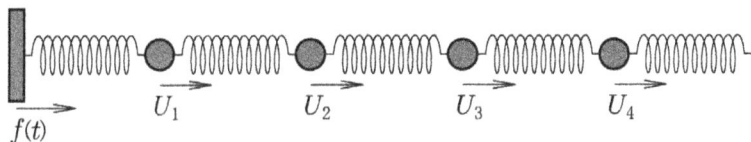

Fig. 6.6 One-dimensional lattice: case of a forced-vibration end

As a boundary condition, we set[7]

$$U_0(t) = 0, \quad U_{N+1}(t) = 0. \tag{6.3.53}$$

And we set the initial condition that the displacement and the velocity are all zero,

$$U_n(0) = 0, \quad \dot{U}_n(0) = 0, \quad n = 1, 2, \ldots, N. \tag{6.3.54}$$

The displacement vector $U(t)$ and the coefficient matrix P which are defined by expression (6.3.4) in the earlier case are used to write equation (6.3.52). In addition, the vector C is defined as

$$C = {}^{t}(1, 0, \ldots, 0). \tag{6.3.55}$$

By taking into account these notations, equation (6.3.52) is rewritten in the matrix format

$$m \frac{d^2}{dt^2} U(t) = -kPU(t) + kf(t)C. \tag{6.3.56}$$

Solution to the equation

Now we consider the eigenvalue equation of the matrix P to diagonalize it,

$$PV = \lambda V, \quad V = {}^{t}(V_1, V_2, \ldots, V_N). \tag{6.3.57}$$

Since the method is the same as in the earlier case, we will not repeat it here, and only the result will be cited. From this eigenvalue equation, N eigenvalues $\lambda^{(\ell)}$, $(\ell = 1, 2, \ldots, N)$ and the n-th component of the ℓ-th-order normalized eigenvector $V_n^{(\ell)}$ are determined as

$$\lambda^{(\ell)} = 4 \sin^2 \left(\frac{\ell \pi}{2(N+1)} \right), \quad V_n^{(\ell)} = \sqrt{\frac{2}{N+1}} \sin \left(\frac{n\ell \pi}{N+1} \right). \tag{6.3.58}$$

The eigenvector $V^{(\ell)}$ arranged horizontally with respect to ℓ is defined as the matrix Λ, just as in (6.3.21). The matrix P of equation (6.3.56) is diagonalized by this Λ, as the form $\Lambda^\dagger P \Lambda$ being a diagonal matrix with $\lambda^{(\ell)}$ on the diagonal part.

---

[7]There is another way to set $U_0(t) = f(t)$. However, since it will be confusing when we use the matrix formalism as shown below, this method is not adopted here. This will be discussed again at the end of this section.

Here this is also the same as before, by multiplying $\Lambda^\dagger$ from the left side of equation (6.3.56), and replacing the variable $U(t)$ with $W(t)$,

$$\Lambda^\dagger U(t) = W(t), \quad W = {}^t(W^{(1)}, W^{(2)}, \dots, W^{(N)}), \qquad (6.3.59)$$

we have the separated inhomogeneous equations,

$$m\frac{d^2}{dt^2}W^{(\ell)}(t) = -k\lambda^{(\ell)}W^{(\ell)}(t) + kf(t)V_1^{(\ell)}, \quad \ell = 1, 2, \dots, N. \qquad (6.3.60)$$

It is assumed here that $f(t)$ which is a forced displacement, is given by Fourier-sine transformation,

$$f(t) = \int_0^\infty g(\omega)\sin(\omega t)d\omega, \qquad (6.3.61)$$

and this inverse transformation is

$$g(\omega) = \frac{2}{\pi}\int_0^\infty f(t')\sin(\omega t')dt'. \qquad (6.3.62)$$

First, we find the special solution of this inhomogeneous equation (6.3.60). For this purpose, $W^{(\ell)}$ can also be written in terms of Fourier-sine transform,

$$W^{(\ell)}(t) = \int_0^\infty h^{(\ell)}(\omega)\sin(\omega t)d\omega. \qquad (6.3.63)$$

Substituting this expression and $f(t)$ of (6.3.61) into equation (6.3.60), we have

$$\int_0^\infty (\omega^{(\ell)\,2} - \omega^2)h^{(\ell)}(\omega)\sin(\omega t)d\omega = \omega_0^2 V_1^{(\ell)}\int_0^\infty g(\omega)\sin(\omega t)d\omega. \qquad (6.3.64)$$

Here, we define the angular frequency determined by the spring and weight ω_0, and another angular frequency $\omega^{(\ell)}$ in the ℓ mode of this vibration system,

$$\omega_0 = \sqrt{\frac{k}{m}}, \quad \omega^{(\ell)} = \omega_0\sqrt{\lambda^{(\ell)}} = 2\omega_0\sin\left(\frac{\ell\pi}{2(N+1)}\right). \qquad (6.3.65)$$

To satisfy equation (6.3.64),

$$h^{(\ell)}(\omega) = \frac{\omega_0^2 V_1^{(\ell)} g(\omega)}{\omega^{(\ell)\,2} - \omega^2}. \qquad (6.3.66)$$

From this, a particular solution of the inhomogeneous equation (6.3.60) is found

$$W^{(\ell)}(t) = \omega_0^2 V_1^{(\ell)} \int_0^\infty \frac{g(\omega)\sin(\omega t)}{\omega^{(\ell)\,2} - \omega^2} d\omega. \tag{6.3.67}$$

When $\omega = \omega^{(\ell)}$ here, that is, when the frequency of the forced vibration ω and the frequency of the ℓ mode $\omega^{(\ell)}$ are equal, the integrand shows the existence of a resonance state. The integration over ω should be defined by taking the **Cauchy principal value** at this point.

Furthermore, the general solution of equation (6.3.60) can be obtained by adding to the general solution of the homogeneous part,

$$W^{(\ell)}(t) = A\sin\left(\omega^{(\ell)}t\right) + B\cos\left(\omega^{(\ell)}t\right) + \omega_0^2 V_1^{(\ell)} \int_0^\infty \frac{g(\omega)\sin(\omega t)}{\omega^{(\ell)\,2} - \omega^2} d\omega, \tag{6.3.68}$$

where A and B are arbitrary constants. These are the quantities given by the initial values $W^{(\ell)}(0)$, $\dot{W}^{(\ell)}(0)$. In this case, as shown by expression (6.3.54), since $U_n(0)$, $\dot{U}_n(0)$ are all zero, these linear combinations $W^{(\ell)}(0)$, $\dot{W}^{(\ell)}(0)$ are also all zero. Considering these conditions, the constants A and B are determined as

$$A = -\frac{\omega_0^2 V_1^{(\ell)}}{\omega^{(\ell)}} \int_0^\infty \frac{\omega g(\omega)}{\omega^{(\ell)\,2} - \omega^2} d\omega, \quad B = 0. \tag{6.3.69}$$

Thus the solution of equation (6.3.60) is

$$W^{(\ell)}(t) = -\frac{\omega_0^2 V_1^{(\ell)}}{\omega^{(\ell)}} \left[\int_0^\infty \frac{\omega g(\omega)}{\omega^{(\ell)\,2} - \omega^2} d\omega \right] \sin\left(\omega^{(\ell)}t\right)$$

$$+ \omega_0^2 V_1^{(\ell)} \int_0^\infty \frac{g(\omega)\sin(\omega t)}{\omega^{(\ell)\,2} - \omega^2} d\omega. \tag{6.3.70}$$

We must substitute the $g(\omega)$ of expression (6.3.62), which is the inverse transformation of forced displacement $f(t)$, into this equation. Since the expression is long, we will partially calculate it. First, the ω integral of this second term becomes

$$\int_0^\infty \frac{g(\omega)\sin(\omega t)}{\omega^{(\ell)\,2} - \omega^2} d\omega = \frac{2}{\pi} \int_0^\infty \left[\int_0^\infty \frac{\sin(\omega t)\sin(\omega t')}{\omega^{(\ell)\,2} - \omega^2} d\omega \right] f(t') dt'. \tag{6.3.71}$$

To perform this ω integral, the product-to-sum formula of trigonometric functions, and the integration formula[8]

$$\int_0^\infty \frac{\cos(ax)}{b^2 - x^2}\,dx = \frac{\pi}{2b}\sin(ab), \quad a,\ b > 0 \tag{6.3.72}$$

are used. Here at the zero point of the denominator of the integrand, the Cauchy principal value of the integral should be taken. By using this formula, (6.3.71) becomes

$$(6.3.71) = \frac{1}{2\omega^{(\ell)}}\int_0^\infty \left[\sin\left(\omega^{(\ell)}|t - t'|\right) - \sin\left(\omega^{(\ell)}(t + t')\right)\right]f(t')dt'. \tag{6.3.73}$$

On the other hand, the ω integral of the first term of (6.3.70) becomes

$$\int_0^\infty \frac{\omega g(\omega)}{\omega^{(\ell)\,2} - \omega^2}\,d\omega = \frac{2}{\pi}\int_0^\infty \left[\int_0^\infty \frac{\omega \sin(\omega t')}{\omega^{(\ell)\,2} - \omega^2}\,d\omega\right]f(t')dt'$$

$$= -\frac{2}{\pi}\int_0^\infty f(t')\partial_{t'}\left[\int_0^\infty \frac{\cos(\omega t')}{\omega^{(\ell)\,2} - \omega^2}\,d\omega\right]dt'. \tag{6.3.74}$$

Also using formula (6.3.72), we have

$$(6.3.74) = -\frac{1}{\omega^{(\ell)}}\int_0^\infty f(t')\partial_{t'}\sin(\omega^{(\ell)}t')dt' = -\int_0^\infty \cos(\omega^{(\ell)}t')f(t')dt'. \tag{6.3.75}$$

By substituting these results into expression (6.3.70), $W^{(\ell)}(t)$ becomes

$$W^{(\ell)}(t) = \frac{\omega_0^2 V_1^{(\ell)}}{2\omega^{(\ell)}}\int_0^\infty \left[2\sin\left(\omega^{(\ell)}t\right)\cos\left(\omega^{(\ell)}t'\right)\right.$$

$$\left. + \sin\left(\omega^{(\ell)}|t - t'|\right) - \sin\left(\omega^{(\ell)}(t + t')\right)\right]f(\iota')d\iota', \tag{6.3.76}$$

and furthermore, by using the product-to-sum formula of trigonometric functions, we obtain the solution

$$W^{(\ell)}(t) = \frac{\omega_0^2 V_1^{(\ell)}}{\omega^{(\ell)}}\int_0^t \sin\left(\omega^{(\ell)}(t - t')\right)f(t')dt'. \tag{6.3.77}$$

This result is surprisingly simple. Notice that the part of $t' > t$ vanishes, therefore, the range of t' integral becomes from 0 to t.

---

[8] S. Moriguchi, K. Udagawa and S. Hitotsumatsu (Eds), *Mathematical Formula I*, (1956) Iwanami shoten (in Japanese), p. 255.

Now we have $W^{(\ell)}(t)$. From now on, we use the inverse transformation of relation (6.3.59) $U(t) = \Lambda W(t)$ to convert it back to $U(t)$,

$$U_n(t) = \sum_{\ell=1}^{N} V_n^{(\ell)} W^{(\ell)}(t) = \omega_0^2 \sum_{\ell=1}^{N} \frac{V_1^{(\ell)} V_n^{(\ell)}}{\omega^{(\ell)}} \int_0^t \sin\left(\omega^{(\ell)}(t - t')\right) f(t') dt'.$$
(6.3.78)

Now, we just have $U(t)$. For our convenience later, we convert the part of the sine function to the integral form and we rewrite this expression as

$$U_n(t) = \omega_0^2 \sum_{\ell=1}^{N} V_1^{(\ell)} V_n^{(\ell)} \int_0^t \left[\int_{t'}^{t} \cos\left(\omega^{(\ell)}(t - t'')\right) dt'' \right] f(t') dt'. \quad (6.3.79)$$

Finally, by substituting the definition of $\omega^{(\ell)}$ (6.3.65) and $V_n^{(\ell)}$ (6.3.58), we rewrite it further to obtain the final result

$$U_n(t) = \frac{2\omega_0^2}{N+1} \sum_{\ell=1}^{N} \sin\left(\frac{\ell\pi}{N+1}\right) \sin\left(\frac{n\ell\pi}{N+1}\right)$$

$$\times \int_0^t \left[\int_{t'}^{t} \cos\left(2\omega_0(t - t'') \sin\left(\frac{\ell\pi}{2(N+1)}\right)\right) dt'' \right] f(t') dt'.$$
(6.3.80)

Case of $N \to \infty$

We consider here the case of the limit $N \to \infty$ in solution (6.3.80). In this case, we replace the summation over ℓ by the integration introducing the continuous variable x, just as in expression (6.3.32),

$$\frac{\ell\pi}{2(N+1)} \mapsto x, \quad \frac{\pi}{2(N+1)} \mapsto dx. \quad (6.3.81)$$

Then expression (6.3.80) becomes

$$U_n(t) = \frac{4\omega_0^2}{\pi} \int_0^{\pi/2} dx \sin(2x) \sin(2nx)$$

$$\times \int_0^t \left[\int_{t'}^{t} \cos\left(2\omega_0(t - t'') \sin x\right) dt'' \right] f(t') dt'. \quad (6.3.82)$$

Here, by using the product-to-sum formula of trigonometric functions, and the integral representation formula of the Bessel function (6.3.34), this $U_n(t)$

is represented in the following form by using the Bessel function:

$$U_n(t) = \omega_0^2 \int_0^t \left[\int_{t'}^t \Big(J_{2n-2}\big(2\omega_0(t-t'')\big) - J_{2n+2}\big(2\omega_0(t-t'')\big) \Big) dt'' \right] f(t')dt'.$$

(6.3.83)

Furthermore, by using the indefinite integral formula of the Bessel function

$$\int J_\nu(z)dz = 2\sum_{r=0}^\infty J_{\nu+2r+1}(z),$$

(6.3.84)

this t'' integration is carried out, and the displacement $U_n(t)$ becomes

$$U_n(t) = \omega_0 \int_0^t \left[J_{2n-1}\big(2\omega_0(t-t')\big) + J_{2n+1}\big(2\omega_0(t-t')\big) \right] f(t')dt'.$$

(6.3.85)

At this time, notice that the infinite sum disappears, leaving these two terms. Thus, the result is in the form of a **convolution integral** of the Bessel function $J_{2n\pm1}$ and the forced displacement f. This is a natural result that shows that the displacement at time t is determined based on the information of the forced displacement in the past.

Finally, let us check whether the resulting solution (6.3.85) satisfies equation (6.3.52). By using the differential formula

$$\frac{d}{dz}J_\nu(z) = \frac{1}{2}\big(J_{\nu-1}(z) - J_{\nu+1}(z)\big),$$

(6.3.86)

the argument $2\omega_0(t-t')$ of the Bessel function will be omitted altogether, and we get

$$\frac{d}{dt}U_n(t) = \omega_0^2 \int_0^t \big(J_{2n-2} - J_{2n+2}\big) f(t')dt',$$

(6.3.87)

$$\frac{d^2}{dt^2}U_n(t) = \omega_0^2 f(t)\delta_{n,1} + \omega_0^3 \int_0^t \big(J_{2n-3} - J_{2n-1} - J_{2n+1} + J_{2n+3}\big) f(t')dt'.$$

(6.3.88)

Here, we used $J_0(0) = 1$, $J_n(0) = 0$, for $n \neq 0$. Furthermore, we have

$$-U_{n-1}(t) + 2U_n(t) - U_{n+1}(t)$$

$$= -\omega_0 \int_0^t \big(J_{2n-3} - J_{2n-1} - J_{2n+1} + J_{2n+3}\big) f(t')dt'. \quad (6.3.89)$$

Substituting these expressions into equation (6.3.52) with $N \to \infty$, we can confirm that this equation certainly holds.

6.3.5 *Energy formula*

Of course, the total energy is not generally conserved because the model here gives a forced displacement.

Example 6.3.1. We consider the case where the forced displacement $f(t)$ is given by

$$f(t) = f_0 \theta(t), \tag{6.3.90}$$

where θ is a unit step function. In other words, $f(t)$ jumps to a constant value f_0 at $t = 0$. We substitute this $f(t)$ into expression (6.3.85) and carry out the integration by using the indefinite integration formula (6.3.84) to get

$$U_n(t) = f_0 \sum_{r=0}^{\infty} \Big[J_{2(n+r)}(2\omega_0 t) + J_{2(n+r+1)}(2\omega_0 t) \Big]. \tag{6.3.91}$$

Furthermore, by using the summation formula for the Bessel function

$$J_0(z) + 2 \sum_{r=1}^{\infty} J_{2r}(z) = 1, \tag{6.3.92}$$

and using the formula for sum

$$\sum_{r=1}^{\infty} J_{2r}(z) - \sum_{r=1}^{n-1} J_{2r}(z) = \sum_{r=n}^{\infty} J_{2r}(z) = \sum_{r=0}^{\infty} J_{2(n+r)}(z), \tag{6.3.93}$$

we can convert the infinite sum to a finite one,

$$U_n(t) = f_0 \Big[1 - 2 \sum_{r=1}^{n-1} J_{2r}(2\omega_0 t) - J_0(2\omega_0 t) - J_{2n}(2\omega_0 t) \Big]. \tag{6.3.94}$$

In particular, when $n = 1$, it becomes

$$U_1(t) = f_0 \Big[1 - J_0(2\omega_0 t) - J_2(2\omega_0 t) \Big]. \tag{6.3.95}$$

We can see from this expression that as t increases, $U_1(t)$ approaches f_0.

Let the energy of the spring and weight at time t be

$$E(t) = \frac{1}{2}k\big(U_1 - f(t)\big)^2 + \frac{1}{2}k \sum_{n=1}^{\infty} (U_{n+1} - U_n)^2 + \frac{1}{2}m \sum_{n=1}^{\infty} \dot{U}_n^2. \tag{6.3.96}$$

Substitute expression (6.3.94) into this. Since it will be long, we will partially calculate. Omitting the argument of the Bessel function, we can obtain

$$U_{n+1} - U_n = -f_0(J_{2n} + J_{2(n+1)}), \tag{6.3.97}$$

$$\dot{U}_n = \omega_0 f_0(J_{2n-1} + J_{2n+1}). \tag{6.3.98}$$

From now on, the sum of the second and third terms on the right-hand side of expression (6.3.96) becomes

$$\frac{1}{2}k\sum_{n=1}^{\infty}(U_{n+1} - U_n)^2 + \frac{1}{2}m\sum_{n=1}^{\infty}\dot{U}_n^2$$

$$= \frac{1}{2}kf_0^2\sum_{n=1}^{\infty}\left[(J_{2n} + J_{2n+2})^2 + (J_{2n-1} + J_{2n+1})^2\right]$$

$$= \frac{1}{2}kf_0^2\sum_{r=1}^{\infty}(J_r + J_{r+2})^2$$

$$= \frac{1}{2}kf_0^2\left[2\sum_{r=1}^{\infty}J_r^2 - J_1^2 - J_2^2 + 2\sum_{r=1}^{\infty}J_rJ_{r+2}\right]. \tag{6.3.99}$$

We use the summation formula of the Bessel function (6.3.44), which we repeat here:

$$\sum_{r=-\infty}^{\infty}J_r^2(z) = 1, \quad \sum_{r=-\infty}^{\infty}J_r(z)J_{r+m}(z) = 0, \quad m \text{ is positive integer.}$$

$$\tag{6.3.100}$$

By using these formulas, we have

$$J_0^2 + 2\sum_{r=1}^{\infty}J_r^2 = 1, \quad -J_1^2 + 2J_0J_2 + 2\sum_{r=1}^{\infty}J_rJ_{r+2} = 0. \tag{6.3.101}$$

Therefore, expression (6.3.99) becomes

$$(6.3.99) = \frac{1}{2}kf_0^2\left[1 - (J_0 + J_2)^2\right]. \tag{6.3.102}$$

On the other hand, since U_1 is given by (6.3.95) and $f(t) = f_0$ at $t > 0$, the first term of the energy (6.3.96) becomes

$$\text{the first term of (6.3.96)} = \frac{1}{2}kf_0^2(J_0 + J_2)^2. \tag{6.3.103}$$

By adding these two results (6.3.102) and (6.3.103), the Bessel function disappears, and the energy $E(t)$ becomes a constant value,

$$E(t) = \frac{1}{2}kf_0^2. \tag{6.3.104}$$

This value is the energy given to the system when the diaphragm first moves momentarily. After that, the diaphragm does not move, so this value is maintained eternally.

Example 6.3.2. We consider the case where the displacement $f(t)$ of the forced vibration is given by the Bessel function,

$$f(t) = f_0 J_\nu(2\omega_0 t). \qquad (6.3.105)$$

By using the formula for convolution between Bessel functions

$$\int_0^x J_\mu(x-t)J_\nu(t)dt = 2\sum_{r=0}^{\infty}(-1)^r J_{\mu+\nu+2r+1}(x), \qquad (6.3.106)$$

we can carry out the integral of (6.3.85), and obtain the displacement $U_n(t)$ in a surprisingly simple form,

$$U_n(t) = f_0 J_{2n+\nu}(2\omega_0 t). \qquad (6.3.107)$$

Substitute this into the energy expression (6.3.96), and write it as

$$E(t,\nu) = \frac{1}{2}kf_0^2 \sum_{r=0}^{\infty}\left(J_{\nu+r} - J_{\nu+r+2}\right)^2. \qquad (6.3.108)$$

This should be equal to the work $W(t)$ that the diaphragm did on the spring and weight system by time t. This work is obtained as follows. At time t', the force with which the diaphragm pushes the leftmost spring is $k\left[f(t') - U_1(t')\right]$. Multiplying this by the length of the movement of the diaphragm $\dot{f}(t')dt'$ during the infinitesimal time dt', and summing up these values from 0 to t for t', we obtain

$$W(t) = \int_0^t k\left[f(t') - U_1(t')\right]\dot{f}(t')dt'. \qquad (6.3.109)$$

Substituting $f(t)$ of (6.3.105) and $U_n(t)$ of (6.3.107) with $n = 1$, and writing $W(t)$ as $W(t,\nu)$ to clarify the ν dependency,

$$W(t,\nu) = kf_0^2\omega_0 \int_0^t \left[J_\nu(2\omega_0 t') - J_{\nu+2}(2\omega_0 t')\right]$$

$$\times \left[J_{\nu-1}(2\omega_0 t') - J_{\nu+1}(2\omega_0 t')\right]dt'. \qquad (6.3.110)$$

On the other hand, $E(t,\nu)$ of expression (6.3.108) is differentiated by time, and omitting the argument of the Bessel function

$$\frac{d}{dt}E(t,\nu) = kf_0^2\omega_0\left(J_\nu - J_{\nu+2}\right)\left(J_{\nu-1} - J_{\nu+1}\right). \qquad (6.3.111)$$

From this result, $W(t, \nu)$ of (6.3.110) is rewritten as

$$W(t, \nu) = \int_0^t \frac{d}{dt'} E(t', \nu) dt' = E(t, \nu) - E(0, \nu). \tag{6.3.112}$$

Because $E(0, \nu) = 0$ for $\nu > 0$ from (6.3.108), it becomes $E(t, \nu) = W(t, \nu)$. Thus the total energy of the system is certainly equal to the work done by the diaphragm.

By the way, it is difficult to find the sum of expression (6.3.108) for general ν. However, if ν is an integer, the summation can be carried out with the aid of formula (6.3.101). For example, if $\nu = 0$, it becomes

$$E(t, 0) = \frac{1}{2} k f_0^2 \left[1 - 2 J_1^2 (2\omega_0 t) \right]. \tag{6.3.113}$$

In the case of $\nu = 1$, we have

$$E(t, 1) = \frac{1}{2} k f_0^2 \left[1 - \left(J_0(2\omega_0 t) - J_2(2\omega_0 t) \right)^2 - 2 J_1^2 (2\omega_0 t) \right]. \tag{6.3.114}$$

If $\nu = 0$, then $E(0, 0) = \frac{1}{2} k f_0^2$. This is $f(0) = f_0$ from (6.3.105); because it jumps from zero to f_0 at the moment $t = 0$, the energy of $\frac{1}{2} k f_0^2$ is acquired at that moment. Thereafter, the energy acquired by time t is $W(t, 0)$. As a result, the energy at time t is $E(t, 0) = E(0, 0) + W(t, 0)$. In the $\nu = 1$ case, we then have $E(0, 1) = 0$, and hence $E(t, 1) = W(t, 1)$.

6.3.6 *Two remarks*

Remark 6.3.1. The most important conclusion in this analysis is that the solution $U(t)$ has been obtained in the form of the convolution integral (6.3.85), which we quote again here:

$$U_n(t) = \omega_0 \int_0^t \left[J_{2n-1}\left(2\omega_0(t - t')\right) + J_{2n+1}\left(2\omega_0(t - t')\right) \right] f(t') dt'. \tag{6.3.115}$$

The process to reach this conclusion required a considerable amount of calculation. Nevertheless, the resulting expression has a simple form. Therefore, it seems that there are other shortcuts to reach this conclusion. The differential operator formalism described in PART I may make it easy to obtain the solution. Only the essential points are described here.

First, in equation (6.3.52), we decided to write the function U_n, where the inhomogeneous part is a delta function, as $\hat{U}_n$:

$$\frac{d^2}{dt^2}\hat{U}_n = -\omega_0^2(-\hat{U}_{n-1} + 2\hat{U}_n - \hat{U}_{n+1}) + \delta(t)\delta_{n,1}. \tag{6.3.116}$$

Here, we used ω_0 of (6.3.65). Also notice that this $\hat{U}_n$ has a dimension of time. If this solution $\hat{U}_n$ can be found, the original solution U_n is obtained as the convolution of this $\hat{U}_n$ and the function $\omega_0^2 f(t)$

$$U_n(t) = \omega_0^2 \int_0^t \hat{U}_n(t - t')f(t')dt'. \tag{6.3.117}$$

Therefore, equation (6.3.116) should be solved by using the operator calculus.

In general, we consider the second-order inhomogeneous linear differential equation in which the coefficient of highest derivative is 1, and the inhomogeneous term is a delta function, like this equation. At this time, it is known that the fundamental solution is equivalent to the solution of the homogeneous equation. However, this solution shall be identically zero at $t < 0$ and the initial value at $t = 0$ is that the 1st derivative $= 1$, 0th derivative $= 0$. Therefore, this is the solution in (6.3.35), that is,

$$U_n(t) = \sum_{r=1}^{\infty} \int_0^t dt' \left[\dot{U}_r(0) + U_r(0)\delta(t - t')\right]\left(J_{2(n-r)}(2\omega_0 t') - J_{2(n+r)}(2\omega_0 t')\right), \tag{6.3.118}$$

where we set the conditions

$$\dot{U}_1(0) = 1, \quad \dot{U}_r(0) = 0 \quad (r = 2, 3, \ldots), \quad U_r(0) = 0 \quad (r = 1, 2, 3, \ldots), \tag{6.3.119}$$

to get the solution of $\hat{U}_n(t)$,

$$\hat{U}_n(t) = \int_0^t \left(J_{2(n-1)}(2\omega_0 t') - J_{2(n+1)}(2\omega_0 t')\right)dt'. \tag{6.3.120}$$

This integration is easily performed by the integration formula (6.3.84) to give

$$\hat{U}_n(t) = \frac{1}{\omega_0}\left[J_{2n-1}(2\omega_0 t) + J_{2n+1}(2\omega_0 t)\right]. \tag{6.3.121}$$

Substituting this expression into the convolution (6.3.117), we can obtain the solution (6.3.115). Thus, it is very easy to find a solution by using the differential operator formalism.

Remark 6.3.2. What happens to $U_0(t)$? As shown in (6.3.53), we initially set $U_0(t) = 0$. When the eigenvalue equation of the coefficient matrix P is solved, according to this fact, the zeroth component of the eigenvector is set to $V_0 = 0$. However, in reality, the value of $U_0(t)$ has never been used until the final conclusion (6.3.85) is obtained. Therefore, it does not matter if $U_0(t)$ is zero or not. If we set $n = 0$ in expression (6.3.85), we have

$$U_0(t) = \omega_0 \int_0^t \left[J_{-1}\big(2\omega_0(t - t')\big) + J_1\big(2\omega_0(t - t')\big)\right] f(t')dt'. \quad (6.3.122)$$

Since $J_{-1}(z) = -J_1(z)$, it becomes $U_0(t) = 0$. However, as shown in Example 6.3.2, if we put $n = 0$ in solution (6.3.107), it becomes $U_0(t) = f(t)$ by using (6.3.105). This is clearly a contradiction. Also, from Fig. 6.6 and equation (6.3.52), it is certainly more natural and easier to understand that the condition $U_0(t) = f(t)$ is set from the beginning.

As for the argument of the Bessel function, because only the zero or positive part is used, so the Bessel function is treated in the form multiplied by a step function $\theta(x)$ as $J_\nu(x)\theta(x)$. For this purpose, by using the series expansion formula for the Bessel function

$$J_\nu(x) = \sum_{m=0}^{\infty} \frac{(-1)^m}{m!\,\Gamma(\nu + m + 1)} \left(\frac{x}{2}\right)^{\nu+2m}, \quad (6.3.123)$$

and the definition of the generalized function Y introduced in PART I,

$$Y_\nu(x) = \begin{cases} \dfrac{x^{\nu-1}}{\Gamma(\nu)}\theta(x), & \nu \neq -n \\[2mm] \delta^{(n)}(x), & \nu = -n \ (n \text{ is non-negative integer}). \end{cases} \quad (6.3.124)$$

Analytic continuation with respect to ν implies

$$J_\nu(x)\theta(x) = \sum_{m=0}^{\infty} \frac{(-1)^m \, x^m}{m! \, 2^{\nu+2m}} Y_{\nu+m+1}(x). \quad (6.3.125)$$

If we set $\nu = -1$ in this formula, because $Y_0(x) = \delta(x)$, then the term $m = 0$ that would normally disappear remains as a correction term,

$$J_{-1}(x)\theta(x) = 2\delta(x) - J_1(x)\theta(x). \quad (6.3.126)$$

If we regard the Bessel function in expression (6.3.122) as that with the step function $\theta(x)$,

$$U_0(t) = \omega_0 \int_0^\infty \Big[J_{-1}\big(2\omega_0(t-t')\big)\theta\big(2\omega_0(t-t')\big)$$

$$+ J_1\big(2\omega_0(t-t')\big)\theta\big(2\omega_0(t-t')\big)\Big] f(t')dt'$$

$$= 2\omega_0 \int_0^\infty \delta(2\omega_0(t-t'))f(t')dt' = f(t). \qquad (6.3.127)$$

Thus we will obtain a consistent result. This resolves the mystery, owing to the fact that the generalized function plays an important role in this calculation.

The existence of the delta function in formula (6.3.126) seems strange. In the definition formula (6.3.123) of the Bessel function, the factor $x^{-1}/\Gamma(0)$ is entered when $\nu = -1$, $m = 0$. This factor can be considered to be zero for $x \neq 0$, but when $x = 0$, it becomes indeterminate.

Here, we have made it possible to correctly evaluate this difficult situation by using the generalized function Y. In equation (6.3.126), changing the sign of x to $-x$, because the $J_{-1}(x)$, $J_1(x)$ is odd-function, and $\delta(x)$ is even-function, it becomes

$$J_{-1}(x)\theta(-x) = -2\delta(x) - J_1(x)\theta(-x). \qquad (6.3.128)$$

When this is summed with the original equation (6.3.126), the delta function part disappears, and we can obtain the commonly-used formula $J_{-1}(x) = -J_1(x)$. It should be noted that the expanded evaluation formula of $J_{-n}(x)\theta(x)$ for the natural number n is obtained by one of the authors N. Nakanishi. (See §2.6.7 on Bessel Functions.)

6.4 Bessel Function and Characteristic Oscillations of Keyboard Percussion

The keyboard percussion instrument is a generic name for percussion instruments that are arranged in a pattern similar to a piano keyboard. These include the xylophone, metallophone, marimba, vibraphone, etc. These instruments are made of wooden or metallic rods of various lengths, and sounds are made by hitting these rods with something called a mallet. Striking a rod produces a sound with the **fundamental oscillation** of the lowest frequency, and at the same time, a sound called the **higher harmonics** is also produced. However, for a simple-shaped rod, the frequency

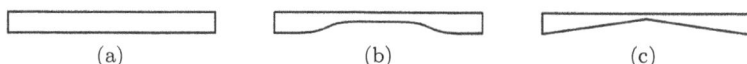

Fig. 6.7 A rod with (a) uniform thickness, (b) a center that is thinner than both ends, (c) linearly-changed thickness

of this higher harmonic is not an integral multiple of the fundamental frequency. Therefore, a dissonant sound (disharmony) is produced, and this is not suitable for a melody instrument. To remedy this, the rods used in these instruments are not simple rods; the center of each rod is made thinner than both ends of the rod. Figure 6.7 shows the side views of 3 types of rods: (a) a rod with uniform thickness, (b) a rod with a center that is thinner than both ends, and (c) a rod with linearly-changed thickness.

In this section, we analyze the characteristic oscillation of a keyboard percussion instrument. It is much too complicated to analyze the oscillation of a rod with a center that is thinner than both ends (Fig. 6.7(b)). Instead, we will analyze the case where the thickness of the rod changes linearly from the center to both ends as shown in Fig. 6.7(c). This model will be quite different from that of a rod in an actual keyboard, but it will be a valuable one in which an analytical solution can be clearly obtained. In fact, the solution of this model can be solved by using the Bessel function. As a result, the frequencies of not only the first and second harmonics, but up to the 9th harmonic can be calculated as almost an integral multiple of the fundamental frequency. Therefore, this shape is superior to that in an actual keyboard. However, we must be prepared for the long tedious calculations.

6.4.1 *Introduction to the equation, and the solution*

Introduction to the equation

Let the length of the rod be 2ℓ and the width be a. When the x axis is taken in the length direction with the center of the rod as the origin, the thickness $h(x)$ for the shape in Fig. 6.7(c) is given by

$$h(x) = h_0\left(1 + \alpha\frac{|x|}{\ell}\right). \qquad (6.4.1)$$

Here, α is a dimensionless parameter for determining the changing ratio in thickness. That is, the thickness becomes h_0 at the center of the rod and it becomes $h_0(1 + \alpha)$ at both ends.

The wave equation used here is that of the Bernoulli–Euler beam. When the displacement in the direction perpendicular to the surface of the rod at the point x and time t is $V(x,t)$, the wave equation is given by

$$\rho\, ah(x)V_{tt}(x,t) = -\partial_x^2\big(EI(x)\,V_{xx}(x,t)\big), \quad I(x) = \frac{ah(x)^3}{12}. \tag{6.4.2}$$

Here, the subscript attached to the displacement V represents the derivative of that variable, for example, $V_x \equiv \partial_x V \equiv \partial V/\partial x$. The meanings of the symbols used here are as follows: ρ is the **volume density**, E is the **Young's modulus** of the rod and $I(x)$ is the **moment of inertia of the area** being defined by this second equation.

In this equation, both ends of the rod ($x = \pm\ell$) are regarded as completely free ends. Therefore we impose the boundary conditions that the **bending moment** $M = EIV_{xx}$ is zero and **shear force** $Q = \partial_x(EIV_{xx})$ is also zero,

$$M = EI(x)V_{xx}\Big|_{x=\pm\ell} = 0, \quad Q = \partial_x\big(EI(x)V_{xx}\big)\Big|_{x=\pm\ell} = 0. \tag{6.4.3}$$

Also since the point of $x = 0$ is a connection point in the equation, we set the connection condition that the rod is not broken or bent

$$\lim_{x\to+0} V(x,t) = \lim_{x\to-0} V(x,t), \quad \lim_{x\to+0} V_x(x,t) = \lim_{x\to-0} V_x(x,t). \tag{6.4.4}$$

Furthermore, we impose the connection condition that the bending moment M and shear force Q are continuous at this point

$$\lim_{x\to+0} EI(x)V_{xx}(x,t) = \lim_{x\to-0} EI(x)V_{xx}(x,t),$$

$$\lim_{x\to+0} \partial_x\big(EI(x)V_{xx}(x,t)\big) = \lim_{x\to-0} \partial_x\big(EI(x)V_{xx}(x,t)\big). \tag{6.4.5}$$

Before solving the equation, we make the variables dimensionless to simplify the equation. First, we introduce c which has the dimension of velocity and τ which has the dimension of time,

$$c = \sqrt{\frac{E}{12\rho}}, \quad \tau = \frac{\ell}{c}. \tag{6.4.6}$$

Then, we set ℓ as the unit of length and τ as the unit of time, and we replace x, h_0, V, t again:

$$\frac{x}{\ell} \mapsto x, \quad \frac{h_0}{\ell} \mapsto h_0, \quad \frac{V}{\ell} \mapsto V, \quad \frac{t}{\tau} \mapsto t. \tag{6.4.7}$$

This replacement makes equation (6.4.2) dimensionless,

$$h(x)V_{tt}(x,t) = -\partial_x^2\left(h(x)^3 V_{xx}(x,t)\right). \tag{6.4.8}$$

Here, instead of (6.4.1), a newly-defined dimensionless thickness $h(x)$ is

$$h(x) = h_0(1 + \alpha|x|). \tag{6.4.9}$$

With this dimensionless, the boundary condition (6.4.3) is changed to

$$V_{xx}(x,t)\Big|_{x=\pm 1} = 0, \quad \partial_x\left(h(x)^3 V_{xx}(x,t)\right)\Big|_{x=\pm 1} = 0$$

$$\Rightarrow \quad V_{xxx}(x,t)\Big|_{x=\pm 1} = 0. \tag{6.4.10}$$

Here, the first and second expressions are directly derived from expression (6.4.3). Note that this second expression is transformed into the third one by using the first expression. Furthermore, the connection condition (6.4.4) remains the same, but condition (6.4.5) changes to

$$\lim_{x\to+0} V_{xx}(x,t) = \lim_{x\to-0} V_{xx}(x,t),$$

$$\lim_{x\to+0} \partial_x\left(h(x)^3 V_{xx}(x,t)\right) = \lim_{x\to-0} \partial_x\left(h(x)^3 V_{xx}(x,t)\right). \tag{6.4.11}$$

Note that this second expression does not become $\lim_{x\to+0} V_{xxx} = \lim_{x\to-0} V_{xxx}$ even if the first expression is used, because it is $\lim_{x\to+0} \partial_x h(x) \neq \lim_{x\to-0} \partial_x h(x)$.

Eigenvalues and eigenfunctions
In solving equation (6.4.8), we assume that the displacement $V(x,t)$ is the separation of variables, and the time-dependent part is the trigonometric function,

$$V(x,t) = X(x)\sin(\omega t + \delta), \tag{6.4.12}$$

where ω is the dimensionless angular frequency and δ is the phase angle. Then equation (6.4.8) becomes the one containing only one variable x,

$$h(x)\omega^2 X(x) = \frac{d^2}{dx^2}\left(h(x)^3 X_{xx}(x)\right), \quad h(x) = h_0(1 + \alpha|x|). \tag{6.4.13}$$

This equation must be solved under the boundary condition derived from (6.4.10),

$$X_{xx}(x)\Big|_{x=\pm 1} = 0, \quad X_{xxx}(x)\Big|_{x=\pm 1} = 0, \tag{6.4.14}$$

and under the connecting condition derived from (6.4.4) and (6.4.11),

$$\lim_{x \to +0} X(x) = \lim_{x \to -0} X(x), \quad \lim_{x \to +0} X_x(x) = \lim_{x \to -0} X_x(x),$$

$$\lim_{x \to +0} X_{xx}(x) = \lim_{x \to -0} X_{xx}(x),$$

$$\lim_{x \to +0} \frac{d}{dx}\left(h(x)^3 X_{xx}(x)\right) = \lim_{x \to -0} \frac{d}{dx}\left(h(x)^3 X_{xx}(x)\right). \tag{6.4.15}$$

Equation (6.4.13) is a fourth-order differential equation, but it can be solved by using the Bessel function. First, in this equation, we carry out the change of variable from x to y,

$$y = \frac{\omega}{h_0 \alpha^2}(1 + \alpha|x|), \tag{6.4.16}$$

and the equation is transformed to

$$yX = \frac{d^2}{dy^2}\left(y^3 \frac{d^2}{dy^2} X\right). \tag{6.4.17}$$

At this time, note that the equation is converted to the same expression regardless of whether x is positive or negative, because the original equation (6.4.13) is invariant to the sign change of x.

On the other hand, the equation satisfied by the order-1 Bessel function and the order-1 modified Bessel function are written, in terms of the independent variable z, the dependent variable $f_{(\pm)}$,

$$\left(\frac{d^2}{dz^2} + \frac{1}{z}\frac{d}{dz} - \frac{1}{z^2} \pm 1\right) f_{(\pm)}(z) = 0. \tag{6.4.18}$$

Here, the double sign is in the common arrangement; plus sign and minus sign represent the normal Bessel function and the modified Bessel function, respectively. If we convert the dependent variable of this expression from $f_{(\pm)}(z)$ to $f_{(\pm)}(z)/z$,

$$\left(\frac{d^2}{dz^2} + \frac{3}{z}\frac{d}{dz} \pm 1\right)\frac{f_{(\pm)}(z)}{z} = 0, \tag{6.4.19}$$

and furthermore, if we convert the independent variable from z to y,

$$z = 2\sqrt{y}, \tag{6.4.20}$$

the equation is transformed to

$$\left(y\frac{d^2}{dy^2} + 2\frac{d}{dy} \pm 1\right)\frac{f_{(\pm)}(2\sqrt{y})}{2\sqrt{y}} = 0. \tag{6.4.21}$$

Furthermore, by making a product of the above two (upper and lower signs) differential operators, we have

$$\left[\left(y\frac{d^2}{dy^2} + 2\frac{d}{dy}\right)^2 - 1\right]\frac{f_{(\pm)}(2\sqrt{y})}{2\sqrt{y}} = 0. \tag{6.4.22}$$

Now, we rewrite the part of this differential operator

$$\left(y\frac{d^2}{dy^2} + 2\frac{d}{dy}\right)^2 = y^2\frac{d^4}{dy^4} + 6y\frac{d^3}{dy^3} + 6\frac{d^2}{dy^2} = \frac{1}{y}\frac{d^2}{dy^2}\left(y^3\frac{d^2}{dy^2}\right). \tag{6.4.23}$$

Then equation (6.4.22) is rewritten as

$$y \cdot \left(\frac{f_{(\pm)}(2\sqrt{y})}{2\sqrt{y}}\right) = \frac{d^2}{dy^2}\left[y^3\frac{d^2}{dy^2}\left(\frac{f_{(\pm)}(2\sqrt{y})}{2\sqrt{y}}\right)\right]. \tag{6.4.24}$$

This exactly reproduces equation (6.4.17) for X, and therefore, four independent solutions are obtained by using the first and second kind Bessel functions of order-1 J_1, N_1, and the first and second kind modified Bessel functions of order-1 I_1, K_1,

$$X = \frac{J_1(2\sqrt{y})}{2\sqrt{y}}, \quad \frac{N_1(2\sqrt{y})}{2\sqrt{y}}, \quad \frac{I_1(2\sqrt{y})}{2\sqrt{y}}, \quad \frac{K_1(2\sqrt{y})}{2\sqrt{y}}. \tag{6.4.25}$$

As mentioned earlier, equation (6.4.13), the boundary condition (6.4.14), and the connection condition (6.4.15) are all invariant for the sign change of x, and therefore, solution X can be an even function or an odd function. For simplification, we make use of the relationship between the variables x and z as it is, that is, from (6.4.16) and (6.4.20),

$$z = 2\sqrt{y} = 2\sqrt{\frac{\omega}{h_0\alpha^2}(1 + \alpha|x|)}, \tag{6.4.26}$$

and we will use this z as the argument of the Bessel function. Then, the solution of the even function $F(x)$ is defined by the linear combination of expression (6.4.25), with A, B, C, D as appropriate coefficients,

$$F(x) = A\frac{J_1(z)}{z} + B\frac{N_1(z)}{z} + C\frac{I_1(z)}{z} + D\frac{K_1(z)}{z}, \tag{6.4.27}$$

and the solution of odd function $G(x)$ is defined by

$$G(x) = \text{sgn}(x)\left[A\frac{J_1(z)}{z} + B\frac{N_1(z)}{z} + C\frac{I_1(z)}{z} + D\frac{K_1(z)}{z}\right]. \tag{6.4.28}$$

Here, the symbol "sgn" is a sign function. We also used the same coefficients A, B, C, D in these two expressions, to avoid unnecessarily increasing the number of the variables involved, though, of course, they are generally different.

For these two expressions $F(x)$ and $G(x)$, we imposed two boundary conditions of (6.4.14) and four connection conditions of (6.4.15); a total of six conditions were imposed. Among these, the two boundary conditions of (6.4.14) are necessary conditions, but all of the four connection conditions of (6.4.15) are not always necessary. For example, $F(x)$ and F_{xx} are even functions from the definition (6.4.27), so they are already continuous at $x = 0$. On the other hand, F_x and $(d/dx)[h^3 F_{xx}]$ are odd functions, therefore, these values must be zero at $x = 0$ to be continuous there. Similarly, for $G(x)$, because G_x and $(d/dx)[h^3 G_{xx}]$ are even functions, these are already continuous at $x = 0$. But, since G and G_{xx} are odd functions, these values must be zero at $x = 0$ to be continuous there.

To summarize the above, the conditions for the even function $F(x)$ are[9]

$$F_{xx}(1) = 0, \quad F_{xxx}(1) = 0, \quad F_x(0) = 0, \quad \frac{d}{dx}\left[h(x)^3 F_{xx}\right]\Big|_{x=0} = 0.$$

$$(6.4.29)$$

Also, the conditions for the odd function $G(x)$ are

$$G_{xx}(1) = 0, \quad G_{xxx}(1) = 0, \quad G(0) = 0, \quad G_{xx}(0) = 0. \qquad (6.4.30)$$

Here, since the functions F and G have a fixed even/odd parity, the conditions for positive and negative x are practically required to impose only positive x.

In what follows, we will apply these conditions to $F(x)$ and $G(x)$ defined in (6.4.27) and (6.4.28), respectively. For this purpose, some differential formulas for the Bessel function are required,

$$\left(\frac{1}{z}\frac{d}{dz}\right)^n \left(\frac{J_\nu(z)}{z^\nu}\right) = (-1)^n \frac{J_{\nu+n}(z)}{z^{\nu+n}}, \quad \left(\frac{1}{z}\frac{d}{dz}\right)^n \left(\frac{N_\nu(z)}{z^\nu}\right) = (-1)^n \frac{N_{\nu+n}(z)}{z^{\nu+n}},$$

$$\left(\frac{1}{z}\frac{d}{dz}\right)^n \left(\frac{I_\nu(z)}{z^\nu}\right) = \frac{I_{\nu+n}(z)}{z^{\nu+n}}, \quad \left(\frac{1}{z}\frac{d}{dz}\right)^n \left(\frac{K_\nu(z)}{z^\nu}\right) = (-1)^n \frac{K_{\nu+n}(z)}{z^{\nu+n}}.$$

$$(6.4.31)$$

---

[9]We want to decompose the derivative that appears in the last expression of (6.4.29), using the distributive law of differentiation. Then, the later calculation becomes rather cumbersome.

We use the case of $\nu = 1$, $n = 1, 2$. In addition, we must apply the last condition expressed in (6.4.29),

$$\left(\frac{1}{z}\frac{d}{dz}\right)^n \left(z^\nu J_\nu(z)\right) = z^{\nu-n} J_{\nu-n}(z),$$

$$\left(\frac{1}{z}\frac{d}{dz}\right)^n \left(z^\nu N_\nu(z)\right) = z^{\nu-n} N_{\nu-n}(z),$$

$$\left(\frac{1}{z}\frac{d}{dz}\right)^n \left(z^\nu I_\nu(z)\right) = z^{\nu-n} I_{\nu-n}(z),$$

$$\left(\frac{1}{z}\frac{d}{dz}\right)^n \left(z^\nu K_\nu(z)\right) = (-1)^n z^{\nu-n} K_{\nu-n}(z).$$

(6.4.32)

Here, we use the case of $\nu = 3$, $n = 1$. Further, when applying the last condition in (6.4.29), the $h(x)$ must be represented by z using (6.4.9), (6.4.16) and (6.4.20),

$$h(x) = \frac{(h_0 \alpha)^2}{4\omega} z^2.$$

(6.4.33)

After that, we can use the relationship between the x derivative and the z derivative, in the range of $x > 0$,

$$\frac{d}{dx} = \frac{dy}{dx}\frac{d}{dy} = \frac{\omega}{h_0 \alpha}\frac{d}{dy} = \frac{\omega}{h_0 \alpha}\frac{dz}{dy}\frac{d}{dz} = \frac{2\omega}{h_0 \alpha}\left(\frac{1}{z}\frac{d}{dz}\right).$$

(6.4.34)

This is exactly the form in which the differential formulas of (6.4.31) and (6.4.32) can be used. Before writing the result, the values of z at $x = 0$ and $x = 1$ are defined by using (6.4.26) to simplify the expressions,

$$z(0) = 2\sqrt{\frac{\omega}{h_0 \alpha^2}} \equiv k, \quad z(1) = \gamma k, \quad (\gamma \equiv \sqrt{1+\alpha}).$$

(6.4.35)

Next, we apply the condition (6.4.29) to $F(x)$ of (6.4.27),

$$AJ_3(\gamma k) + BN_3(\gamma k) + CI_3(\gamma k) + DK_3(\gamma k) = 0,$$

$$AJ_4(\gamma k) + BN_4(\gamma k) - CI_4(\gamma k) + DK_4(\gamma k) = 0,$$

$$AJ_2(k) + BN_2(k) - CI_2(k) + DK_2(k) = 0,$$

$$AJ_2(k) + BN_2(k) + CI_2(k) - DK_2(k) = 0.$$

(6.4.36)

Similarly, we apply the condition (6.4.30) to $G(x)$ of (6.4.28),

$$AJ_3(\gamma k) + BN_3(\gamma k) + CI_3(\gamma k) + DK_3(\gamma k) = 0,$$
$$AJ_4(\gamma k) + BN_4(\gamma k) - CI_4(\gamma k) + DK_4(\gamma k) = 0,$$
$$AJ_1(k) + BN_1(k) + CI_1(k) + DK_1(k) = 0,$$
$$AJ_3(k) + BN_3(k) + CI_3(k) + DK_3(k) = 0.$$
(6.4.37)

Because these expressions are **homogeneously** linear simultaneous equations, the value of the coefficients determinant must be zero for the unknown numbers A, B, C, D to be not zero. That is, the coefficients determinant defined from (6.4.36) for F,

$$\Lambda_F(k) = \begin{vmatrix} J_3(\gamma k) & N_3(\gamma k) & I_3(\gamma k) & K_3(\gamma k) \\ J_4(\gamma k) & N_4(\gamma k) & -I_4(\gamma k) & K_4(\gamma k) \\ J_2(k) & N_2(k) & -I_2(k) & K_2(k) \\ J_2(k) & N_2(k) & I_2(k) & -K_2(k) \end{vmatrix},$$
(6.4.38)

must vanish

$$\Lambda_F(k) = 0.$$
(6.4.39)

From now on, assuming that $\gamma = \sqrt{1+\alpha}$ is given, the value of k is obtained as an **eigenvalue**. Once the value of k is determined, the eigenvalue of the angular frequency ω is also determined from the first equation of (6.4.35).

Similarly, the determinant defined from (6.4.37) for G,

$$\Lambda_G(k) = \begin{vmatrix} J_3(\gamma k) & N_3(\gamma k) & I_3(\gamma k) & K_3(\gamma k) \\ J_4(\gamma k) & N_4(\gamma k) & -I_4(\gamma k) & K_4(\gamma k) \\ J_1(k) & N_1(k) & I_1(k) & K_1(k) \\ J_3(k) & N_3(k) & I_3(k) & K_3(k) \end{vmatrix},$$
(6.4.40)

must vanish

$$\Lambda_G(k) = 0.$$
(6.4.41)

Similarly, from this equation, the values of k and ω are determined as eigenvalues.

These eigenvalues for k should be discrete, because the zero point of an analytic function is isolated. Thus the eigenvalues obtained from equations (6.4.39) and (6.4.41) are successively named k_n ($n = 0, 1, 2, 3, \ldots$) in increasing order. For each k_n, we set $A = 1$ in equation (6.4.36) or (6.4.37), while the remaining coefficients B, C, D can be determined from any three of these four equations. The function $F(x)$ or $G(x)$ obtained in this way is called the **eigenfunction** $F(x, k_n)$ or $G(x, k_n)$ to specify the eigenvalue k_n dependency. However, these are not normalized eigenfunctions.

Normalization of the eigenfunction

Let the two eigenvalues be k_n, $k_{n'}$. If the even/odd parity of n and n' is different, the even/odd parity of the original functions is also considered to be different. Here, for example, when n is even and n' is odd, the integral of the product of their even function $F(x, k_n)$ and odd function $G(x, k_{n'})$ multiplied by the weight $h(x)$ becomes

$$\int_{-1}^{1} h(x)F(x, k_n)G(x, k_{n'})dx = 0. \tag{6.4.42}$$

This is obvious because $h(x)$ is an even function. Therefore, when calculating the orthogonality of eigenfunctions, it is enough to consider the integral between even functions or between odd functions. In the following, we consider the case of a pair of two even functions.

We consider here two k's, k and k', which are not necessarily eigenvalues, and let the functions belonging to them be $F(x, k)$ and $F(x, k')$, respectively. These functions satisfy the equation (6.4.13),

$$\omega^2 h(x)F(x, k) = \frac{d^2}{dx^2}\left(h(x)^3 F_{xx}(x, k)\right),$$
$$\omega'^2 h(x)F(x, k') = \frac{d^2}{dx^2}\left(h(x)^3 F_{xx}(x, k')\right), \tag{6.4.43}$$

where k and ω, or k' and ω' are related by the first equation (6.4.35). First, multiplying the above first equation by $F(x, k')$ and multiplying the second equation by $F(x, k)$, we write down the difference of both equations:

$$(\omega^2 - \omega'^2)h(x)F(x, k)F(x, k')$$

$$= \frac{d}{dx}\left[F(x, k')\frac{d}{dx}\left(h(x)^3 F_{xx}(x, k)\right) - F(x, k)\frac{d}{dx}\left(h(x)^3 F_{xx}(x, k')\right)\right.$$

$$\left. + h(x)^3 F_x(x, k)F_{xx}(x, k') - h(x)^3 F_x(x, k')F_{xx}(x, k)\right]. \tag{6.4.44}$$

Next, we integrate both sides of this equation over x from -1 to 1. Then, since the function F has different definitions depending on the sign of x, as seen in expressions (6.4.26) and (6.4.27), we need to divide the integral into two regions,

$$\int_{-1}^{1} = \int_{-1}^{-0} + \int_{+0}^{1}. \tag{6.4.45}$$

If we assume that the connection condition (6.4.15) is satisfied, then each of F, F_x, F_{xx}, $(d/dx)[h^3 F_{xx}]$ becomes continuous at $x = 0$. Therefore the boundary values are canceled here and it is integral smooth from -1 to 1. The result is

$$(\omega^2 - \omega'^2) \int_{-1}^{1} h(x) F(x, k) F(x, k') dx$$

$$= \Big[F(x, k') \frac{d}{dx} \big(h(x)^3 F_{xx}(x, k) \big) - F(x, k) \frac{d}{dx} \big(h(x)^3 F_{xx}(x, k') \big)$$

$$+ h(x)^3 F_x(x, k) F_{xx}(x, k') - h(x)^3 F_x(x, k') F_{xx}(x, k) \Big]_{-1}^{1}. \tag{6.4.46}$$

Further, we assume that the first boundary condition of (6.4.29) is satisfied, that is, $F_{xx}(\pm 1, k) = F_{xx}(\pm 1, k') = 0$, and we have

$$\int_{-1}^{1} h(x) F(x, k) F(x, k') dx$$

$$= \frac{h(1)^3}{\omega^2 - \omega'^2} \Big[F(x, k') F_{xxx}(x, k) - F(x, k) F_{xxx}(x, k') \Big]_{-1}^{1}. \tag{6.4.47}$$

Here, we have used the relation $(d/dx)\big(h(x)^3 F_{xx}(x) \big) \big|_{x=\pm 1} = h(1)^3 F_{xxx}(\pm 1)$. Furthermore, since the inside of the square brackets of this expression is an odd function, it becomes

$$\int_{-1}^{1} h(x) F(x, k) F(x, k') dx$$

$$= \frac{2h(1)^3}{\omega^2 - \omega'^2} \Big[F(1, k') F_{xxx}(1, k) - F(1, k) F_{xxx}(1, k') \Big]. \tag{6.4.48}$$

If the last condition, namely the second boundary condition of (6.4.29) is satisfied, the parameters k and k' take eigenvalues k_n and $k_{n'}$, respectively. Thus under this condition, the inside of the brackets on the right-hand side becomes zero. Therefore, if $k_n \neq k_{n'}$, we are led to the orthogonality relation between the eigenfunctions belonging to the different eigenvalues,

$$\int_{-1}^{1} h(x) F(x, k_n) F(x, k_{n'}) dx = 0, \quad n \neq n'. \tag{6.4.49}$$

When the same eigenvalues are substituted, the right-hand side of expression (6.4.48) becomes $0/0$, so we set $k' = k_n$ first, and then take the limit of $k \to k_n$. That is, ω is represented by k using the first expression of (6.4.35), and we can obtain the orthogonality relation of the eigenfunctions by using l'Hôpital's rule,

$$\int_{-1}^{1} h(x)F(x, k_n)F(x, k_{n'})dx = N_{F,n}^2 \delta_{n,n'},$$

$$N_{F,n}^2 = \frac{8h_0(1+\alpha)^3}{\alpha^4 k_n^3} \left[\partial_k F_{xxx}(1, k)\right]_{k=k_n} F(1, k_n), \quad (6.4.50)$$

where $N_{F,n}$ is a normalization constant, and $F(x, k_n)/N_{F,n}$ is the normalized eigenfunction.

Similarly, the orthogonality relation between odd functions $G(x, k_n)$ is obtained,

$$\int_{-1}^{1} h(x)G(x, k_n)G(x, k_{n'})dx = N_{G,n}^2 \delta_{n,n'},$$

$$N_{G,n}^2 = \frac{8h_0(1+\alpha)^3}{\alpha^4 k_n^3} \left[\partial_k G_{xxx}(1, k)\right]_{k=k_n} G(1, k_n). \quad (6.4.51)$$

$N_{G,n}$ is the normalization constant.

Numerical calculation

In the following, we solve the eigenvalue equations $\Lambda_F(k) = 0$, $\Lambda_G(k) = 0$ defined in (6.4.39) and (6.4.41), by using the formula manipulation software, *Maxima*. The solutions obtained from $\Lambda_F(k) = 0$ are added even numbers $k_0, k_2, k_4, \ldots$ from the smallest, and the solutions obtained from $\Lambda_G(k) = 0$ are added odd numbers $k_1, k_3, k_5, \ldots$ from the smallest. Although not proved, the magnitude relation of the eigenvalues obtained in this way is $k_0 < k_1 < k_2 < k_3 < k_4 < \cdots$.

The parameter α of expression (6.4.9) which determines the thickness of the rod is positive when the edges are thicker than the center of the rod. On the contrary, when the edges are thinner than the center, it becomes negative. This value cannot be equal to zero because it appears in a denominator of expression (6.4.16). After all, the possible range of α is greater than -1 and non-zero.

Here, for the five values of $\alpha = -0.50, -0.75, 0.25, 0.50, 0.75$, the eigenvalues k_n are obtained by using the *Maxima* software. The results are listed with precision up to 3 digits after the decimal point, for six values of $n = 0, 1, 2, 3, 4, 5$,

$$\alpha = -0.50, \quad k_n = \quad 9.343, \quad 14.138, \quad 19.455, \quad 24.592, \quad 29.956, \quad 35.195,$$

$$\alpha = -0.75, \quad k_n = \quad 6.280, \quad 8.846, \quad 11.947, \quad 14.823, \quad 17.950, \quad 20.939,$$

$$\alpha = 0.25, \quad k_n = 19.086, \quad 32.777, \quad 46.190, \quad 59.650, \quad 72.951, \quad 86.342,$$

$$\alpha = 0.50, \quad k_n = \quad 9.626, \quad 17.018, \quad 24.126, \quad 31.250, \quad 38.234, \quad 45.297,$$

$$\alpha = 0.75, \quad k_n = \quad 6.471, \quad 11.735, \quad 16.733, \quad 21.718, \quad 26.579, \quad 31.510.$$

$$(6.4.52)$$

As can be seen from these numbers, for $\alpha < 0$, as $|\alpha|$ increases, the value of k_n decreases. On the other hand, for $\alpha > 0$, as α increases, the value of k_n first increases sharply and subsequently decreases slowly. This is because the mass of the vibrating part increases as the value of α increases. As a result, it is thought that it becomes difficult to vibrate.

When the eigenvalue k_n is determined, the angular frequency ω_n is also determined from the first expression of (6.4.35).

$$\omega_n = \frac{h_0 \alpha^2}{4} k_n^2. \tag{6.4.53}$$

From this, the ratio of the harmonic frequency to the fundamental wave is determined,

$$\frac{\omega_n}{\omega_0} = \left(\frac{k_n}{k_0}\right)^2. \tag{6.4.54}$$

Since k_n is determined by only one parameter α, the value of this ratio is also determined only by α. The actual values of this ratio are found from (6.4.52),

$$\alpha = -0.50, \quad \omega_n/\omega_0 = 1, \quad 2.289, \quad 4.335, \quad 6.927, \quad 10.279, \quad 14.188,$$

$$\alpha = -0.75, \quad \omega_n/\omega_0 = 1, \quad 1.983, \quad 3.618, \quad 5.569, \quad 8.167, \quad 11.114,$$

$$\alpha = 0.25, \quad \omega_n/\omega_0 = 1, \quad 2.949, \quad 5.856, \quad 9.767, \quad 14.608, \quad 20.464,$$

$$\alpha = 0.50, \quad \omega_n/\omega_0 = 1, \quad 3.125, \quad 6.280, \quad 10.538, \quad 15.774, \quad 22.141,$$

$$\alpha = 0.75, \quad \omega_n/\omega_0 = 1, \quad 3.289, \quad 6.686, \quad 11.263, \quad 16.870, \quad 23.711.$$

$$(6.4.55)$$

What is important here is that the frequency of the harmonics is an integral multiple of the lowest fundamental wave. If this is not the case, there will occur a dissonance from the rod. If we seek the value which is close to an integral multiple, the frequency of the first harmonic when $\alpha = -0.75$ is almost twice that of the fundamental wave. Furthermore, the frequencies of the first and second harmonics of $\alpha = 0.25$ are almost 3 times and 6 times the fundamental wave.

If we change the value of α in small steps, we may get a value closer to an integral multiple. Therefore, if we try to solve the eigenvalue equation when $\alpha = 1/3 = 0.333\ldots$, the result is, for $n = 0, 1, 2, 3, 4, 5$,

$$k_n = 14.356, \quad 24.905, \quad 35.168, \quad 45.469, \quad 55.616, \quad 65.849,$$

$$\frac{\omega_n}{\omega_0} = 1, \quad 3.009, \quad 6.000, \quad 10.030, \quad 15.006, \quad 21.037. \tag{6.4.56}$$

In fact, from the 1st harmonic to the 5th harmonic, the values are almost 3-times, 6-times, 10-times, 15-times, and 21-times which are extremely close to integers. In particular, the vibration of the first harmonic is 3-times that of the fundamental wave; this is the same as in the case of the clarinet.

Here, let us find the size of the rod to make the sound of the center [La] (440 Hz) on a piano. Here, only the fundamental wave is considered, so we set $n = 0$ in relation (6.4.53). As for the parameter α, we use the value $\alpha = 1/3$ with which expression (6.4.56) was derived. Then, the relation between k_0 and ω_0 becomes from (6.4.53),

$$\omega_0 = \frac{h_0}{36} k_0^2. \tag{6.4.57}$$

This value ω_0 is a dimensionless angular frequency, so to restore it, first, let h_0 be the original h_0/ℓ, and then divide it by the time unit $\tau = \ell/c$ defined in expression (6.4.6). Furthermore, the frequency ν is obtained by dividing by 2π. That is, the frequency ν is given by

$$\nu = \frac{h_0 c}{36 \times 2\pi \ell^2} k_0^2. \tag{6.4.58}$$

From now on, we find ℓ and multiply it by 2 to make the length of the rod,

$$2\ell = \frac{k_0}{3} \sqrt{\frac{h_0 c}{2\pi \nu}}. \tag{6.4.59}$$

After that, we have to find h_0 and c. Here, the thickness h_0 at the thinnest part of the center of the rod is adopted appropriately,

$$h_0 = 1\,\mathrm{cm} = 0.01\,\mathrm{m}. \tag{6.4.60}$$

Further, we assume that the rod is made of pine wood, and adopt the values of Young's modulus E and density ρ as found on the Internet,

$$E = 10\,790\,\text{MPa}, \quad \rho = 0.63\,\text{g/cm}^3 = 630\,\text{kg/m}^3. \tag{6.4.61}$$

By using these values, the velocity c defined by (6.4.6) becomes

$$c = \sqrt{\frac{E}{12\rho}} = 1194.675\,\text{m/s}. \tag{6.4.62}$$

These values of h_0, c and the frequency $\nu = 440\,\text{Hz}$, $k_0 = 14.356$ are substituted into expression (6.4.59). Thus, we can obtain the rod length 2ℓ as a plausible value,

$$2\ell = \frac{14.356}{3}\sqrt{\frac{0.01 \times 1194.675}{2\pi \times 440}} = 0.31458\,\text{m} = 31.458\,\text{cm}. \tag{6.4.63}$$

We set the thickness of the rod to be 1 cm at the thinnest part in the center. For the thickest part of both ends, we obtain from expression (6.4.1),

$$h(\ell) = h_0(1+\alpha) = 1.333\cdots\,\text{cm}. \tag{6.4.64}$$

Finally, let us find the eigenfunctions numerically. We set $A = 1$ in (6.4.36) or (6.4.37), and let k be the n-th eigenvalue k_n. The remaining coefficients B, C, D are determined by arbitrarily using three of the four equations. These calculations are carried out by using the *Maxima* software. As a result, we can obtain the eigenfunctions $F(x, k_n)$ and $G(x, k_n)$. After that, the normalization coefficients $N_{F,n}$, $N_{G,n}$ (6.4.50) and (6.4.51) are obtained and made into a normalized eigenfunction.

The graph of eigenfunctions obtained here is shown in Fig. 6.8. This graph was drawn by using four types of Bessel functions J_1, N_1, I_1, K_1, with 99% contributions from J_1 and N_1 and very little from I_1 and K_1.

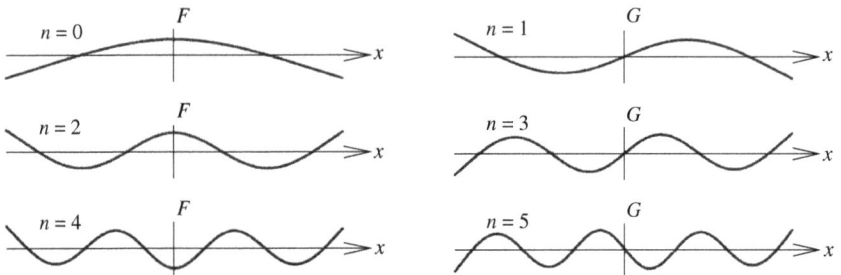

Fig. 6.8　$F(x, k_n)(n = 0, 2, 4)$, $G(x, k_n)(n = 1, 3, 5)$, case of $\alpha = 1/3$

Mysterious Phenomena in Nature

The best result of this analysis is that the frequency of the harmonics becomes almost an integral multiple of the fundamental wave as shown in expression (6.4.56) for $\alpha = 1/3$. It may not make sense, but in the continuation of this analysis, we will try to solve the case of $n = 6, 7, 8, 9$. The result is

$$k_n = 75.986, \quad 86.198, \quad 96.330, \quad 106.524,$$

$$\frac{\omega_n}{\omega_0} = 28.012, \quad 36.044, \quad 45.019, \quad 55.052.$$

It can also be seen that the ratio ω_n/ω_0 is an almost integral multiple. Moreover, if we start from $n = 0$, the ratio becomes 1-time, 3-times, 6-times, 10-times, 15-times, 21-times, 28-times, 36-times, 45-times, and 55-times, and its difference sequence becomes 2, 3, 4, 5, 6, 7, 8, 9, 10. Therefore, the general term of this ratio is represented by the **triangular number** $\omega_n/\omega_0 \simeq (n+1)(n+2)/2$.

On another subject, in the Bethe–Salpeter equation which describes the bound state of two elementary particles, there is a model called the Wick–Cutkosky model which is known to have an analytically exact solution. Moreover, the eigenvalue of this model is also curiously represented by using the triangular number.

There are two phenomena that seem to have nothing to do with each other at first glance. For example, there are the phenomenon of an apple falling from a tree and the phenomenon of a planet rotating around the sun. These two phenomena were successfully combined by Newtonian mechanics. It might not always be the case that the phenomenon dealt with here and the phenomenon of the Wick–Cutkosky model are not linked. At this point, we wonder whether there is any undiscovered mysterious law in nature. What is it exactly?

Going back to the topic at hand, the shape of a rod in an actual keyboard is as shown in Fig. 6.7(b) as mentioned before. However, if it is shaped like in Fig. 6.7(c), the tone of the sound produced should definitely be better. If at all possible, I would wish that some company produces this type of rods for our keyboard percussion instruments.

Chapter 7

Potential Problems in Quantum Mechanics

Regarding potential problems in quantum mechanics, the most familiar ones may be the harmonic oscillator problem, which is solvable by using the Hermite polynomial, and the Coulomb potential problem, which is solvable by using the Laguerre polynomial. However, we will leave these familiar problems to other books; here we discuss more difficult problems.

First, we discuss the Aharonov–Bohm effect in quantum mechanics. This is a problem of what happens to the orbit when an electron runs in a vector potential. We can see how the Bessel function works here as well. Next, we deal with a slightly difficult potential problem called the Rosen–Morse potential in which Gauss' hypergeometric function plays an active role. Finally, we deal with two periodic potential problems in quantum mechanics. One is the potential problem where the delta functions are arranged periodically and the other is the serrated (saw blade) shape potential problem. The delta function problem can be solved by an elementary function, but for the serrated shape potential problem, the Airy function described in the previous chapter plays an active role.

7.1 Aharonov–Bohm Effect

The **Aharonov–Bohm effect** is a mysterious phenomenon. In classical mechanics, if an electron passes through a place where there is no magnetic field, its movement is not affected. However, in quantum mechanics, even in places where there is no magnetic field, as long as that vector potential exists, the movement of electrons is affected.

When this treatise was first published in 1959, many doubts arose and its reliability was questioned. In this treatise, Yakir Aharonov and David Bohm, assuming a solenoid that is infinitely long with a thickness that is extremely close to zero, solved the problem of electron waves incident perpendicular to this solenoid by using the Bessel function. However, this received a lot of criticism. Firstly, infinitely-long solenoids do not really exist, and secondly, if its thickness is zero, the magnetic flux density there becomes a delta function. Therefore, this was thought to be just a mathematical trick. However, in 1986, 27 years after this treatise was published, the real existence of this effect was definitely confirmed by Tonomura *et al.*[1] by an experiment using a ring-shaped solenoid.

7.1.1 *Introduction to the equation*

We consider here an infinitely-long solenoid with a finite radius of r_0, and assume that a **magnetic flux** of constant magnitude Φ penetrates this solenoid. The z-axis of Cartesian coordinates (x, y, z) is taken along the central axis of the solenoid. The distance from the solenoid r, in two-dimension, is

$$r = \sqrt{x^2 + y^2}. \tag{7.1.1}$$

Then, the **vector potential** $\boldsymbol{A}(\boldsymbol{r})$ at point $\boldsymbol{r}(x, y, z)$ is given by

$$\boldsymbol{A}(\boldsymbol{r}) = \begin{cases} \left(-\dfrac{y\Phi}{2\pi r^2}, \dfrac{x\Phi}{2\pi r^2}, 0 \right), & r \geq r_0, \\[4mm] \left(-\dfrac{y\Phi}{2\pi r_0{}^2}, \dfrac{x\Phi}{2\pi r_0{}^2}, 0 \right), & r < r_0. \end{cases} \tag{7.1.2}$$

Then, the **magnetic flux density** $\boldsymbol{B}(\boldsymbol{r})$ is derived as

$$\boldsymbol{B}(\boldsymbol{r}) = \boldsymbol{\nabla} \times \boldsymbol{A}(\boldsymbol{r}) = \begin{cases} (0, 0, 0), & r \geq r_0, \\[4mm] \left(0, 0, \dfrac{\Phi}{\pi r_0{}^2} \right), & r < r_0. \end{cases} \tag{7.1.3}$$

From these expressions, we can see that although the vector potential exists outside the solenoid, the magnetic flux density becomes zero. Therefore, classically speaking, the passage of electrons outside the solenoid should

---

[1]M. Peshkin and A. Tonomura, *The Aharonov–Bohm Effect*, (1989) Springer-Verlag Berlin Heidelberg.

have no effect. We will clarify that this is not the case in quantum theory from now.

When an electron with mass μ and charge e is incident from the positive direction of the x-axis toward a solenoid on the z-axis, we analyze what kind of scattering occurs by using the time-independent **Schrödinger equation**

$$H\Psi = E\Psi, \tag{7.1.4}$$

where Hamiltonian H of the electron moving in the vector potential $\boldsymbol{A}$ is given by

$$H = -\frac{\hbar^2}{2\mu}\left(\boldsymbol{\nabla} - \frac{ie}{c\hbar}\boldsymbol{A}\right)^2, \tag{7.1.5}$$

where c is the velocity of light. When this Hamiltonian is converted to two-dimensional polar coordinates (r, θ) by using the vector potential $\boldsymbol{A}$ of (7.1.2), we have

$$H = -\frac{\hbar^2}{2\mu}\left[\frac{\partial^2}{\partial r^2} + \frac{1}{r}\frac{\partial}{\partial r} + \frac{1}{r^2}\left(\frac{\partial}{\partial \theta} + i\alpha\right)^2\right], \tag{7.1.6}$$

where we define the dimensionless constant α by

$$\alpha = -\frac{e\Phi}{2\pi\hbar c}. \tag{7.1.7}$$

Now the effect of the vector potential is summed up to the one parameter α.

Next, it is necessary to determine the incident wave. Before doing that, let us define flow vector in quantum theory. By using the time-dependent Schrödinger equation

$$H\Psi = i\hbar\frac{\partial\Psi}{\partial t}, \tag{7.1.8}$$

we can derive the **equation of continuity**

$$\frac{\partial(\Psi^*\Psi)}{\partial t} + \boldsymbol{\nabla}\cdot\boldsymbol{j} = 0, \tag{7.1.9}$$

where the **flow vector** $\boldsymbol{j}$ is defined as

$$\begin{aligned}
\boldsymbol{j} &= \frac{\hbar}{2i\mu}\left[\Psi^*\left(\overrightarrow{\boldsymbol{\nabla}} - \frac{ie}{c\hbar}\boldsymbol{A}\right)\Psi - \Psi^*\left(\overleftarrow{\boldsymbol{\nabla}} - \frac{ie}{c\hbar}\boldsymbol{A}\right)^*\Psi\right] \\
&= \frac{\hbar}{2i\mu}\Psi^*\left(\overrightarrow{\boldsymbol{\nabla}} - \overleftarrow{\boldsymbol{\nabla}}\right)\Psi - \frac{e}{\mu c}\boldsymbol{A}\Psi^*\Psi.
\end{aligned} \tag{7.1.10}$$

Next, we introduce the incident wave Ψ_{inc} using a positive constant k as a **wave number**

$$\Psi_{\text{inc}} = e^{-ikx - i\alpha\theta}, \quad |\theta| < \pi. \tag{7.1.11}$$

By using this incident wave to find the flow vector (7.1.10), we have

$$\boldsymbol{j}_{\text{inc}} = (-\hbar k/\mu, 0, 0). \tag{7.1.12}$$

This represents a wave traveling from the positive direction to the negative direction of the x-axis at speed $\hbar k/\mu$. If there is no vector potential, the incident wave can be e^{-ikx}; however, when the vector potential exists, it is noted that the incident wave needs to have a α dependency as shown in expression (7.1.11) to cancel the vector potential term of (7.1.10)

If we set here the electron energy E moving at the velocity $\hbar k/\mu$

$$E = \frac{\hbar^2}{2\mu} k^2, \tag{7.1.13}$$

the equation to be solved is from (7.1.4) and (7.1.6)

$$\left[\frac{\partial^2}{\partial r^2} + \frac{1}{r} \frac{\partial}{\partial r} + \frac{1}{r^2} \left(\frac{\partial}{\partial \theta} + i\alpha \right)^2 + k^2 \right] \Psi = 0. \tag{7.1.14}$$

Caution is required here. It is possible to eliminate the parameter α derived from the vector potential. In fact, if the wave function is phase-transformed $\Psi \mapsto e^{-i\alpha\theta}\Psi$, this parameter α in (7.1.14) apparently disappears. Instead, the wave function becomes a multi-valued function unless α is an integer. Conversely speaking, if α is an integer, the influence of the vector potential can be completely eliminated, maintaining the single-valuedness of the wave function. Thus, for the general case where α is not necessarily an integer, we let the Gauss symbol of α be $n = [\alpha]$, and transform the Ψ and α:

$$\Psi \mapsto e^{-in\theta}\Psi, \quad \alpha \mapsto n + \alpha. \tag{7.1.15}$$

The range of this new α is $0 \le \alpha < 1$. It is noted that equation (7.1.14) is invariant for this transformation, and the incident wave Ψ_{inc} defined in (7.1.11) is also invariant. Therefore, in the following, the Ψ and α are assumed to be redefined ones. Also, this incident wave is no longer unique at $\lim_{\theta \to \pm\pi} \Psi_{\text{inc}}$. Since this is not a solution according to equation (7.1.14), it can be adopted as an incident wave. What must be single-valued is the solution of equation (7.1.14), which is the sum of the incident wave and the scattering wave.

7.1.2 *Solution of the scattering problem*

After preparing so far, let us solve equation (7.1.14). First, the wave function Ψ is separated into variables using the integer m

$$\Psi(r,\theta) = R(r)e^{im\theta}. \qquad (7.1.16)$$

Then the equation becomes

$$\left[\frac{d^2}{dr^2} + \frac{1}{r}\frac{d}{dr} - \frac{1}{r^2}(m+\alpha)^2 + k^2\right]R = 0. \qquad (7.1.17)$$

There are four types of solutions to this equation: the Bessel function $J_{|m+\alpha|}(kr)$; the Neumann function $N_{|m+\alpha|}(kr)$, and their linear combinations; and the first and second kind Hankel functions $H^{(1)}_{|m+\alpha|}(kr)$, $H^{(2)}_{|m+\alpha|}(kr)$. Here, $J_{|m+\alpha|}(kr)$ and $H^{(1)}_{|m+\alpha|}(kr)$ are adopted as two independent solutions. Therefore, the solution of R is represented with A_m, B_m being arbitrary constants,

$$R(r) = A_m J_{|m+\alpha|}(kr) + B_m H^{(1)}_{|m+\alpha|}(kr). \qquad (7.1.18)$$

In the following, in order to be able to carry out later calculations smoothly, it is assumed that the wave number k has a positive small imaginary part ϵ

$$k \mapsto k + i\epsilon. \qquad (7.1.19)$$

This is a kind of **adiabatic approximation**.

Next, it is assumed that the surface of the solenoid is shielded so that the electron wave does not penetrate the inside. Therefore, the wave function becomes zero at $r = r_0$, that is,

$$A_m J_{|m+\alpha|}(kr_0) + B_m H^{(1)}_{|m+\alpha|}(kr_0) = 0. \qquad (7.1.20)$$

Eliminating the coefficient B_m from this, expression (7.1.18) is rewritten as

$$R(r) = A_m\left[J_{|m+\alpha|}(kr) - \frac{J_{|m+\alpha|}(kr_0)}{H^{(1)}_{|m+\alpha|}(kr_0)} H^{(1)}_{|m+\alpha|}(kr)\right]. \qquad (7.1.21)$$

It is noticed here that the Hankel function does not have zero-point on the real axis.

Now, the general solution of equation (7.1.14) is obtained as a superposition over the integer m,

$$\Psi = \sum_{m=-\infty}^{\infty} A_m e^{im\theta} \left[J_{|m+\alpha|}(kr) - \frac{J_{|m+\alpha|}(kr_0)}{H_{|m+\alpha|}^{(1)}(kr_0)} H_{|m+\alpha|}^{(1)}(kr) \right]. \quad (7.1.22)$$

The remaining work is to decompose this solution into the incident and the scattering wave. By using the formula for delta function

$$\frac{1}{2\pi} \sum_{m=-\infty}^{\infty} e^{im(\theta-\theta')} = \delta(\theta - \theta'), \quad |\theta|, \ |\theta'| < \pi, \quad (7.1.23)$$

the incident wave of (7.1.11) Ψ_{inc}, in which x is $r\cos\theta$, is deformed into

$$\Psi_{\text{inc}} = \int_{-\pi}^{\pi} \delta(\theta - \theta') e^{-ikr\cos\theta' - i\alpha\theta'} d\theta'$$

$$= \frac{1}{\pi} \sum_{m=-\infty}^{\infty} e^{im\theta} \int_0^{\pi} \cos[(m+\alpha)\theta'] e^{-ikr\cos\theta'} d\theta'. \quad (7.1.24)$$

This is considered to be an equation obtained by expanding the incident wave into a Fourier series.[2] Here, the integral representation formula of the Bessel function is quoted

$$J_\nu(z) = \frac{e^{i\nu\pi/2}}{\pi} \left[\int_0^{\pi} e^{-iz\cos t} \cos(\nu t)dt - \sin(\nu\pi) \int_0^{\infty} e^{-\nu t + iz\cosh t}dt \right],$$

$$0 < \arg(z) < \pi. \quad (7.1.25)$$

This formula with $z = kr$ can be used as a prescription to resolve into the incident and scattering waves, because the wave number k has a positive imaginary part according to equation (7.1.19). This integral formula is applied only to $J_{|m+\alpha|}(kr)$ in the wave function of (7.1.22), and thus

---

[2]In the case of $\alpha = 0$, this formula is a two-dimensional version of the formula that expands plane wave by spherical waves: $e^{ikr\cos\theta} = J_0(kr) + 2\sum_{m=1}^{\infty} i^m J_m(kr)\cos(m\theta)$.

we have

$$\Psi = \sum_{m=-\infty}^{\infty} \frac{e^{i|m+\alpha|\pi/2} A_m}{\pi} e^{im\theta} \int_0^{\pi} \cos(|m+\alpha|t) e^{-ikr\cos t} dt$$

$$- \sum_{m=-\infty}^{\infty} e^{i|m+\alpha|\pi/2} A_m e^{im\theta} \left[\frac{\sin(|m+\alpha|\pi)}{\pi} \int_0^{\infty} e^{-|m+\alpha|t+ikr\cosh t} dt \right.$$

$$\left. + e^{-i|m+\alpha|\pi/2} \frac{J_{|m+\alpha|}(kr_0)}{H^{(1)}_{|m+\alpha|}(kr_0)} H^{(1)}_{|m+\alpha|}(kr) \right]. \tag{7.1.26}$$

By comparing this expression with the incident wave of (7.1.24), if we put for all m

$$A_m = e^{-i|m+\alpha|\pi/2}, \tag{7.1.27}$$

it can be seen that the first term of the right-hand side of (7.1.26) is just the incident wave. Therefore, the remaining second term is the scattering wave, so we write it as Ψ_{scat},

$$\Psi_{\text{scat}} = - \sum_{m=-\infty}^{\infty} e^{im\theta} \left[\frac{\sin(|m+\alpha|\pi)}{\pi} \int_0^{\infty} e^{-|m+\alpha|t+ikr\cosh t} dt \right.$$

$$\left. + A_m \frac{J_{|m+\alpha|}(kr_0)}{H^{(1)}_{|m+\alpha|}(kr_0)} H^{(1)}_{|m+\alpha|}(kr) \right]. \tag{7.1.28}$$

Next, we calculate the sum over m included in this expression. It is difficult to handle the second term in the square brackets, so we will take the sum of the first term only. We assume here that the order of integration and sum can be exchanged, and extract only the summation part. Furthermore, with the summation being divided into parts from $-\infty$ to -1 and from 0 to ∞, we find the sum, noticing that $0 \le \alpha < 1$,

$$\sum_{m=-\infty}^{\infty} \sin(|m+\alpha|\pi) e^{im\theta-|m+\alpha|t} = \sin(\pi\alpha) \left(\frac{e^{-\alpha t}}{1+e^{-t+i\theta}} + \frac{e^{\alpha t}}{1+e^{t+i\theta}} \right). \tag{7.1.29}$$

Therefore, the sum of the first term in the brackets of (7.1.28) becomes

$$\frac{\sin(\pi\alpha)}{\pi} \int_0^{\infty} \left(\frac{e^{-\alpha t}}{1+e^{-t+i\theta}} + \frac{e^{\alpha t}}{1+e^{t+i\theta}} \right) e^{ikr\cosh t} dt$$

$$= \frac{\sin(\pi\alpha)}{\pi} \int_{-\infty}^{\infty} \frac{e^{-\alpha t+ikr\cosh t}}{1+e^{-t+i\theta}} dt. \tag{7.1.30}$$

As a result, the expression (7.1.28) is rewritten,

$$\Psi_{\text{scat}} = -\left[\frac{\sin(\pi\alpha)}{\pi}\int_{-\infty}^{\infty}\frac{e^{-\alpha t+ikr\cosh t}}{1+e^{-t+i\theta}}dt\right.$$

$$\left.+\sum_{m=-\infty}^{\infty}A_m e^{im\theta}\frac{J_{|m+\alpha|}(kr_0)}{H_{|m+\alpha|}^{(1)}(kr_0)}H_{|m+\alpha|}^{(1)}(kr)\right].\qquad(7.1.31)$$

Finally, the differential cross-section in scattering is obtained from this expression. Therefore, it is necessary to find the asymptotic form where r is sufficiently large. As shown in expression (7.1.19), the wave number k should have a positive small imaginary part. Therefore, when r is large enough, the imaginary part of ikr also becomes large and its real part is negative. Then, the part $e^{ikr\cosh t}$ in the integral of the first term becomes rapidly zero as the absolute value of t increases. Therefore, since the part that contributes to this integration is considered to be only near zero of t, this integral is approximated as follows,

$$\int_{-\infty}^{\infty}\frac{e^{-\alpha t+ikr\cosh t}}{1+e^{-t+i\theta}}dt\approx\frac{1}{1+e^{i\theta}}\int_{-\infty}^{\infty}e^{-\alpha t+ikr\cosh t}dt=\frac{2}{1+e^{i\theta}}K_\alpha(-ikr),$$

$$(7.1.32)$$

where K_α in this last expression is the second-kind modified Bessel function, and we used the formula

$$K_\nu(z)=\frac{1}{2}\int_{-\infty}^{\infty}e^{-\nu t-z\cosh t}dt.\qquad(7.1.33)$$

It is cited here the asymptotic forms of the modified Bessel function, the Bessel function and the Hankel function, when z is large enough,

$$K_\nu(z)\approx\sqrt{\frac{\pi}{2z}}e^{-z},\quad J_\nu(z)\approx\sqrt{\frac{2}{\pi z}}\cos\left(z-\frac{2\nu+1}{4}\pi\right),$$

$$H_\nu^{(1)}(z)\approx\sqrt{\frac{2}{\pi z}}e^{i(z-(2\nu+1)\pi/4)}.\qquad(7.1.34)$$

By using these asymptotic forms of $K_\nu(z)$, $H_\nu^{(1)}(z)$, we can derive the asymptotic form of the scattering wave of expression (7.1.31) when r is large enough,

$$\Psi_{\text{scat}}\approx-\sqrt{\frac{2}{\pi k}}e^{-\pi i/4}\left[i\frac{\sin(\pi\alpha)}{1+e^{i\theta}}+\sum_{m=-\infty}^{\infty}(A_m)^2 e^{im\theta}\frac{J_{|m+\alpha|}(kr_0)}{H_{|m+\alpha|}^{(1)}(kr_0)}\right]\frac{e^{ikr}}{\sqrt{r}}.$$

$$(7.1.35)$$

If we convert this to the standard form of a two-dimensional scattering wave, we get

$$\Psi_{\text{scat}} \approx \frac{f(\theta)}{\sqrt{r}} e^{ikr}, \tag{7.1.36}$$

where the angle-dependent part $f(\theta)$ becomes

$$f(\theta) = -\sqrt{\frac{2}{\pi k}} e^{-\pi i/4} \left[i \frac{\sin(\pi\alpha)}{1 + e^{i\theta}} + \sum_{m=-\infty}^{\infty} (A_m)^2 e^{im\theta} \frac{J_{|m+\alpha|}(kr_0)}{H^{(1)}_{|m+\alpha|}(kr_0)} \right]. \tag{7.1.37}$$

From now on, the scattering **differential cross-section** $\sigma(\theta)$ is obtained as

$$\sigma(\theta) \equiv |f(\theta)|^2 = \frac{2}{\pi k} \left| i \frac{\sin(\pi\alpha)}{1 + e^{i\theta}} + \sum_{m=-\infty}^{\infty} (A_m)^2 e^{im\theta} \frac{J_{|m+\alpha|}(kr_0)}{H^{(1)}_{|m+\alpha|}(kr_0)} \right|^2. \tag{7.1.38}$$

This is the differential cross-section of electron scattering in the case of an infinitely-long solenoid with a finite radius r_0.

Here, if the thickness of the solenoid is made infinitely small and taking the limit of $r_0 \to 0$, the second term in the absolute value disappears, because $\lim_{r_0 \to 0} H^{(1)}_{|m+\alpha|}(kr_0) = \infty$. Therefore, when $r_0 = 0$, we have the cross-section

$$\sigma(\theta) = \frac{\sin^2(\pi\alpha)}{2\pi k \cos^2(\theta/2)}. \tag{7.1.39}$$

From now on, if $\alpha = 0$, the scattering will not occur, but if the value of $\alpha > 0$, even if just little, the scattering will occur. This is the first result which was given by Aharonov and Bohm.

7.1.3 *Numerical calculation*

The sum over m in the differential cross-section (7.1.38) has not yet been executed. It seems that the only way to find this summation is a numerical calculation. Before that, we clarify the α-dependence of this differential cross-section. Denoting this as $\sigma(\theta, \alpha)$, we present the behavior of $\sigma(\theta, \alpha)$ concerning α

$$\sigma(-\theta, 0) = \sigma(\theta, 0), \quad \sigma(-\theta, 1 - \alpha) = \sigma(\theta, \alpha). \tag{7.1.40}$$

This first expression is easily derived by changing the sign of the sum-variable m. As for the second expression, after performing this conversion, we notice $\sin\big((1-\alpha)\pi\big) = \sin(\alpha\pi)$ and multiply the denominator and numerator by $e^{i\theta}$ in the first term. In the second term, we change the sign of the sum-variable m and then replace m with $m+1$. Although an extra overall factor $e^{i\theta}$ arises, this factor disappears by taking the absolute value and therefore the original definition formula is reproduced. From this conversion formula, it can be seen that when $\alpha = 0,\ 1/2$, the differential cross-section is an even function of θ.

When performing numerical calculations, it is better to use as few parameters as possible. Since the $\sigma(\theta)/r_0$ obtained from (7.1.38) contains only two parameters α and kr_0, in the following, this $\sigma(\theta)/r_0$ will be calculated numerically.

Another problem is how large the range of the sum of m should be. If we fix z, the Bessel function $J_\nu(z)$ rapidly approaches zero when the value of ν nearly exceeds z. On the contrary, the Hankel function $H_\nu^{(1)}(z) = J_\nu(z) + iN_\nu(z)$ grows rapidly, because the Neumann function is included in its imaginary part. We take here the range of the sum of m: by adding kr_0 to 10 and converting it to an integer value as $M = [kr_0 + 10]$, we set the range of the sum of m from $-M$ to $+M$.

Figures 7.1 and 7.2 show the case of $kr_0 = 1$ and $\alpha = 1/4$ and $1/2$, respectively. In these figures, the curves drawn with solid lines correspond to each value of these parameters. For comparison, in the case of $\alpha = 0$, that is, when the Aharonov–Bohm effect does not occur, the curve is drawn with a dotted line. Figures 7.3 and 7.4 are graphs for $kr_0 = 0.1$ and 10, respectively, and $\alpha = 1/4$.

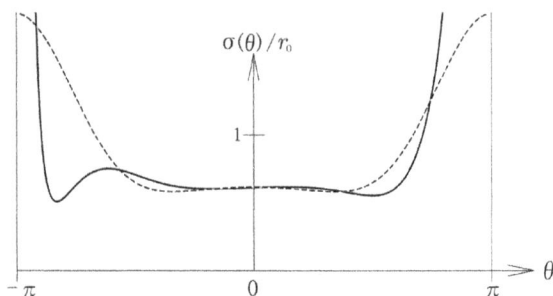

Fig. 7.1 $kr_0 = 1,\ \ \alpha = 1/4$

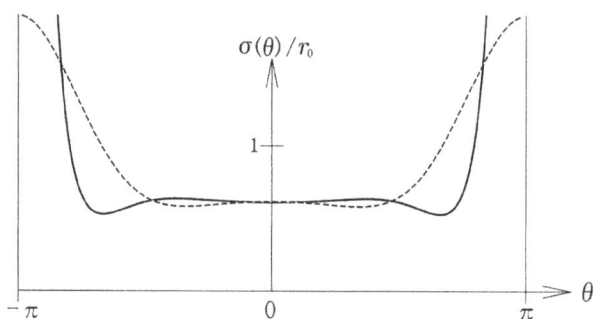

Fig. 7.2 $kr_0 = 1, \quad \alpha = 1/2$

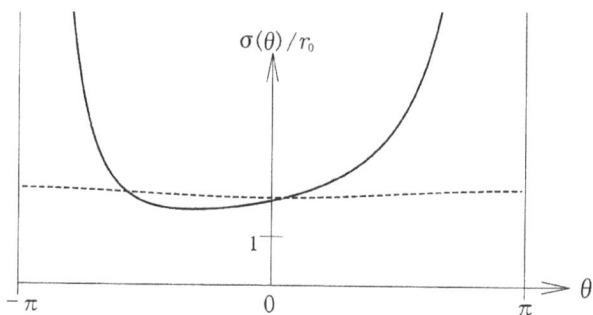

Fig. 7.3 $kr_0 = 0.1, \quad \alpha = 1/4$

As shown in relation (7.1.40), in the case of Fig. 7.1 where $\alpha = 1/4$, it becomes asymmetric with respect to θ, but, in Fig. 7.2 with $\alpha = 1/2$, it becomes symmetrical. Also, if we flip this Fig. 7.1 left and right around $\theta = 0$, the graph becomes for $\alpha = 3/4$.

Figure 7.3 shows the case where the value of kr_0 is taken to be as small as 0.1. It can be seen that the Aharonov–Bohm effect is prominent in this figure, because there is a big difference between the solid line and the dotted line. However, as shown in Fig. 7.4, when $kr_0 = 10$, which is regarded as large enough, this effect only appears in backward scattering where θ is close to $\pm\pi$. This means that as kr_0 increases, the movement of electrons becomes particle-like and approaches classical theory. It is noted that the scale of the vertical-axis is different between Fig. 7.3 and Fig. 7.4.

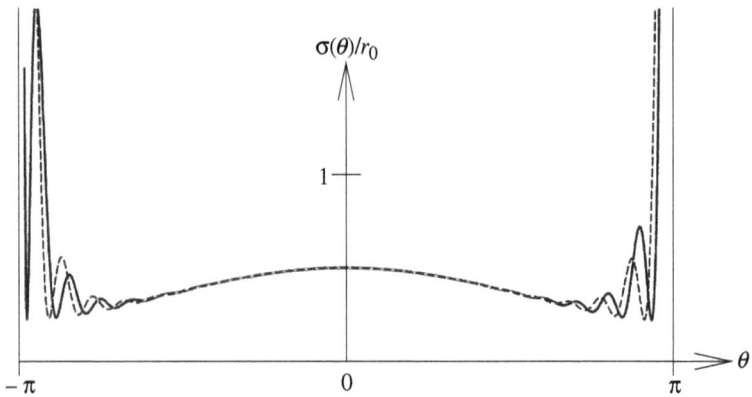

Fig. 7.4 $kr_0 = 10, \quad \alpha = 1/4$

Can Order of Summation and Asymptotic Limit be Exchangeable?

When expression (7.1.38) is obtained at the beginning, it seems to be impossible to take the sum over m as it is. However, since the asymptotic form of this function when kr_0 is large enough can be obtained, so I thought that it would be possible to take the sum. Because the wavelength of the electron is very small in many cases, this condition will be satisfied. Therefore, by using the asymptotic form of $J_\nu(z)$, $H_\nu^{(1)}(z)$ of expression (7.1.34), if we execute the remaining sum part as

$$\sum_{m=-\infty}^{\infty} (A_m)^2 e^{im\theta} \frac{J_{|m+\alpha|}(kr_0)}{H_{|m+\alpha|}^{(1)}(kr_0)}$$

$$\approx e^{-ikr_0} \sum_{m=-\infty}^{\infty} e^{i(m\theta - \frac{1}{2}|m+\alpha|\pi + \frac{1}{4}\pi)} \cos\left(kr_0 - \frac{2|m+\alpha|+1}{4}\pi\right)$$

$$= \pi i e^{-2ikr_0} \delta(\theta) - i\frac{\sin(\alpha\pi)}{1 + e^{i\theta}},$$

surprisingly enough, the second term of the last expression coincides with the opposite sign of the first term of (7.1.38) and therefore the cancellation occurs. As a result, only the delta function part remains. This is clearly false. That is, the sum calculated after taking the asymptotic form of each term and the asymptotic form after taking the sum do not generally coincide. For example, the next calculation is the correct one,

$$\lim_{x\to\infty} \sum_{m=0}^{\infty} \frac{1+(-\log x)^m}{m!} = \lim_{x\to\infty}\left(e + e^{-\log x}\right) = \lim_{x\to\infty}\left(e + \frac{1}{x}\right) = e.$$

However, if we take first the asymptotic form of each term, this sum becomes $1/x$, and when $x \to \infty$, it becomes zero. This situation is the same as when the order of summation and integration cannot be exchanged generically. However, when we think about it, in the asymptotic form of the Bessel function when the variable z is large enough, what is it larger than? It is the same as being asked if the earth is big; if there is nothing to compare, we cannot answer. The asymptotic form of the Bessel function $J_\nu(z)$ can be used where the variable z is large enough compared to the degree ν. The problem here is that it can be correct to take the sum from $-\infty$ to $+\infty$ of the degree for the fixed z. Therefore, it is impossible to use the asymptotic form from the beginning.

At this stage, to obtain the asymptotic form (7.1.35) when r is sufficiently large from the Ψ_{scat} of (7.1.28), for the first term, we first sum up and then find the asymptotic form. On the other hand, since the sum of the second term could not be obtained, the asymptotic form was created before taking the summation. I was wondering if this procedure was correct. Therefore, I tried to confirm whether the same result can be obtained by first making the asymptotic form for this first term and then taking the sum later. That is, we convert the integral part of (7.1.28) to

$$\int_0^\infty e^{-|m+\alpha|t + ikr\cosh t}\,dt \approx \int_0^\infty e^{ikr\cosh t}\,dt$$

$$= \frac{1}{2}\int_{-\infty}^\infty e^{ikr\cosh t}\,dt = K_0(-ikr),$$

and carry out the remaining summation:

$$-\frac{K_0(-ikr)}{\pi}\sum_{m=-\infty}^{\infty} e^{im\theta}\sin(|m+\alpha|\pi) = -\frac{2\sin(\alpha\pi)K_0(-ikr)}{\pi(1+e^{i\theta})}$$

$$\approx -\sqrt{\frac{2}{\pi k}}\frac{\sin(\alpha\pi)e^{\pi i/4}}{1+e^{i\theta}}\frac{e^{ikr}}{\sqrt{r}}.$$

This is exactly the same as the first term in the asymptotic expression (7.1.35).

For the second term, it was impossible to take the sum. As mentioned in the numerical calculation, when the value of kr_0 is fixed, although it is an infinite sum, the range of the sum is actually enough to carry out to the value a little larger than kr_0. Therefore if the value of kr is sufficiently larger than kr_0, there should be no problem.

When I first learned quantum mechanics, I was reluctant to believe the Schrödinger equation. Firstly, why is this equation linear? Is it just to make it easier to solve? Secondly, why does the Schrödinger equation include the energy term from the beginning? In classical mechanics, the energy is introduced secondarily from the basic equation to make it easier to solve. I suspected that there was something more basic hidden deeper, and that the Schrödinger equation should be derived from it secondarily. These are questions that I, and others who learn quantum mechanics, hold. The Aharonov–Bohm effect which contains the vector potential might be able to answer these questions.

The Schrödinger equation was proposed in 1926, while the Aharonov–Bohm effect was presented 33 years later. Furthermore, it took 27 years to confirm it experimentally. It took such a long time because it was difficult both theoretically and experimentally.

7.2 The Rosen–Morse Potential Problem

Here, we explain the **Rosen–Morse potential** problem in quantum mechanics. This potential is difficult in quantum mechanics, but in recent years, it became famous because it is one of candidates for the potential between quark-gluon in quantum chromodynamics (QCD). By using two positive parameters V_1, V_2 having the dimension of energy and one positive parameter x_0 having the dimension of length, the potential $V(x)$ is defined as

$$V(x) = V_1 \tanh(x/x_0) - \frac{V_2}{\cosh^2(x/x_0)}. \qquad (7.2.1)$$

As shown in Fig. 7.5, this potential becomes $V(x) \to \pm V_1$ as $x \to \pm\infty$. Also, when $2V_2 > V_1$, it takes a minimum value $V_{\min} = -V_2 - (V_1{}^2/4V_2)$

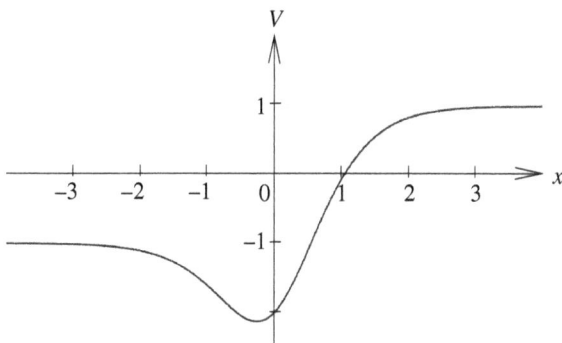

Fig. 7.5   Rosen–Morse potential (case of $V_1 = 1$, $V_2 = 2$)

smaller than $-V_1$ at $x = -x_0 \operatorname{arctanh}(V_1/2V_2)$. In the following, it is required that the condition $2V_2 > V_1$ is satisfied. Let the mass and energy of the electron be m and E, respectively, then the Schrödinger equation for this model is given by

$$\left[-\frac{\hbar^2}{2m}\frac{d^2}{dx^2} + V_1 \tanh(x/x_0) - \frac{V_2}{\cosh^2(x/x_0)} \right] \Psi = E\Psi. \qquad (7.2.2)$$

In the following, to simplify the formula, the coordinate variable x, the potential size V_1, V_2, and the energy E are converted to the dimensionless style,

$$\frac{x}{x_0} \mapsto x, \quad \frac{2mx_0{}^2}{\hbar^2}V_1 \mapsto V_1, \quad \frac{2mx_0{}^2}{\hbar^2}V_2 \mapsto V_2, \quad \frac{2mx_0{}^2}{\hbar^2}E \mapsto E. \tag{7.2.3}$$

Figure 7.5 shows this potential in the case of $V_1 = 1$, $V_2 = 2$ after making this dimensionless. Equation (7.2.2) is rewritten with this dimensionless style

$$\left[\frac{d^2}{dx^2} + E - V_1 \tanh(x) + \frac{V_2}{\cosh^2(x)} \right] \Psi = 0. \qquad (7.2.4)$$

To solve this equation, we transform the wave function from Ψ to Φ by using the constant a

$$\Psi = e^{ax}\Phi, \tag{7.2.5}$$

then equation (7.2.4) is rewritten as

$$\left[\frac{d^2}{dx^2} + 2a\frac{d}{dx} + a^2 + E - V_1\tanh(x) + \frac{V_2}{\cosh^2(x)}\right]\Phi = 0. \tag{7.2.6}$$

Further, we convert the independent variable from x to ξ,

$$\xi = \tanh(x), \tag{7.2.7}$$

and the equation becomes

$$\left[(1-\xi^2)\frac{d^2}{d\xi^2} + 2(a-\xi)\frac{d}{d\xi} + V_2 + \frac{a^2+E-V_1\xi}{1-\xi^2}\right]\Phi = 0. \tag{7.2.8}$$

Then we convert Φ to φ by using the constant b,

$$\Phi = (1-\xi^2)^{b/2}\varphi, \tag{7.2.9}$$

and the equation becomes

$$\left[(1-\xi^2)\frac{d^2}{d\xi^2} + 2(a-\xi-b\xi)\frac{d}{d\xi} + V_2 - b\right.$$

$$\left. + \frac{b^2\xi^2 - (2ab+V_1)\xi + a^2 + E}{1-\xi^2}\right]\varphi = 0. \tag{7.2.10}$$

So far, nothing has been said about the constants a and b, but now, we convert the last term in the brackets into a constant, that is,

$$2ab = -V_1, \quad a^2 + b^2 = -E, \tag{7.2.11}$$

and the equation becomes

$$\left[(1-\xi^2)\frac{d^2}{d\xi^2} + 2(a-\xi-b\xi)\frac{d}{d\xi} + V_2 - b(b+1)\right]\varphi = 0. \tag{7.2.12}$$

We convert the independent variable further,

$$\zeta = \frac{1}{2}(1-\xi), \tag{7.2.13}$$

and the equation becomes

$$\left[\zeta(1-\zeta)\frac{d^2}{d\zeta^2} + \left(-a+b+1-2(b+1)\zeta\right)\frac{d}{d\zeta} + V_2 - b(b+1)\right]\varphi = 0. \tag{7.2.14}$$

This is the same type as the differential equation for **Gauss' hypergeo-metric function** $u = F(\alpha, \beta; \gamma; \zeta)$ which satisfies the equation

$$\left[\zeta(1-\zeta)\frac{d^2}{d\zeta^2} + [\gamma - (\alpha+\beta+1)\zeta]\frac{d}{d\zeta} - \alpha\beta\right]u = 0. \qquad (7.2.15)$$

The relationships of the variables between this equation and (7.2.14) are

$$\alpha + \beta = 2b + 1, \quad \alpha\beta = b(b+1) - V_2, \quad \gamma = -a + b + 1. \qquad (7.2.16)$$

To summarize the above, the wave function Ψ is solved as, excluding the normalization constants[3]

$$\Psi(x) = 2^{-b}e^{ax}\cosh^{-b}(x)F\left(\alpha, \beta; \gamma; \tfrac{1}{2}(1-\tanh x)\right), \qquad (7.2.17)$$

where $a, b, \alpha, \beta, \gamma$ can be obtained from the equations (7.2.11) and (7.2.16).

In the following subsections, we will classify the energy E to find the **eigenvalue** and the **eigenfunction**; first, the case of bound state at $V_{min} < E < -V_1$, second, the case of scattering state at $-V_1 < E < V_1$, and finally, the case of scattering state at $V_1 < E$. We will solve the problem by distinguishing the energy E in the above three cases.

7.2.1 *Case of bound state at $V_{min} < E < -V_1$*

Eigenvalues and eigenfunctions

It is necessary to impose the following conditions on the wave function to find a solution of the bound state,

$$\lim_{x \to \pm\infty} \Psi(x) = 0. \qquad (7.2.18)$$

First, for the case of $x \to \infty$, the hypergeometric function part of (7.2.17) is 1, so the asymptotic form of the wave function becomes

$$\Psi \sim e^{(a-b)x}. \qquad (7.2.19)$$

Because this must approach zero as $x \to \infty$, we have a condition

$$a - b < 0. \qquad (7.2.20)$$

From expression (7.2.11), a and b are different signs. Combined with this result, we have $a < 0 < b$.

---

[3]There is another independent solution to the differential equation of the hypergeometric function (7.2.15). This will be described in §7.2.5.

Next, for the case of $x \to -\infty$, since expression (7.2.17) cannot be estimated as it is, therefore the hypergeometric function part is converted by using **Gauss' transformation formula**,

$$
\Psi = 2^{-b} e^{ax} \cosh^{-b}(x) \Gamma(\gamma) \left[\frac{\Gamma(\alpha + \beta - \gamma)}{\Gamma(\alpha)\Gamma(\beta)} \left(\frac{1 + \tanh(x)}{2} \right)^{\gamma - \alpha - \beta} \right.
$$

$$
\times F\left(\gamma - \alpha, \gamma - \beta; \gamma - \alpha - \beta + 1; \tfrac{1}{2}(1 + \tanh x)\right)
$$

$$
\left. + \frac{\Gamma(\gamma - \alpha - \beta)}{\Gamma(\gamma - \alpha)\Gamma(\gamma - \beta)} F\left(\alpha, \beta; \alpha + \beta - \gamma + 1; \tfrac{1}{2}(1 + \tanh x)\right) \right].
$$

$$
(7.2.21)
$$

We can make here the asymptotic form of $x \to -\infty$ with this expression, the hypergeometric function part also becomes 1, and we have

$$
\Psi \sim \frac{\Gamma(\gamma)\Gamma(a + b)}{\Gamma(\alpha)\Gamma(\beta)} e^{-(a+b)x} + \frac{\Gamma(\gamma)\Gamma(-a - b)}{\Gamma(\gamma - \alpha)\Gamma(\gamma - \beta)} e^{(a+b)x}, \qquad (7.2.22)
$$

where we used $\gamma - \alpha - \beta = -(a + b)$ from equation (7.2.16). Here, the sign of $a + b$ becomes a problem. From equation (7.2.11), $(a + b)^2 = -E - V_1$, but the sign of $a + b$ is not determined. Here, the positive sign is adopted.[4] Then, the second term of (7.2.22) converges to zero as $x \to -\infty$, but, the first term diverges. To prevent this situation, either $\Gamma(\alpha)$ or $\Gamma(\beta)$ must be in the pole position. Since the hypergeometric function which appears in expression (7.2.17) is symmetric under the interchange of α and β, we assume here that $\Gamma(\alpha)$ is in the pole position

$$
\alpha = -n, \quad n = 0, \ 1, \ 2, \dots . \qquad (7.2.23)
$$

Based on the above remarks, we can determine a, b, α, β and γ. Before that, to simplify the formula, we set

$$
V_2 = \nu(\nu + 1), \qquad (7.2.24)
$$

where the positive ν is defined as

$$
\nu = \sqrt{V_2 + \tfrac{1}{4}} - \tfrac{1}{2}, \qquad (7.2.25)
$$

and μ is defined as

$$
\mu = \frac{V_1}{2}. \qquad (7.2.26)
$$

---

[4]The case where the negative sign is adopted is shown in §7.2.5, but the result is exactly the same.

Since α is already determined by (7.2.23), we eliminate β from the first and second relations of (7.2.16) and determine b, noting $b > 0$,

$$b = \nu - n. \tag{7.2.27}$$

β is obtained from the first relation of the same (7.2.16),

$$\beta = 2\nu + 1 - n, \tag{7.2.28}$$

a is obtained from the first relation of (7.2.11),

$$a = -\frac{\mu}{\nu - n}, \tag{7.2.29}$$

and finally, γ is from the third relation of (7.2.16),

$$\gamma = \frac{\mu}{\nu - n} + \nu - n + 1. \tag{7.2.30}$$

Thus the quantized energy E_n is obtained from the second relation of (7.2.11)

$$E_n = -\frac{\mu^2}{(\nu - n)^2} - (\nu - n)^2. \tag{7.2.31}$$

This E_n is proved to be less than or equal to $-V_1$ by using the relation of "arithmetic mean $\geq$ geometric mean".

Since b in expression (7.2.27) is positive, the value of the non-negative integer n must be $n < \nu$. Furthermore, since $a + b > 0$ is set here, the following relation must be held:

$$a + b = -\frac{\mu}{\nu - n} + \nu - n > 0, \tag{7.2.32}$$

and from this relation, the maximum value of n is determined by the relation

$$n < \nu - \sqrt{\mu}. \tag{7.2.33}$$

Substituting these results into expression (7.2.17), and letting the eigenfunction belonging to the energy eigenvalue E_n be denoted by $\Psi_n(x)$, we obtain the expression

$$\Psi_n(x) = 2^{-(\nu-n)} e^{-[\mu/(\nu-n)]x} \cosh^{-(\nu-n)}(x)$$
$$\times F\left(-n, 2\nu + 1 - n, \frac{\mu}{\nu - n} + \nu - n + 1; \tfrac{1}{2}(1 - \tanh x)\right). \tag{7.2.34}$$

In particular, the part of this hypergeometric function becomes a polynomial of order n, and can also be represented by using the Jacobi polynomial.

Orthogonality of eigenfunctions

The orthogonality of eigenfunctions will be described here. By using (7.2.4), the equations satisfied by two eigenfunctions Ψ_n, $\Psi_{n'}$ belonging to two generically different eigenvalues E_n, $E_{n'}$ are written as

$$\left[\frac{d^2}{dx^2} + E_n - V_1\tanh(x) + \frac{V_2}{\cosh^2(x)}\right]\Psi_n = 0,$$

$$\left[\frac{d^2}{dx^2} + E_{n'} - V_1\tanh(x) + \frac{V_2}{\cosh^2(x)}\right]\Psi_{n'} = 0. \tag{7.2.35}$$

Multiplying this first equation by $\Psi_{n'}$ and the second equation by Ψ_n, and subtracting both sides, we have

$$\frac{d}{dx}\left(\Psi_{n'}\frac{d\Psi_n}{dx} - \Psi_n\frac{d\Psi_{n'}}{dx}\right) + (E_n - E_{n'})\Psi_n\Psi_{n'} = 0, \tag{7.2.36}$$

and integrating this with respect to x, we obtain

$$\int_{-\infty}^{\infty}\Psi_n\Psi_{n'}dx = \frac{1}{E_n - E_{n'}}\left(\Psi_n\frac{d\Psi_{n'}}{dx} - \Psi_{n'}\frac{d\Psi_n}{dx}\right)\bigg|_{-\infty}^{\infty}. \tag{7.2.37}$$

The functions Ψ_n, $\Psi_{n'}$ and their derivatives exponentially approach to zero as $x \to \pm\infty$. Therefore, when $E_n \neq E_{n'}$, the right-hand side becomes zero, and the orthogonality between eigenfunctions belonging to different eigenvalues is shown. However, when the same eigenvalue $E_n = E_{n'}$, it is expected that it will be difficult to carry out this integration. Even if this is represented by using the Jacobi polynomial, it is not connected to the orthogonality of the Jacobi polynomial. Another method is to use l'Hôpital's rule to take the limit, but we will not touch on this problem anymore here.

7.2.2 *Case of scattering state*

If we solve a, b from equation (7.2.11)

$$\begin{cases} a = \dfrac{1}{2}(k_1 - k_2), \\ b = \dfrac{1}{2}(k_1 + k_2), \end{cases} \quad \text{or} \quad \begin{cases} a = \dfrac{1}{2}(k_1 + k_2), \\ b = \dfrac{1}{2}(k_1 - k_2). \end{cases} \tag{7.2.38}$$

Other than these, there are cases where the signs of a and b are changed at the same time in these solutions. There are a total of 4 solutions. Here, we define k_1, k_2 as

$$k_1 = \sqrt{-E - V_1}, \quad k_2 = \sqrt{-E + V_1}. \tag{7.2.39}$$

Case of $-V_1 < E < V_1$

In this case, k_2 is real, but k_1 is imaginary, so we rewrite k_1 as

$$k_1 = i\kappa_1, \quad \kappa_1 = \sqrt{E + V_1}. \tag{7.2.40}$$

For a and b in (7.2.38), adopting the first solutions, we solve α, β and γ from equation (7.2.16)

$$\alpha = \tfrac{1}{2}(i\kappa_1 + k_2 + 1 + \lambda), \quad \beta = \tfrac{1}{2}(i\kappa_1 + k_2 + 1 - \lambda), \quad \gamma = k_2 + 1. \tag{7.2.41}$$

We define λ as

$$\lambda = \sqrt{1 + 4V_2} \quad (= 2\nu + 1). \tag{7.2.42}$$

If we write the wave function of this case as $\Psi(x, E)$, from (7.2.17)

$$\Psi(x, E) = 2^{-(i\kappa_1 + k_2)/2} e^{(i\kappa_1 - k_2)x/2} \cosh^{-(i\kappa_1 + k_2)/2}(x)$$
$$\times F\left(\tfrac{1}{2}(i\kappa_1 + k_2 + 1 + \lambda), \tfrac{1}{2}(i\kappa_1 + k_2 + 1 - \lambda); k_2 + 1; \tfrac{1}{2}(1 - \tanh x)\right). \tag{7.2.43}$$

From this expression, the asymptotic form of the wave function for $x \to \infty$ becomes

$$\Psi(x, E) \sim e^{-k_2 x}. \tag{7.2.44}$$

Also, the asymptotic form of $x \to -\infty$ is the same as expression (7.2.22)

$$\Psi(x, E) \sim f(\kappa_1, k_2)e^{i\kappa_1 x} + \overline{f(\kappa_1, k_2)}e^{-i\kappa_1 x}, \tag{7.2.45}$$

where the overline means the complex conjugate, and we define

$$f(\kappa_1, k_2) = \frac{\Gamma(k_2 + 1)\Gamma(-i\kappa_1)}{\Gamma\left(\tfrac{1}{2}(-i\kappa_1 + k_2 + 1 - \lambda)\right)\Gamma\left(\tfrac{1}{2}(-i\kappa_1 + k_2 + 1 + \lambda)\right)}. \tag{7.2.46}$$

It can be seen that the first term of the right-hand side of (7.2.45) is an incident wave propagating from the negative direction to the positive direction, and the second term is the reflected waves. The coefficients of the incident and reflected waves are just complex conjugates, naturally, their absolute values are equal. On the other hand, expression (7.2.44) represents a penetrating wave. In this case, because the energy E is smaller than the potential V_1, so it quickly decreases. As a result, this scattering becomes the total reflection.

Case of $V_1 < E$

In this case, k_2 is also an imaginary number. Here, the second expression of (7.2.38) in which the sign of k_2 is changed is adopted, and we put the previous k_2 as

$$k_2 = -i\kappa_2, \quad \kappa_2 = \sqrt{E - V_1}. \tag{7.2.47}$$

The variables α, β and γ are rewritten from (7.2.41)

$$\alpha = \tfrac{1}{2}(i\kappa_1 - i\kappa_2 + 1 + \lambda), \quad \beta = \tfrac{1}{2}(i\kappa_1 - i\kappa_2 + 1 - \lambda), \quad \gamma = -i\kappa_2 + 1. \tag{7.2.48}$$

Then the wave function becomes

$$\Psi(x, E) = 2^{-i(\kappa_1 - \kappa_2)/2} e^{i(\kappa_1 + \kappa_2)x/2} \cosh^{-i(\kappa_1 - \kappa_2)/2}(x)$$
$$\times F\big(\tfrac{1}{2}(i\kappa_1 - i\kappa_2 + 1 + \lambda), \tfrac{1}{2}(i\kappa_1 - i\kappa_2 + 1 - \lambda); -i\kappa_2 + 1; \tfrac{1}{2}(1 - \tanh x)\big). \tag{7.2.49}$$

Also, from the expressions which correspond to (7.2.44) $\sim$ (7.2.46), we have for $x \to \infty$

$$\Psi(x, E) \sim e^{i\kappa_2 x}, \tag{7.2.50}$$

and for $x \to -\infty$

$$\Psi(x, E) \sim g_1(\kappa_1, \kappa_2)e^{i\kappa_1 x} + g_2(\kappa_1, \kappa_2)e^{-i\kappa_1 x}, \tag{7.2.51}$$

where g_1, g_2 are defined as

$$g_1(\kappa_1, \kappa_2) = \frac{\Gamma(-i\kappa_2 + 1)\Gamma(-i\kappa_1)}{\Gamma\big(\tfrac{1}{2}(-i\kappa_1 - i\kappa_2 + 1 - \lambda)\big)\Gamma\big(\tfrac{1}{2}(-i\kappa_1 - i\kappa_2 + 1 + \lambda)\big)},$$

$$g_2(\kappa_1, \kappa_2) = \frac{\Gamma(-i\kappa_2 + 1)\Gamma(i\kappa_1)}{\Gamma\big(\tfrac{1}{2}(i\kappa_1 - i\kappa_2 + 1 + \lambda)\big)\Gamma\big(\tfrac{1}{2}(i\kappa_1 - i\kappa_2 + 1 - \lambda)\big)}. \tag{7.2.52}$$

The first term of the right-hand side of (7.2.51) is the incident wave propagating from the negative to positive direction of the x-axis, the second term is the reflected wave, and the expression of (7.2.50) is the transmitted wave.

Here, the coefficient ratio of the reflected wave and the transmitted wave to the incident wave are denoted by R and T, respectively

$$R = \frac{g_2(\kappa_1, \kappa_2)}{g_1(\kappa_1, \kappa_2)}$$

$$= \frac{\Gamma\left(\frac{1}{2}(-i\kappa_1 - i\kappa_2 + 1 - \lambda)\right)\Gamma\left(\frac{1}{2}(-i\kappa_1 - i\kappa_2 + 1 + \lambda)\right)\Gamma(i\kappa_1)}{\Gamma\left(\frac{1}{2}(i\kappa_1 - i\kappa_2 + 1 + \lambda)\right)\Gamma\left(\frac{1}{2}(i\kappa_1 - i\kappa_2 + 1 - \lambda)\right)\Gamma(-i\kappa_1)},$$

$$T = \frac{1}{g_1(\kappa_1, \kappa_2)} = \frac{\Gamma\left(\frac{1}{2}(-i\kappa_1 - i\kappa_2 + 1 - \lambda)\right)\Gamma\left(\frac{1}{2}(-i\kappa_1 - i\kappa_2 + 1 + \lambda)\right)}{\Gamma(-i\kappa_2 + 1)\Gamma(-i\kappa_1)}.$$

$$(7.2.53)$$

Calculating $|R|^2$ and $|T|^2$ from these expressions, we have

$$|R|^2 = \frac{\cosh^2\left(\frac{\pi}{2}(\kappa_1 - \kappa_2)\right) - \sin^2(\frac{\pi}{2}\lambda)}{\cosh^2\left(\frac{\pi}{2}(\kappa_1 + \kappa_2)\right) - \sin^2(\frac{\pi}{2}\lambda)},$$

$$(7.2.54)$$

$$|T|^2 = \frac{\kappa_1}{\kappa_2} \cdot \frac{\sinh(\pi\kappa_1)\sinh(\pi\kappa_2)}{\cosh^2\left(\frac{\pi}{2}(\kappa_1 + \kappa_2)\right) - \sin^2(\frac{\pi}{2}\lambda)}.$$

Here, by using the formula

$$\cosh^2\left(\tfrac{1}{2}(x - y)\right) + \sinh(x)\sinh(y) = \cosh^2\left(\tfrac{1}{2}(x + y)\right), \qquad (7.2.55)$$

we obtain the relation of $|R|^2$ and $|T|^2$ from (7.2.54)

$$\kappa_1|R|^2 + \kappa_2|T|^2 = \kappa_1. \qquad (7.2.56)$$

This is an expression for the **conservation of particle number** which is taken as the difference of wave numbers κ_1 and κ_2 as $x \to -\infty$ and $+\infty$, respectively.

7.2.3 *Normalization of wave function*

Here, the wave function obtained by expressions (7.2.43) and (7.2.49) will be normalized. The method is the same as the previous (7.2.35) $\sim$ (7.2.37), but the symbols are different, so we repeat it again. Let the wave functions for two generically different energies E and E' be $\Psi(x, E)$ and $\Psi(x, E')$, respectively. The equation of the form (7.2.4) imply that these wave functions satisfy

$$\left[\frac{d^2}{dx^2} + E - V_1 \tanh(x) + \frac{V_2}{\cosh^2(x)}\right]\overline{\Psi(x, E)} = 0,$$

$$\left[\frac{d^2}{dx^2} + E' - V_1 \tanh(x) + \frac{V_2}{\cosh^2(x)}\right]\Psi(x, E') = 0,$$

(7.2.57)

where the overline of $\overline{\Psi(x, E)}$ means the complex conjugate. Multiplying this first equation by $\Psi(x, E')$ and the second equation by $\overline{\Psi(x, E)}$, and then subtracting both sides, we have

$$\frac{d}{dx}\left(\Psi(x, E')\frac{d\overline{\Psi(x, E)}}{dx} - \overline{\Psi(x, E)}\frac{d\Psi(x, E')}{dx}\right) + (E - E')\overline{\Psi(x, E)}\Psi(x, E')$$
$$= 0.$$

(7.2.58)

Integrating this expression with respect to x, we have

$$\int_{-\infty}^{\infty}\overline{\Psi(x, E)}\Psi(x, E')dx = \frac{1}{E - E'}[F(x)]_{-\infty}^{\infty},$$

(7.2.59)

where the function $F(x)$ is defined as

$$F(x) = \overline{\Psi(x, E)}\frac{d\Psi(x, E')}{dx} - \Psi(x, E')\frac{d\overline{\Psi(x, E)}}{dx}.$$

(7.2.60)

Hereinafter, primed symbols denote the quantities related to E', that is, k_1', k_2', κ_1', κ_2' correspond to k_1, k_2, κ_1, κ_2, respectively, and we will examine this function $F(x)$, classifying the cases according to the two energy ranges.

Case of $-V_1 < E < V_1$

First, from expression (7.2.44), the wave function Ψ is exponentially zero as $x \to \infty$, and its derivative also becomes zero. Therefore, we obtain

$$\lim_{x \to \infty} F(x) = 0.$$

(7.2.61)

Next, we research what happens as $x \to -\infty$, where we use expression (7.2.45) to find the asymptotic form. Here, for simplification of the expression, $f(\kappa_1, k_2)$, $f(\kappa'_1, k'_2)$ are abbreviated as f, f', respectively. We have

$$
\begin{aligned}
F(x) &\sim i\kappa'_1 \left(fe^{i\kappa_1 x} + \overline{f}e^{-i\kappa_1 x}\right)\left(f'e^{i\kappa'_1 x} - \overline{f'}e^{-i\kappa'_1 x}\right) \\
&\quad - i\kappa_1 \left(f'e^{i\kappa'_1 x} + \overline{f'}e^{-i\kappa'_1 x}\right)\left(fe^{i\kappa_1 x} - \overline{f}e^{-i\kappa_1 x}\right) \\
&= i(\kappa_1 + \kappa'_1)\left(\overline{f}f'e^{-i(\kappa_1 - \kappa'_1)x} - f\overline{f'}e^{i(\kappa_1 - \kappa'_1)x}\right) \\
&\quad + i(\kappa_1 - \kappa'_1)\left(\overline{f}\overline{f'}e^{-i(\kappa_1 + \kappa'_1)x} - ff'e^{i(\kappa_1 + \kappa'_1)x}\right). \quad (7.2.62)
\end{aligned}
$$

When this $F(x)$ is substituted into (7.2.59), the term proportional to $e^{\pm i(\kappa_1 + \kappa'_1)x}$ of the second term of (7.2.62) can be regarded as zero for $x \to -\infty$ in the sense of a generalized function. Because $\kappa_1 + \kappa'_1$ is always positive, it can be ignored. As a result, expression (7.2.59) becomes

$$
\begin{aligned}
&\int_{-\infty}^{\infty} \overline{\Psi(x, E)}\Psi(x, E')dx \\
&= - \lim_{x \to -\infty} \frac{i(\kappa_1 + \kappa'_1)}{E - E'}\left[\overline{f}f'e^{-i(\kappa_1 - \kappa'_1)x} - f\overline{f'}e^{i(\kappa_1 - \kappa'_1)x}\right]. \quad (7.2.63)
\end{aligned}
$$

Using the relation

$$
\frac{\kappa_1^2 - \kappa'_1{}^2}{E - E'} = 1, \quad (7.2.64)
$$

which follows from (7.2.40), namely $\kappa_1 = \sqrt{E + V_1}$, and the formula of generalized function for delta function

$$
\lim_{M \to \infty} \frac{e^{\pm iM\kappa}}{\pi\kappa} = \pm i\delta(\kappa), \quad (7.2.65)
$$

we can obtain the orthogonality relation of the eigenfunctions

$$
\int_{-\infty}^{\infty} \overline{\Psi(x, E)}\Psi(x, E')dx = 2\pi\left|f(\kappa_1, k_2)\right|^2 \delta(\kappa_1 - \kappa'_1), \quad (7.2.66)
$$

or equivalently

$$
\int_{-\infty}^{\infty} \overline{\Psi(x, E)}\Psi(x, E')dx = 4\pi\kappa_1\left|f(\kappa_1, k_2)\right|^2 \delta(E - E'). \quad (7.2.67)
$$

In this case, the orthogonality relation is written by using Dirac's delta function, instead of the Kronecker delta, because the eigenvalue spectrum is a continuous one.

Case of $V_1 < E$

The asymptotic form of $F(x)$ for $x \to \infty$ is given by the asymptotic form of Ψ (7.2.50) and the definition of $F(x)$ (7.2.60) is

$$F(x) \sim i(\kappa_2 + \kappa_2')e^{-i(\kappa_2 - \kappa_2')x}. \tag{7.2.68}$$

The asymptotic form for $x \to -\infty$ is from expressions (7.2.51) and (7.2.60):

$$
\begin{aligned}
F(x) &\sim i\kappa_1'\big(\overline{g_1}e^{-i\kappa_1 x} + \overline{g_2}e^{i\kappa_1 x}\big)\big(g_1'e^{i\kappa_1' x} - g_2'e^{-i\kappa_1' x}\big) \\
&\quad + i\kappa_1\big(g_1'e^{i\kappa_1' x} + g_2'e^{-i\kappa_1'}\big)\big(\overline{g_1}e^{-i\kappa_1 x} - \overline{g_2}e^{i\kappa_1 x}\big) \\
&= i(\kappa_1 + \kappa_1')\big(\overline{g_1}g_1'e^{-i(\kappa_1 - \kappa_1')x} - \overline{g_2}g_2'e^{i(\kappa_1 - \kappa_1')x}\big) \\
&\quad - i(\kappa_1 - \kappa_1')\big(\overline{g_2}g_1'e^{i(\kappa_1 + \kappa_1')x} - \overline{g_1}g_2'e^{-i(\kappa_1 + \kappa_1')x}\big). \tag{7.2.69}
\end{aligned}
$$

Substituting these results into expression (7.2.59), the term proportional to $e^{\pm i(\kappa_1 + \kappa_1')x}$ of the second term can be regarded as zero for $x \to -\infty$, just as in (7.2.62). Therefore, we have

$$
\begin{aligned}
\int_{-\infty}^{\infty} \overline{\Psi(x, E)}\Psi(x, E')dx &= \lim_{x \to \infty} \frac{i(\kappa_2 + \kappa_2')}{E - E'}e^{-i(\kappa_2 - \kappa_2')x} \\
&\quad - \lim_{x \to -\infty} \frac{i(\kappa_1 + \kappa_1')}{E - E'}\big(\overline{g_1}g_1'e^{-i(\kappa_1 - \kappa_1')x} - \overline{g_2}g_2'e^{i(\kappa_1 - \kappa_1')x}\big). \tag{7.2.70}
\end{aligned}
$$

Using the expression

$$\frac{\kappa_2^2 - \kappa_2'^2}{E - E'} = 1, \tag{7.2.71}$$

which follows from (7.2.47), namely $\kappa_2 = \sqrt{E - V_1}$, together with (7.2.64) and (7.2.65), we obtain

$$
\begin{aligned}
\int_{-\infty}^{\infty} \overline{\Psi(x, E)}\Psi(x, E')dx &= \pi\delta(\kappa_2 - \kappa_2') + \pi\big(|g_1|^2 + |g_2|^2\big)\delta(\kappa_1 - \kappa_1') \\
&= 2\pi\big[\kappa_2 + \kappa_1\big(|g_1|^2 + |g_2|^2\big)\big]\delta(E - E'). \tag{7.2.72}
\end{aligned}
$$

Furthermore, since we can find $|g_1|^2 + |g_2|^2$ by using (7.2.53), (7.2.54) and (7.2.55), we obtain the orthogonality relation of the eigenfunction:

$$\int_{-\infty}^{\infty} \overline{\Psi(x, E)}\Psi(x, E')dx = 4\pi\kappa_2 \frac{\cosh^2\big(\frac{\pi}{2}(\kappa_1 + \kappa_2)\big) - \sin^2(\frac{\pi}{2}\lambda)}{\sinh(\pi\kappa_1)\sinh(\pi\kappa_2)}\delta(E - E'). \tag{7.2.73}$$

7.2.4 *Case of incidence from positive direction of x-axis*

We have dealt with so far the case where particles are incident from the negative direction of the x-axis to the positive direction. However, when $V_1 < E$, particles can be incident from the positive direction to the negative direction. Here, we will briefly mention this case. In this case, we change the sign of the coordinate x and replace x with $-x$. However, this alone would change equation (7.2.4). So to make it invariant, we should change the sign of V_1 and replace V_1 with $-V_1$. Because if the equation changes, the previous result cannot be used. The variable V_1 is contained in the κ_1 and κ_2 defined in (7.2.40) and (7.2.47) in the form of $\kappa_1 = \sqrt{E + V_1}$ and $\kappa_2 = \sqrt{E - V_1}$. Here, we adopt the method of exchanging κ_1 and κ_2 of the previous result without changing the definitions of them. Let the wave function in this case be $\hat{\Psi}(x, E)$. We have from (7.2.49), (7.2.50) and (7.2.51)

$$\hat{\Psi}(x, E) = 2^{-i(\kappa_2 - \kappa_1)/2} e^{-i(\kappa_1 + \kappa_2)x/2} \cosh^{-i(\kappa_2 - \kappa_1)/2}(x)$$
$$\times F\left(\tfrac{1}{2}(i\kappa_2 - i\kappa_1 + 1 + \lambda), \tfrac{1}{2}(i\kappa_2 - i\kappa_1 + 1 - \lambda), -i\kappa_1 + 1; \tfrac{1}{2}(1 + \tanh x)\right).$$
$$(7.2.74)$$

This solution becomes, as $x \to -\infty$,

$$\hat{\Psi} \sim e^{-i\kappa_1 x}, \tag{7.2.75}$$

and it becomes, at $x \to \infty$,

$$\hat{\Psi} \sim g_1(\kappa_2, \kappa_1)e^{-i\kappa_2 x} + g_2(\kappa_2, \kappa_1)e^{i\kappa_2 x}. \tag{7.2.76}$$

The first term of the right-hand side is the incident wave propagating to the negative direction, the second term is the reflected wave and expression (7.2.75) is a transmitted wave propagating to the negative direction. Let the reflection coefficient and transmission coefficient be $\hat{R}$ and $\hat{T}$, respectively. Then we have

$$\hat{R} = \frac{g_2(\kappa_2, \kappa_1)}{g_1(\kappa_2, \kappa_1)}, \quad \hat{T} = \frac{1}{g_1(\kappa_2, \kappa_1)}, \tag{7.2.77}$$

and the relation for the conservation of the number of particles corresponding to relation (7.2.56) is written as

$$\kappa_2|\hat{R}|^2 + \kappa_1|\hat{T}|^2 = \kappa_2. \tag{7.2.78}$$

Furthermore, the normalization formula corresponding to (7.2.73) becomes

$$\int_{-\infty}^{\infty} \overline{\hat{\Psi}(x,E)}\hat{\Psi}(x,E')dx = 4\pi\kappa_1 \frac{\cosh^2\left(\frac{\pi}{2}(\kappa_1+\kappa_2)\right) - \sin^2(\frac{\pi}{2}\lambda)}{\sinh(\pi\kappa_1)\sinh(\pi\kappa_2)}\delta(E-E').$$

(7.2.79)

7.2.5 *Final comment*

We have presented here the Rosen–Morse potential problem in quantum mechanics. There are other books on this problem about the solution of the bound state, but, little is written about the scattering state. Here, we focused mainly on the solution of the scattering state. Again, we are keenly aware that there is a limit to what can be done analytically.

Another independent solution

There is another independent solution to the differential equation of the hypergeometric function (7.2.15):

$$u = \zeta^{1-\gamma}F(\alpha-\gamma+1,\ \beta-\gamma+1,\ 2-\gamma;\ \zeta).$$ (7.2.80)

Now, let us consider what happens if this solution is adopted. Substituting this into (7.2.17) with $\zeta = \frac{1}{2}(1 - \tanh x)$, by using $1 - \gamma = a - b$ from the third relation of (7.2.16), we have

$$\Psi = 2^{-a}e^{bx}\cosh^{-a}(x)F\left(\alpha-\gamma+1,\ \beta-\gamma+1,\ 2-\gamma;\ \tfrac{1}{2}(1-\tanh x)\right).$$

(7.2.81)

If we compare this with (7.2.17), this expression becomes the one that a and b are interchanged except for the hypergeometric function part. However, in fact, it can be proven that a and b are interchanged in the hypergeometric function part as well. To show this situation, if we replace the variables as

$$\alpha' = \alpha - \gamma + 1, \quad \beta' = \beta - \gamma + 1, \quad \gamma' = 2 - \gamma,$$ (7.2.82)

and using expressions (7.2.16), we have

$$\alpha' + \beta' = 2a + 1, \quad \alpha'\beta' = a(a+1) - V_2, \quad \gamma' = a - b + 1.$$ (7.2.83)

These expressions just mean that a and b are interchanged when the variables α, β, γ are replaced with α', β', γ' in expression (7.2.17). It shows that equation (7.2.11) is symmetric for the interchange of a and b.

Thus, there are two independent solutions in equation (7.2.4), and the general solution becomes a linear combination of these two solutions. Then we require that this solution converge to zero as $x \to \infty$, and the result is $a < b$ as shown in (7.2.20) from expression (7.2.17). On the other hand, from expression (7.2.81), a and b are interchanged, and the result becomes $a > b$. Of course, these two conditions are in contradiction, and they are not satisfied simultaneously. Therefore, for this solution, either expression (7.2.17) or expression (7.2.81) will be adopted. Here, we discussed using expression (7.2.17). However, when expression (7.2.81) is adopted, the same conclusion is reached simply by interchanging the roles of a and b.

Case of $a + b < 0$ in (7.2.22)

We consider the case of $a + b < 0$ in expression (7.2.22). In this case, the exponential function of the second term on the right-hand side diverges as $x \to -\infty$. To prevent this, either $\Gamma(\gamma - \alpha)$ or $\Gamma(\gamma - \beta)$ should be in the pole position. Because it will be the same regardless of which is in the pole position, we assume here that $\Gamma(\gamma - \alpha)$ has the pole, and we put

$$\gamma - \alpha = -n, \quad n = 0, 1, 2, \ldots . \tag{7.2.84}$$

We can derive the following relations from (7.2.16):

$$(\gamma - \alpha) + (\gamma - \beta) = -2a + 1, \quad (\gamma - \alpha)(\gamma - \beta) = a(a - 1) - V_2. \tag{7.2.85}$$

Eliminating $\gamma - \beta$ from these expressions and using (7.2.84), we have

$$a^2 - (2n + 1)a + n(n + 1) - V_2 = 0. \tag{7.2.86}$$

We solve this equation, noting that a is negative, as shown in (7.2.20),

$$a = n - \nu, \tag{7.2.87}$$

where ν is defined by (7.2.25). Thus variable a is obtained, and b is calculated from (7.2.11), using μ of (7.2.26),

$$b = -\frac{\mu}{n - \nu}. \tag{7.2.88}$$

Substituting these expressions of a and b into the second expression of (7.2.11), the energy E is obtained as

$$E_n = -(n - \nu)^2 - \frac{\mu^2}{(n - \nu)^2}. \tag{7.2.89}$$

This is exactly the same as the energy given in expression (7.2.31). The remaining α, β and γ are given from expressions (7.2.16) and (7.2.84) as

$$\alpha = \nu - \frac{\mu}{n-\nu} + 1, \quad \beta = -\nu - \frac{\mu}{n-\nu}, \quad \gamma = \nu - n - \frac{\mu}{n-\nu} + 1.$$

$$(7.2.90)$$

Substituting these expressions into (7.2.17), we have

$$\Psi_n = 2^{\mu/(n-\nu)} e^{(n-\nu)x} \cosh^{\mu/(n-\nu)}(x)$$

$$\times F\left(\nu - \frac{\mu}{n-\nu} + 1, \; -\nu - \frac{\mu}{n-\nu}, \; \nu - n - \frac{\mu}{n-\nu} + 1; \; \tfrac{1}{2}(1 - \tanh x)\right).$$

$$(7.2.91)$$

This is formally different from that obtained by expression (7.2.34). However, if we use **Kummer's formula of the hypergeometric function**

$$F(\alpha, \beta, \gamma; \zeta) = (1 - \zeta)^{\gamma - \alpha - \beta} F(\gamma - \alpha, \gamma - \beta, \gamma; \zeta), \qquad (7.2.92)$$

it can be seen that the result is exactly the same as the previous one including the coefficient. As a result, the sign of $a + b$ gives the same result regardless of whether it is positive or negative.

7.3 Periodic Potential Problems (1)

We discuss here the periodic potential problems in quantum mechanics. That is, we consider a time-independent one-dimensional Schrödinger equation

$$\left[-\frac{\hbar^2}{2m}\frac{d^2}{dx^2} + V(x)\right]\Psi = E\Psi \qquad (7.3.1)$$

in which the potential $V(x)$ has periodicity. In the case of the potential $V(x)$ being represented by trigonometric functions, this equation becomes the **Mathieu equation**. Although the Mathieu equation is the most well-known one with a periodic term, it is also well known that this equation is very complicated to handle analytically.

A periodic potential problem which we can often find in quantum mechanics textbooks is the **Kronig–Penny model**. This potential is a continuous array of square waves. More precisely speaking, if potential $V(x)$ has period ℓ, that is, $V(x + \ell) \equiv V(x)$, it is defined in the range of $[0, \ell]$ as

$$V(x) = V_0\,\theta(x - a), \quad 0 < a < \ell \tag{7.3.2}$$

where θ is a unit step function, V_0 and a are constants having the dimension of potential and length, respectively. In this case all calculations are done within the elementary function. However, we need to solve the equations separately in the ranges $[0, a]$ and $[a, \ell]$, and to connect the solutions at two points $x = a$ and $x = \ell$. This requires a fairly complicated calculation when actually tried. There is literature up to the point of finding the energy eigenvalues, but there is no literature on the normalization of the wave functions. This is probably because the calculation becomes too cumbersome.

Here we will deal with a simplified version of this model. That is, multiplying this potential by $\ell/(\ell - a)$, and we will take the limit value of $a \to \ell$. Since the width of the potential narrows and the height increases, the potential finally becomes the delta functions being periodically arranged at intervals of distance ℓ

$$V(x) = \sum_{n=-\infty}^{\infty} \ell V_0 \delta(x - n\ell). \tag{7.3.3}$$

We will analyze the problem by using this potential. In this case, we only need to connect the solution which is solved in the range of $[0, \ell]$, at one point of $x = \ell$. So it is considerably simpler than the original Kronig–Penny model. We call this model a **periodic delta function model**.

For the simplification of mathematical formulas handled below, we rewrite here the coordinate x, magnitude of the potential V_0 and the energy E in the dimensionless styles

$$\frac{x}{\ell} \mapsto x, \quad \frac{2m\ell^2 V_0}{\hbar^2} \mapsto V_0, \quad \frac{2m\ell^2 E}{\hbar^2} \mapsto E. \tag{7.3.4}$$

The potential becomes a periodic function of period 1 by this transformation, and the wave equation becomes

$$\left[\frac{d^2}{dx^2} + E - \sum_{n=-\infty}^{\infty} V_0 \delta(x - n) \right] \Psi = 0. \tag{7.3.5}$$

7.3.1 *Floquet's theorem*

First, we explain **Floquet's theorem** in second-order linear differential equations which have the periodic terms. Here we give the equation generality, we set

$$\left[\frac{d^2}{dx^2} + F(x)\right]\Phi = 0, \tag{7.3.6}$$

where function $F(x)$ is assumed to have a periodicity such that $F(x+1) \equiv F(x)$. Let the two linearly-independent solutions of this equation be $\Phi_1(x)$ and $\Phi_2(x)$. Then, $\Phi_1(x+1)$, $\Phi_2(x+1)$ are also solutions from the periodicity of the equation. However, because there are only two independent solutions, these two solutions should be linearly dependent on $\Phi_1(x)$ and $\Phi_2(x)$, and so we have

$$\begin{pmatrix} \Phi_1(x+1) \\ \Phi_2(x+1) \end{pmatrix} = A \begin{pmatrix} \Phi_1(x) \\ \Phi_2(x) \end{pmatrix}, \tag{7.3.7}$$

where the coefficient matrix A is written as

$$A = \begin{pmatrix} a_{11} & a_{12} \\ a_{21} & a_{22} \end{pmatrix}. \tag{7.3.8}$$

If we combine this expression and its x derivative, we have

$$\begin{pmatrix} \Phi_1{}'(x+1) & \Phi_1(x+1) \\ \Phi_2{}'(x+1) & \Phi_2(x+1) \end{pmatrix} = A \begin{pmatrix} \Phi_1{}'(x) & \Phi_1(x) \\ \Phi_2{}'(x) & \Phi_2(x) \end{pmatrix}, \tag{7.3.9}$$

where the prime means the differential with respect to x. Furthermore, if we take the determinant of this expression

$$W(x+1) = (\det A)W(x), \tag{7.3.10}$$

where the **Wronskian** $W(x)$ is defined as

$$W(x) = \begin{vmatrix} \Phi_1{}'(x) & \Phi_1(x) \\ \Phi_2{}'(x) & \Phi_2(x) \end{vmatrix}. \tag{7.3.11}$$

By using equation (7.3.6), we can easily derive the relation

$$\frac{dW(x)}{dx} = 0. \tag{7.3.12}$$

Therefore, this Wronskian $W(x)$ must be a constant. Moreover, because Φ_1, Φ_2 are independent solutions, it should be a non-zero constant. Therefore, we have from (7.3.10)

$$\det A = 1. \tag{7.3.13}$$

Next, we try to diagonalize the coefficient matrix A. For this purpose, we consider the eigenvalue equation with the eigenvalue λ and the eigenvector $^t(u_1, \ u_2)$

$$\begin{pmatrix} a_{11} & a_{12} \\ a_{21} & a_{22} \end{pmatrix} \begin{pmatrix} u_1 \\ u_2 \end{pmatrix} = \lambda \begin{pmatrix} u_1 \\ u_2 \end{pmatrix}. \tag{7.3.14}$$

As is well known, the eigenvalue λ is the solution of the following equation:

$$\begin{vmatrix} a_{11} - \lambda & a_{12} \\ a_{21} & a_{22} - \lambda \end{vmatrix} = 0. \tag{7.3.15}$$

Taking into account (7.3.13), this equation becomes

$$\lambda^2 - (a_{11} + a_{22})\lambda + 1 = 0. \tag{7.3.16}$$

If it is limited to the case in which two independent solutions $\Phi_1(x)$, $\Phi_2(x)$ can be obtained by real numbers, then each element of A of (7.3.8) is also a real number. The quadratic equation (7.3.16) can be solved as

$$\lambda_{1,\ 2} = \frac{1}{2}\left[a_{11} + a_{22} \pm \sqrt{(a_{11} + a_{22})^2 - 4}\right], \tag{7.3.17}$$

where the plus and minus signs correspond to λ_1 and λ_2, respectively. Basing on the above result, we classify this solution depending on the value of $a_{11} + a_{22}$, as follows:

$$\begin{cases} a_{11} + a_{22} < -2, & -1 < \lambda_1 < 0, & \lambda_2 < -1, \\ a_{11} + a_{22} = -2, & \lambda_1 = \lambda_2 = -1, \\ -2 < a_{11} + a_{22} < 2, & |\lambda_1| = |\lambda_2| = 1, & \text{and } \overline{\lambda_1} = \lambda_2, \\ a_{11} + a_{22} = 2, & \lambda_1 = \lambda_2 = 1, \\ 2 < a_{11} + a_{22}, & 1 < \lambda_1, 0 < \lambda_2 < 1. \end{cases} \tag{7.3.18}$$

Next, the normalized eigenvectors belonging to the eigenvalues λ_1 and λ_2 obtained here are $^t(u_{11}, \ u_{21})$ and $^t(u_{12}, \ u_{22})$, respectively. And we define the matrix U which is made from these eigenvectors as

$$U = \begin{pmatrix} u_{11} & u_{12} \\ u_{21} & u_{22} \end{pmatrix}. \tag{7.3.19}$$

Multiplying from the left the inverse matrix U^{-1} of this U to expression (7.3.7), we have

$$U^{-1}\begin{pmatrix}\Phi_1(x+1)\\\Phi_2(x+1)\end{pmatrix} = U^{-1}AUU^{-1}\begin{pmatrix}\Phi_1(x)\\\Phi_2(x)\end{pmatrix}. \qquad (7.3.20)$$

Here we define the transformation from ${}^t(\Phi_1,\ \Phi_2)$ to ${}^t(\Psi_1,\ \Psi_2)$

$$\begin{pmatrix}\Psi_1(x)\\\Psi_2(x)\end{pmatrix} = U^{-1}\begin{pmatrix}\Phi_1(x)\\\Phi_2(x)\end{pmatrix}, \qquad (7.3.21)$$

and if $U^{-1}AU$ is a diagonal matrix, this expression becomes[5]

$$\begin{pmatrix}\Psi_1(x+1)\\\Psi_2(x+1)\end{pmatrix} = \begin{pmatrix}\lambda_1 & 0\\0 & \lambda_2\end{pmatrix}\begin{pmatrix}\Psi_1(x)\\\Psi_2(x)\end{pmatrix}. \qquad (7.3.22)$$

Furthermore, repeating the above procedure n times, we have

$$\begin{pmatrix}\Psi_1(x+n)\\\Psi_2(x+n)\end{pmatrix} = \begin{pmatrix}\lambda_1{}^n & 0\\0 & \lambda_2{}^n\end{pmatrix}\begin{pmatrix}\Psi_1(x)\\\Psi_2(x)\end{pmatrix}. \qquad (7.3.23)$$

As shown in the solution λ (7.3.18), there is the case where λ_i becomes $|\lambda_i| < 1$ or $|\lambda_i| > 1$. In either case, $\Psi(x+n)$ diverges as $n \to -\infty$ or $n \to \infty$ from expression (7.3.23). As a result, these cases are irrelevant as the wave function of quantum mechanics. It is the case of $|\lambda_i| = 1$ that is compatible. Using here the real number K where $0 \le K \le \pi$, we put

$$\lambda_1 = e^{iK}, \quad \lambda_2 = e^{-iK}. \qquad (7.3.24)$$

Then the periodicity of the function Ψ is given by expression (7.3.22)

$$\Psi_1(x+1) = e^{iK}\Psi_1(x), \quad \Psi_2(x+1) = e^{-iK}\Psi_2(x). \qquad (7.3.25)$$

It can be seen from this that the phase e^{iK} or e^{-iK} is added each time the period goes up by one. Furthermore, we express Ψ_1 as

$$\Psi(x) = e^{iKx}P(x), \qquad (7.3.26)$$

where the subscript 1 has been omitted. Substituting this into (7.3.25), we have

$$P(x+1) = P(x). \qquad (7.3.27)$$

---

[5]More generally, when λ becomes a multiple solution, it might be a triangular matrix (Jordan's normal form) without diagonalization. Even in this case, either $\Psi_1(x)$ or $\Psi_2(x)$, it can be written in the form $\Psi_i(x+1) = \lambda_i\Psi_i(x)$.

Thus, the function $P(x)$ is a periodic function. The same applies to Ψ_2. This theory is called Floquet's theorem.

We have analyzed here one-dimensional space. A study to describe the wave motion of electrons in metal crystals was done by Bloch. The three-dimensional version of expression (7.3.26) is called the Bloch function. Then the K becomes a three-dimensional vector, and is called the **propagation vector**. Since this is a one-dimensional analysis here, let us call K the **propagation number**.

7.3.2 *Solution to the equation*

Case of $V_0 < 0$, $E < 0$

Let us go back to the first problem and try to solve equation (7.3.5). First, we consider the case of $V_0 < 0$, $E < 0$. Normally, this describes the motion of electrons in bound state. Since it is a periodic potential here, we do not know what would happen. To simplify the following expressions, we put

$$\kappa = \sqrt{-E}. \tag{7.3.28}$$

Then equation (7.3.5) can be easily solved within the range of $0 < x < 1$, and the result is written with A and B as arbitrary constants:

$$\Psi(x) = A\cosh(\kappa x) + B\sinh(\kappa x). \tag{7.3.29}$$

Also, $\Psi(x)$ in the range of $1 < x < 2$ is given by the first expression of (7.3.25) which is described in Floquet's theorem:

$$\Psi(x) = e^{iK}\left[A\cosh\left(\kappa(x-1)\right) + B\sinh\left(\kappa(x-1)\right)\right]. \tag{7.3.30}$$

Of course, there is also the second expression of (7.3.25) in which the sign of K is changed. However, for the time being, we will proceed with the first expression only.

Here, the two solutions of (7.3.29) and (7.3.30) must be connected "smoothly" at $x = 1$. First, from the continuity of the function itself, we obtain

$$A\cosh\kappa + B\sinh\kappa = e^{iK}A. \tag{7.3.31}$$

What we do next is the connection of the derivatives. However, they cannot be continuous, because the potential includes the delta function. With ε as a positive small quantity, we integrate equation (7.3.5) from $1 - \varepsilon$ to $1 + \varepsilon$,

$$\frac{d\Psi}{dx}\bigg|_{x=1+\varepsilon} - \frac{d\Psi}{dx}\bigg|_{x=1-\varepsilon} = V_0 \Psi(1). \tag{7.3.32}$$

It can be seen that there is a gap in the derivatives. Substituting (7.3.29) and (7.3.30) into this expression, we obtain

$$e^{iK}\kappa B - \kappa(A\sinh\kappa + B\cosh\kappa) = V_0 e^{iK} A. \tag{7.3.33}$$

Summarizing these two expressions (7.3.31) and (7.3.33), we have

$$\begin{pmatrix} \cosh\kappa - e^{iK} & \sinh\kappa \\ \dfrac{V_0 e^{iK}}{\kappa} + \sinh\kappa & \cosh\kappa - e^{iK} \end{pmatrix} \begin{pmatrix} A \\ B \end{pmatrix} = 0. \tag{7.3.34}$$

Because the coefficients A and B must not be both zero, the value of the coefficient determinant must be zero. Therefore, we have

$$\cos K = \cosh\kappa + \frac{V_0}{2\kappa} \sinh\kappa. \tag{7.3.35}$$

From this equation, there is a relationship between the propagation number K and the κ. For this condition to be satisfied, the absolute value of the right-hand side must be 1 or less. As shown in the following numerical calculation, the value of κ that satisfies this equation is not discrete; it has a width, a so-called band structure. Also, from (7.3.28), κ is determined by the energy E, and this expression determines the relationship between the energy E and the propagation number K. In this sense, expression (7.3.35) is called (in a broad sense) the **eigenvalue equation**. Furthermore, since the left-hand side of this expression is an even function of K, it is determined that there are two positive and negative values of K for one E. This is the reason why there are two wave functions which have different signs for K in expression (7.3.25). In this sense, the wave function is degenerate. In particular, when $K = 0$ or π, because these two functions e^{iK} and e^{-iK} become the same, therefore, it must be another wave function which is made by partially differentiating with respect to K of the original one. However, this case is not treated here as an exception. Therefore, in the following, the range of K should be $0 < K < \pi$.

Here, the coefficients A and B are selected so that expression (7.3.31) is satisfied,

$$A = \sinh\kappa, \quad B = e^{iK} - \cosh\kappa. \tag{7.3.36}$$

Hereafter, the wave function in this case would be written as $\Psi(x, \kappa)$ specifying the κ dependency. From relation (7.3.25), for the x in the interval $n < x < n+1$ with n as an arbitrary integer, the wave function is written as

$$\Psi(x, \kappa) = e^{iKn}\left[\sinh \kappa \cosh\left(\kappa(x-n)\right) + (e^{iK} - \cosh \kappa)\sinh\left(\kappa(x-n)\right)\right],$$
(7.3.37)

or, by using the addition theorem of hyperbolic functions, it is rewritten as

$$\Psi(x, \kappa) = e^{iKn}\left[e^{iK}\sinh\left(\kappa(x-n)\right) - \sinh\left(\kappa(x-n-1)\right)\right]. \quad (7.3.38)$$

However, this is not yet normalized.

Next, let us examine what happens to the normalization of this wave function $\Psi(x, \kappa)$. We consider two generically different κ, κ', and let the corresponding propagation number be K, K'. By using expression (7.3.38), the integral for the normalization becomes

$$\int_{-\infty}^{\infty} \overline{\Psi(x, \kappa)}\Psi(x, \kappa')dx = \lim_{M\to\infty} \sum_{n=-M}^{M} \int_{n}^{n+1} \overline{\Psi(x, \kappa)}\Psi(x, \kappa')dx$$

$$= \left(\lim_{M\to\infty} \sum_{n=-M}^{M} e^{-i(K-K')n}\right)\int_{0}^{1} \overline{\Psi(x, \kappa)}\Psi(x, \kappa')dx.$$
(7.3.39)

Here, the overline means a complex conjugate. Now, let us execute this integral part first. Putting $E = -\kappa^2$, or $-\kappa'^2$ in equation (7.3.5), we write the equations which $\overline{\Psi(x, \kappa)}$, or $\Psi(x, \kappa')$ satisfies

$$\left[\frac{d^2}{dx^2} - \kappa^2 - \sum_{n=-\infty}^{\infty} V_0\delta(x-n)\right]\overline{\Psi(x, \kappa)} = 0,$$

$$\left[\frac{d^2}{dx^2} - \kappa'^2 - \sum_{n=-\infty}^{\infty} V_0\delta(x-n)\right]\Psi(x, \kappa') = 0.$$
(7.3.40)

Multiplying these first and second equations by $\Psi(x, \kappa')$ and $\overline{\Psi(x, \kappa)}$, respectively, and subtracting both sides, we have

$$\frac{d}{dx}\left[\Psi(x, \kappa')\frac{d\overline{\Psi(x, \kappa)}}{dx} - \overline{\Psi(x, \kappa)}\frac{d\Psi(x, \kappa')}{dx}\right] - (\kappa^2 - \kappa'^2)\overline{\Psi(x, \kappa)}\Psi(x, \kappa') = 0.$$
(7.3.41)

Integrating this expression from 0 to 1, we obtain

$$\int_0^1 \overline{\Psi(x,\kappa)}\Psi(x,\kappa')dx = \frac{1}{\kappa^2 - \kappa'^2}\left[\Psi(x,\kappa')\frac{\overline{d\Psi(x,\kappa)}}{dx} - \overline{\Psi(x,\kappa)}\frac{d\Psi(x,\kappa')}{dx}\right]_0^1.$$

(7.3.42)

Furthermore, the right-hand side is calculated by using (7.3.38) with $n = 0$,

$$\int_0^1 \overline{\Psi(x,\kappa)}\Psi(x,\kappa')dx$$

$$= \frac{1}{\kappa^2 - \kappa'^2}\left[\left(1 + e^{-i(K-K')}\right)(\kappa\sinh\kappa'\cosh\kappa - \kappa'\sinh\kappa\cosh\kappa')\right.$$

$$\left. + \left(e^{-iK} + e^{iK'}\right)(\kappa'\sinh\kappa - \kappa\sinh\kappa')\right].$$

(7.3.43)

Next, the sum part of expression (7.3.39) can be calculated by using the formula for the delta function,

$$\lim_{M\to\infty}\sum_{n=-M}^{M} e^{-i(K-K')n} = 2\pi\delta(K-K'), \quad 0 < K, \ K' < \pi.$$

(7.3.44)

As a result, expression (7.3.39) becomes

$$\int_{-\infty}^{\infty} \overline{\Psi(x,\kappa)}\Psi(x,\kappa')dx = \frac{4\pi}{\kappa^2 - \kappa'^2}\left[\kappa\sinh\kappa'\cosh\kappa - \kappa'\sinh\kappa\cosh\kappa'\right.$$

$$\left. + \cos(K)(\kappa'\sinh\kappa - \kappa\sinh\kappa')\right]\delta(K-K').$$

(7.3.45)

Now, we return to the eigenvalue equation (7.3.35). As shown in the following numerical calculation, there is only one band structure in this area of $V_0 < 0$, $E < 0$. Therefore, when the value of K is determined, the corresponding κ is uniquely determined. In other words, the relation $K = K'$ just means $\kappa = \kappa'$. Therefore, since the part of expression (7.3.45) excluding the delta function is an indeterminate form of $0/0$, so we calculate the limit value of $\kappa \to \kappa'$. The result is

$$\int_{-\infty}^{\infty} \overline{\Psi(x,\kappa)}\Psi(x,\kappa')dx$$

$$= \frac{2\pi}{\kappa}\left[-\kappa + \sinh\kappa\cosh\kappa - \cos(K)(\sinh\kappa - \kappa\cosh\kappa)\right]\delta(K-K').$$

(7.3.46)

Next, if we remove V_0 from the eigenvalue equation (7.3.35) and its derivative with respect to κ, we have

$$-\kappa + \sinh\kappa\cosh\kappa - \cos(K)(\sinh\kappa - \kappa\cosh\kappa) = \kappa\sinh\kappa\frac{d\cos(K)}{d\kappa}.$$
$$(7.3.47)$$

Then we rewrite (7.3.46) by using this relation, and replace $\delta(K - K')$ by $\delta(\kappa - \kappa')$. Finally, we obtain the orthogonality relation of the wave function,

$$\int_{-\infty}^{\infty}\overline{\Psi(x,\kappa)}\Psi(x,\kappa')dx = N^2(\kappa)\delta(\kappa - \kappa'), \quad N^2(\kappa) \equiv 2\pi\sinh(\kappa)\sin(K).$$
$$(7.3.48)$$

When deriving this formula, as shown in the following numerical calculation, we noticed that the $dK/d\kappa$ becomes negative. Also, here, the condition $0 < K < \pi$ is set, so the result of this integration is a positive definite value. $N(\kappa)$ is the normalization constant, and $\Psi(x,\kappa)/N(\kappa)$ is the normalized wave function. If the sign of K is changed, the part of $\sin(K)$ becomes negative. However, in this case, $dK/d\kappa$ becomes positive on the contrary. Therefore, it is noted that the result of this normalization integral is also a positive definite value.

Case of $E > 0$

We analyze here the case of $E > 0$ where the sign of V_0 can be either positive or negative. For simplification of the expression, we put

$$k = \sqrt{E}. \tag{7.3.49}$$

We could proceed as before, but instead, we put

$$\kappa = ik, \tag{7.3.50}$$

and we borrow the previous result as it is. The formula that determines the relationship between K and k is (7.3.35), that is, the eigenvalue equation in this case becomes

$$\cos K = \cos k + \frac{V_0}{2k}\sin k. \tag{7.3.51}$$

The wave function which is written as $\Psi(x,k)$ is derived from expression (7.3.38), in the interval $n < x < n+1$,

$$\Psi(x,k) = ie^{iKn}\left[e^{iK}\sin\left(k(x-n)\right) - \sin\left(k(x-n-1)\right)\right]. \tag{7.3.52}$$

So far, we can borrow from the previous result. However, at the next stage of the normalization of the wave function, it is noted that the analytic continuation of (7.3.50) would be impossible, because it takes a complex conjugate.

The following calculation is the same as for the expressions from (7.3.39) to (7.3.42), but we obtain the integral form in this case,

$$\int_0^1 \overline{\Psi(x,k)}\Psi(x,k')dx = -\frac{1}{k^2-k'^2}\left[\Psi(x,k')\frac{d\overline{\Psi(x,k)}}{dx} - \overline{\Psi(x,k)}\frac{d\Psi(x,k')}{dx}\right]_0^1.$$

$$(7.3.53)$$

Substituting expression (7.3.52) with $n = 0$ on the right-hand side, and using formula (7.3.44), we can obtain the normalization integral,

$$\int_{-\infty}^{\infty} \overline{\Psi(x,k)}\Psi(x,k')dx = -\frac{4\pi}{k^2-k'^2}\left[k\sin k'\cos k - k'\sin k\cos k'\right.$$

$$\left. + \cos(K)(k'\sin k - k\sin k')\right]\delta(K-K'),$$

$$(7.3.54)$$

as an alternative to (7.3.45). One problem arises here. When $V_0 < 0$, $E < 0$, as discussed in the previous subsection, since there was only one band structure, the κ was uniquely determined by K. However, in the present case, even if the propagation number K is determined, k is not uniquely determined, because one k is determined for each band. Therefore, even if $K = K'$, it does not necessarily mean that $k = k'$. However, the bracket part of (7.3.54) is calculated by eliminating $\cos k$, $\cos k'$ by using the eigenvalue equation (7.3.51) as

$$k\sin k'\cos k - k'\sin k\cos k' + \cos(K)(k'\sin k - k\sin k')$$

$$= k'\sin k\left(\cos(K) - \cos(K')\right).$$

$$(7.3.55)$$

Therefore, when this is multiplied by $\delta(K-K')$, it is always zero. This depends on the periodic boundary condition of the wave function. Therefore, when k and k' belong to different bands, $k \neq k'$, the integral value of this (7.3.54) becomes zero. This means the orthogonality relation between different bands. If k and k' belong to the same band, then it becomes $k = k'$. Then, since the denominator of (7.3.54) is also zero, we take the limit value by using (7.3.55),

$$\int_{-\infty}^{\infty} \overline{\Psi(x,k)}\Psi(x,k')dx = -2\pi\sin k\frac{d\cos(K)}{dk}\delta(K-K').$$

$$(7.3.56)$$

Now we express $\delta(K - K')$ in terms of $\delta(k - k')$. As shown in the numerical calculation below, the sign of dK/dk changes from band to band. With n being a non-negative integer, its sign in the interval $n\pi < k < (n+1)\pi$ becomes $(-1)^n$. This indicates that the sign of dK/dk just matches the sign of $\sin k$. Considering these results, the orthogonality relation of the wave function is obtained

$$\int_{-\infty}^{\infty} \overline{\Psi(x,k)}\Psi(x,k')dx = N^2(k)\delta(k-k'), \quad N^2(k) \equiv 2\pi|\sin k|\sin(K),$$

$$(7.3.57)$$

where the $N(k)$ is the normalization constant, and $\Psi(x,k)/N(k)$ is the normalized wave function. In the case of another solution in which the sign of K is changed, the sign of $\sin(K)$ changes, but the sign of $dK/d\kappa$ also changes; as a result, we have the same orthogonal relation.

We also investigate here the stream function. The dimensionless variables in (7.3.4) are extended for the time t as

$$\frac{\hbar}{2m\ell^2}t \mapsto t, \qquad (7.3.58)$$

and using this expression, the time-dependent Schrödinger equation becomes

$$\left[-\frac{\partial^2}{\partial x^2} + \sum_{n=-\infty}^{\infty} V_0\delta(x-n)\right]\Psi = i\frac{\partial\Psi}{\partial t}. \qquad (7.3.59)$$

From this, the **continuity equation** is derived as follows:

$$\frac{\partial(\overline{\Psi}\Psi)}{\partial t} + \frac{\partial j}{\partial x} = 0, \quad j = i\overline{\Psi}(\overleftarrow{\partial_x} - \overrightarrow{\partial_x})\Psi, \qquad (7.3.60)$$

where the **stream function** j is defined by the second equation. And this is obtained by using the Ψ of (7.3.52) as

$$j = 2k\sin k\sin(K). \qquad (7.3.61)$$

By considering that the sign of $\sin k$ changes from band to band, and that $0 < K < \pi$, the direction of flow repeats positive and negative values for each band in the increasing order of energy; conversely, in the case of a solution in which the sign of K is changed, the direction of flow repeats negative and positive values for each band.

7.3.3 *Numerical analysis*

Some graphs obtained by numerical calculation are displayed here. First, by using the eigenvalue equation (7.3.35) for $E < 0$, the propagation number K obtained as a function of κ is shown in Fig. 7.6. Similarly, by using equation (7.3.51) for $E > 0$, the propagation number K obtained as a function of k is shown in Fig. 7.7. Figure 7.6 shows the case of $V_0 = -5$, while Fig. 7.7 shows the case of $V_0 = 5$. In any of these cases, the absolute value of the right-hand side of these equations must be 1 or less, for the propagation number K to be determined by the real number. The area in which the value of κ or k satisfies this condition is called the **allowed band**, and the area in which κ or k does not satisfy the condition is called the **forbidden band**. It will form a so-called band structure.

As we can see from these figures, when it is $E < 0$, then it becomes $dK/d\kappa < 0$, and when $E > 0$, the sign of dK/dk repeats the positive and negative values for each band in the increasing order of k.

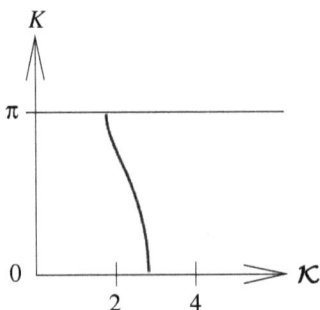

Fig. 7.6 K and κ $(E < 0,\ V_0 = -5)$

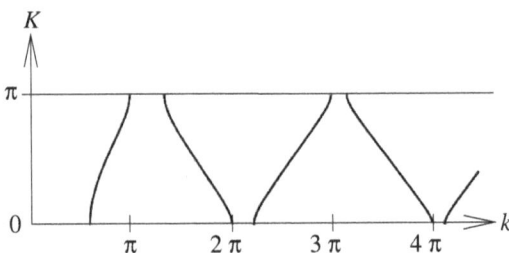

Fig. 7.7 K and k $(E > 0,\ V_0 = 5)$

Since Figs. 7.6 and Fig. 7.7 show only the case of a specific value of V_0, we cannot see the whole structure. Therefore, we solve the potential magnitude V_0 from the eigenvalue equations (7.3.35) and (7.3.51), and write the two equations together:

$$V_0 = \begin{cases} \dfrac{2\sqrt{-E}}{\sinh\sqrt{-E}}\left(\cos K - \cosh\sqrt{-E}\right), & \text{for } E < 0, \\[2ex] \dfrac{2\sqrt{E}}{\sin\sqrt{E}}\left(\cos K - \cos\sqrt{E}\right), & \text{for } E > 0, \end{cases} \qquad (7.3.62)$$

where κ and k are written as $\sqrt{-E}$ and $\sqrt{E}$, respectively, according to equations (7.3.28) and (7.3.49). We consider here the magnitude of the potential V_0 being a function of the energy E. We show this function $V_0(E)$ in Fig. 7.8 in which the value of $\cos K$ is treated as a parameter moving from -1 to 1.

In this figure, when the value of $\cos K$ is -1, it is drawn in cyan. The color continuously changes as the value of $\cos K$ increases, and it turns magenta when $\cos K$ is 1. From this figure, when V_0 is fixed to a positive or negative constant value, some regions having a certain width appear discretely with respect to the energy E. It can be seen that a so-called band structure is formed. It is noted that the value of $\cos K$ goes from -1 to 1 in every band. Also, in this graph, the place where the band rises discontinuously corresponds to $\sqrt{E} = n\pi$, $(n = 1, 2, \ldots)$.

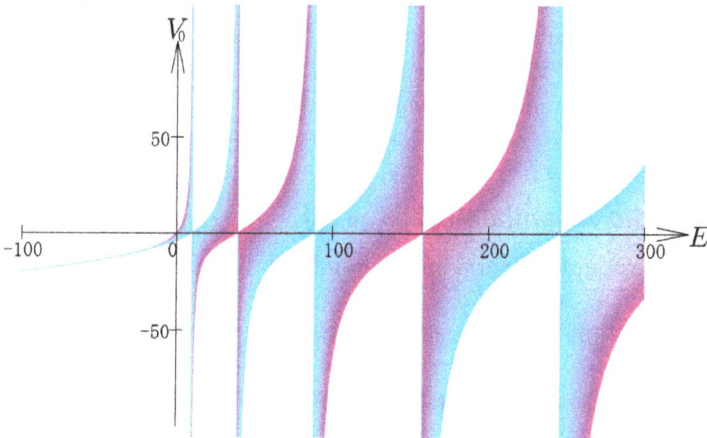

Fig. 7.8 Relation of V_0 and E

Periodic Potential Problems are Very Difficult and Profound

The periodic potential problem dealt with here was a field that I was not specialized in, and it was a problem that was difficult to understand. I searched various textbooks on quantum mechanics, but many books wrote only on the generalities. There were none which gave a concrete potential model or dealt with the normalization of the wave function.

Before starting this calculation, I had doubts on the outcome of the orthogonality between different bands or the orthogonality within the same band. In other books, these two orthogonalities are explained by using two delta functions. However, this is a mistake. In my calculation here, all of these results come from a single delta function, so I decided to solve this problem without ambiguity. I am reminded that before understanding the general theory, it is necessary to understand some concrete examples which are completely solvable. In this respect, since the model where the delta functions are arranged periodically is the easiest one to analyze, it should be helpful to understand this kind of periodic potential problem.

In the following, there is one point to note. For equation (7.3.16) of Floquet's theorem, I wrote that the two independent solutions are limited to those that can be obtained by real numbers. However, this is not a necessary condition for deriving Floquet's theorem. This description follows other books. If we choose $\Phi(x)$ and $\overline{\Phi(x)}$ as two independent solutions, we can derive the relation $a_{22} = \overline{a_{11}}$ and the factor $a_{11} + a_{22}$ which appears in equation (7.3.16) becomes a real number, so the following discussion holds as it is.

7.4 Periodic Potential Problems (2)

As another example of periodic potential, we discuss the following: let the potential $V(x)$ be a periodic function $V(x+\ell) \equiv V(x)$ of the period ℓ. For x in the range $[n\ell, (n+1)\ell]$ with an integer n, we define

$$V(x) = \frac{V_0}{\ell}(x - n\ell). \tag{7.4.1}$$

This is the **potential of a serrated (saw blade) shape**. Here, V_0 is a positive constant with an energy dimension. Alternatively, using brackets [ ] for the Gauss symbol, it can also be represented in a smarter form,

$$V(x) = V_0 \left(\frac{x}{\ell} - \left[\frac{x}{\ell} \right] \right). \tag{7.4.2}$$

In this case, the Schrödinger equation with electron mass m and energy E is written as

$$\left[-\frac{\hbar^2}{2m} \frac{d^2}{dx^2} + V_0 \left(\frac{x}{\ell} - \left[\frac{x}{\ell} \right] \right) \right] \Psi = E\Psi. \tag{7.4.3}$$

To simplify the expression, we make the coordinate x, the magnitude of the potential V_0 and the energy E dimensionless, by replacing them as

$$\frac{x}{\ell} \mapsto x, \quad \frac{2m\ell^2}{\hbar^2} V_0 \mapsto V_0, \quad \frac{2m\ell^2}{\hbar^2} E \mapsto E. \tag{7.4.4}$$

The equation is rewritten as, in terms of these dimensionless quantities,

$$\left[\frac{d^2}{dx^2} - V_0(x - [x]) + E \right] \Psi = 0. \tag{7.4.5}$$

It is noted that the potential becomes a periodic function with the period 1.

7.4.1 *Solution to the equation*

First, we will solve the equation in the range $0 \leq x < 1$

$$\left[\frac{d^2}{dx^2} - V_0 x + E \right] \Psi = 0. \tag{7.4.6}$$

The following variables are introduced to simplify the expression:

$$\mu = V_0^{1/3}, \quad \kappa = E V_0^{-2/3} \left(= E\mu^{-2} \right). \tag{7.4.7}$$

Then, the equation is rewritten as

$$\left[\frac{d^2}{dx^2} - \mu^3 x + \mu^2 \kappa \right] \Psi = 0. \tag{7.4.8}$$

Furthermore, we convert the variable from x to z

$$z = \mu x - \kappa, \tag{7.4.9}$$

then the equation becomes

$$\left[\frac{d^2}{dz^2} - z \right] \Psi = 0. \tag{7.4.10}$$

Since this is the Airy differential equation, its general solution is expressed by using the first and second kind Airy functions $\mathrm{Ai}(z)$ and $\mathrm{Bi}(z)$, and arbitrary constants C, D:

$$\Psi = C\mathrm{Ai}(z) + D\mathrm{Bi}(z). \tag{7.4.11}$$

The Airy function $\mathrm{Ai}(z)$ was described in §6.2, but we did not describe $\mathrm{Bi}(z)$, so we will just write the definition formula again here. These functions are defined by using the Bessel function of order $\pm 1/3$, as follows:

$$\mathrm{Ai}(z) = \begin{cases} \dfrac{1}{\pi}\sqrt{\dfrac{z}{3}}K_{1/3}\left(\tfrac{2}{3}z^{3/2}\right), & z \geq 0, \\[2ex] \dfrac{1}{3}\sqrt{-z}\left[J_{-1/3}\left(\tfrac{2}{3}(-z)^{3/2}\right) + J_{1/3}\left(\tfrac{2}{3}(-z)^{3/2}\right)\right], & z < 0, \end{cases} \tag{7.4.12}$$

$$\mathrm{Bi}(z) = \begin{cases} \sqrt{\dfrac{z}{3}}\left[I_{-1/3}\left(\tfrac{2}{3}z^{3/2}\right) + I_{1/3}\left(\tfrac{2}{3}z^{3/2}\right)\right], & z \geq 0, \\[2ex] \sqrt{-\dfrac{z}{3}}\left[J_{-1/3}\left(\tfrac{2}{3}(-z)^{3/2}\right) - J_{1/3}\left(\tfrac{2}{3}(-z)^{3/2}\right)\right], & z < 0, \end{cases} \tag{7.4.13}$$

where J is the first kind Bessel function, and I and K are the first and second kind modified Bessel functions, respectively. These expressions are defined in apparently different ways depending on whether z is positive or negative. However, they are entire functions. In fact, when they are expanded by using the series expansion formulas of the Bessel functions, the positive and negative parts of z give exactly the same expressions:

$$\mathrm{Ai}(z) = \frac{1}{3^{2/3}}\sum_{n=0}^{\infty}\frac{z^{3n}}{n!\Gamma(n+\tfrac{2}{3})3^{2n}} - \frac{1}{3^{4/3}}\sum_{n=0}^{\infty}\frac{z^{3n+1}}{n!\Gamma(n+\tfrac{4}{3})3^{2n}}, \tag{7.4.14}$$

$$\mathrm{Bi}(z) = \frac{1}{3^{1/6}}\sum_{n=0}^{\infty}\frac{z^{3n}}{n!\Gamma(n+\tfrac{2}{3})3^{2n}} + \frac{1}{3^{5/6}}\sum_{n=0}^{\infty}\frac{z^{3n+1}}{n!\Gamma(n+\tfrac{4}{3})3^{2n}}.$$

The asymptotic forms of these functions are obtained from those of the Bessel functions. When the variable z is positive and large enough, we obtain

$$\mathrm{Ai}(z) \sim \frac{e^{-\frac{2}{3}z^{3/2}}}{2\sqrt{\pi}\,z^{1/4}}, \quad \mathrm{Bi}(z) \sim \frac{e^{\frac{2}{3}z^{3/2}}}{\sqrt{\pi}\,z^{1/4}}. \tag{7.4.15}$$

Also, the asymptotic forms when z is negative and large enough are obtained as

$$\mathrm{Ai}(z) \sim \frac{\sin\left(\frac{2}{3}(-z)^{3/2} + \frac{\pi}{4}\right)}{\sqrt{\pi}\,(-z)^{1/4}}, \quad \mathrm{Bi}(z) \sim \frac{\cos\left(\frac{2}{3}(-z)^{3/2} + \frac{\pi}{4}\right)}{\sqrt{\pi}\,(-z)^{1/4}}. \qquad (7.4.16)$$

These expressions give a good approximation in the area $|z| > 2$.

Here, the solution of equation (7.4.11) in $0 \le x < 1$ is represented by the original variable x:

$$\Psi(x) = C\mathrm{Ai}(\mu x - \kappa) + D\mathrm{Bi}(\mu x - \kappa). \qquad (7.4.17)$$

Just because the potential is periodic, the wave function does not always become periodic. According to Floquet's theorem mentioned in the previous section, it is known that for each cycle increase, the wave function changes to the one of the previous cycle multiplied by the phase e^{iK} or e^{-iK}, where K is the propagation number $(0 < K < \pi)$. The positive and negative signs in this phase indicate that the wave function is degenerate. The calculation method will be the same regardless of which sign is adopted. In the following, we will proceed with the positive phase e^{iK}. According to Floquet's theorem, the wave function $\Psi(x)$ in the next period $1 \le x < 2$ of solution (7.4.17) is written as

$$\Psi(x) = e^{iK}\left[C\mathrm{Ai}\big(\mu(x-1) - \kappa\big) + D\mathrm{Bi}\big(\mu(x-1) - \kappa\big)\right]. \qquad (7.4.18)$$

What should be done next is to connect these two solutions (7.4.17) and (7.4.18), that is, the functions themselves and their derivatives become continuous at $x = 1$. The result is

$$\begin{aligned}
\left[\mathrm{Ai}(\mu - \kappa) - e^{iK}\mathrm{Ai}(-\kappa)\right]C + \left[\mathrm{Bi}(\mu - \kappa) - e^{iK}\mathrm{Bi}(-\kappa)\right]D &= 0, \\
\left[\mathrm{Ai}'(\mu - \kappa) - e^{iK}\mathrm{Ai}'(-\kappa)\right]C + \left[\mathrm{Bi}'(\mu - \kappa) - e^{iK}\mathrm{Bi}'(-\kappa)\right]D &= 0,
\end{aligned} \qquad (7.4.19)$$

where the Ai$'$ and Bi$'$ with prime represent the derivatives of Ai and Bi functions. If these are expressed in a matrix form, we have

$$\begin{pmatrix} \mathrm{Ai}(\mu - \kappa) - e^{iK}\mathrm{Ai}(-\kappa) & \mathrm{Bi}(\mu - \kappa) - e^{iK}\mathrm{Bi}(-\kappa) \\ \mathrm{Ai}'(\mu - \kappa) - e^{iK}\mathrm{Ai}'(-\kappa) & \mathrm{Bi}'(\mu - \kappa) - e^{iK}\mathrm{Bi}'(-\kappa) \end{pmatrix} \begin{pmatrix} C \\ D \end{pmatrix} = \begin{pmatrix} 0 \\ 0 \end{pmatrix}. \qquad (7.4.20)$$

Because the C and D must not both be zero, the value of the coefficient determinant must be zero. Before analyzing that, we define the Wronskian made from the Airy functions Ai and Bi:

$$W(z) = \begin{vmatrix} \text{Ai}(z) & \text{Bi}(z) \\ \text{Ai}'(z) & \text{Bi}'(z) \end{vmatrix}. \tag{7.4.21}$$

This $W(z)$ should be a non-zero constant, because $\text{Ai}(z)$ and $\text{Bi}(z)$ are independent solutions satisfying equation (7.4.10). In fact, using the $z \geq 0$ part in expressions (7.4.12) and (7.4.13) for $\text{Ai}(z)$ and $\text{Bi}(z)$, and using the formula of the modified Bessel function

$$I_\alpha'(z)I_{-\alpha}(z) - I_\alpha(z)I_{-\alpha}'(z) = \frac{2\sin(\alpha\pi)}{\pi z}, \tag{7.4.22}$$

we can derive the result

$$W(z) \equiv \frac{1}{\pi}. \tag{7.4.23}$$

By using this relation, the value of the coefficient determinant of (7.4.20) being set to zero becomes

$$\cos(K) = \frac{\pi}{2}\Big[\text{Ai}(-\kappa)\text{Bi}'(\mu - \kappa) + \text{Ai}(\mu - \kappa)\text{Bi}'(-\kappa)$$
$$- \text{Ai}'(-\kappa)\text{Bi}(\mu - \kappa) - \text{Ai}'(\mu - \kappa)\text{Bi}(-\kappa)\Big]. \tag{7.4.24}$$

From this expression, the absolute value of the right-hand side must be 1 or less, because K must be solved as a real number. Since $\cos(K)$ is an even function of K, two positive and negative K values are obtained. This corresponds to the existence of the two phases $e^{\pm iK}$ as mentioned above. This determines the range that μ and κ should satisfy. Furthermore, the range of energy E for fixed V_0 from (7.4.7), that is, the band structure of energy, is determined. Therefore, this equation is the eigenvalue equation in this case.

If the constants C and D included in the wave function are set to the first expression of (7.4.19),

$$C = \text{Bi}(\mu - \kappa) - e^{iK}\text{Bi}(-\kappa), \quad D = -\text{Ai}(\mu - \kappa) + e^{iK}\text{Ai}(-\kappa), \tag{7.4.25}$$

then the wave function in the range $0 \leq x < 1$ is rewritten as

$$\Psi(x) = \big(\text{Bi}(\mu - \kappa) - e^{iK}\text{Bi}(-\kappa)\big)\text{Ai}(\mu x - \kappa)$$
$$- \big(\text{Ai}(\mu - \kappa) - e^{iK}\text{Ai}(-\kappa)\big)\text{Bi}(\mu x - \kappa). \tag{7.4.26}$$

Of course, this is not normalized. According to Floquet's theorem, every time the period goes up by one, the phase e^{iK} is attached. Therefore, the wave function $\Psi(x)$ for the range $n \leq x < n+1$, n being an arbitrary integer, is written as

$$\Psi(x, E) = e^{inK} \Big[\big(\text{Bi}(\mu - \kappa) - e^{iK} \text{Bi}(-\kappa) \big) \text{Ai}\big(\mu(x - n) - \kappa \big)$$

$$- \big(\text{Ai}(\mu - \kappa) - e^{iK} \text{Ai}(-\kappa) \big) \text{Bi}\big(\mu(x - n) - \kappa \big) \Big]. \quad (7.4.27)$$

In the following, the wave function is written as $\Psi(x, E)$ to clarify the energy dependence.

7.4.2 *Normalization of eigenfunction*

We consider two generically different energies E and E'. For κ and K belonging to these values, we put κ, κ' and K, K', respectively. Then the normalization integral, by considering expression (7.4.27), is written as

$$\int_{-\infty}^{\infty} \overline{\Psi(x, E)} \Psi(x, E') dx = \sum_{n=-\infty}^{\infty} \int_{n}^{n+1} \overline{\Psi(x, E)} \Psi(x, E') dx$$

$$= \Big(\sum_{n=-\infty}^{\infty} e^{-i(K-K')n} \Big) \int_{0}^{1} \overline{\Psi(x, E)} \Psi(x, E') dx. \quad (7.4.28)$$

For this sum part, the same generalized function formula as in (7.3.44) can be used,

$$\lim_{M \to \infty} \sum_{n=-M}^{M} e^{-i(K-K')n} = 2\pi \delta(K - K'), \quad 0 < K, \ K' < \pi. \quad (7.4.29)$$

So orthogonality in terms of the delta function is obtained. Next, to execute the integral part, we write the equation (7.4.8) corresponding to E and E' as

$$\Big(\frac{d^2}{dx^2} - \mu^3 x + \mu^2 \kappa \Big) \overline{\Psi(x, E)} = 0, \quad \Big(\frac{d^2}{dx^2} - \mu^3 x + \mu^2 \kappa' \Big) \Psi(x, E') = 0,$$

$$(7.4.30)$$

where the equation corresponding to E should be the complex conjugate. Next we multiply the first equation by $\Psi(x, E')$, the second equation by $\overline{\Psi(x, E)}$, and then subtract both sides to get

$$\frac{d}{dx} \Big(\Psi(x, E') \frac{d\overline{\Psi(x, E)}}{dx} - \overline{\Psi(x, E)} \frac{d\Psi(x, E')}{dx} \Big)$$

$$+ \mu^2 (\kappa - \kappa') \overline{\Psi(x, E)} \Psi(x, E') = 0. \quad (7.4.31)$$

By integrating this from 0 to 1, it becomes

$$\int_0^1 \overline{\Psi(x, E)}\Psi(x, E')dx$$

$$= \frac{1}{\mu^2(\kappa - \kappa')}\left(\overline{\Psi(x, E)}\frac{d\Psi(x, E')}{dx} - \Psi(x, E')\frac{d\overline{\Psi(x, E)}}{dx}\right)\Bigg|_0^1. \quad (7.4.32)$$

Expression (7.4.26) should be substituted into this right-hand side, but this calculation becomes quite cumbersome. When it is substituted, 64 terms appear, but half of them are canceled, leaving 32 terms. In addition, if we use relation (7.4.23) of Wronskian, the term proportional to e^{-iK} and $e^{iK'}$ can be reduced by half, and finally, 20 terms remain. Furthermore, to summarize the results well, we will introduce a factor $S(\kappa)$ that plays an important role later,

$$S(\kappa) = \text{Ai}(-\kappa)\text{Bi}(\mu - \kappa) - \text{Ai}(\mu - \kappa)\text{Bi}(-\kappa). \quad (7.4.33)$$

By using this function, expression (7.4.32) is expressed in a little long form

$$\int_0^1 \overline{\Psi(x, E)}\Psi(x, E')dx$$

$$= \frac{1}{\mu(\kappa - \kappa')}\Bigg[S(\kappa)\big(\text{Ai}(\mu - \kappa')\text{Bi}'(-\kappa') - \text{Ai}'(-\kappa')\text{Bi}(\mu - \kappa')\big)$$

$$- S(\kappa')\big(\text{Ai}(\mu - \kappa)\text{Bi}'(-\kappa) - \text{Ai}'(-\kappa)\text{Bi}(\mu - \kappa)\big)$$

$$+ e^{i(K'-K)}S(\kappa)\big(\text{Ai}(-\kappa')\text{Bi}'(\mu - \kappa') - \text{Ai}'(\mu - \kappa')\text{Bi}(-\kappa')\big)$$

$$- e^{i(K'-K)}S(\kappa')\big(\text{Ai}(-\kappa)\text{Bi}'(\mu - \kappa) - \text{Ai}'(\mu - \kappa)\text{Bi}(-\kappa)\big)$$

$$+ \frac{1}{\pi}\big(e^{iK'} + e^{-iK}\big) \times \big(S(\kappa') - S(\kappa)\big)\Bigg]. \quad (7.4.34)$$

It is noted that when $\kappa \to \kappa'$ in this expression, it becomes an indeterminate form of $0/0$. Further, if we deform this expression using the eigenvalue equation (7.4.24),

$$\int_0^1 \overline{\Psi(x, E)}\Psi(x, E')dx$$

$$= \frac{1}{\mu(\kappa - \kappa')}\Big[\big(e^{i(K'-K)} - 1\big)S(\kappa)\big(\text{Ai}(-\kappa')\text{Bi}'(\mu - \kappa') - \text{Ai}'(\mu - \kappa')\text{Bi}(-\kappa')\big)$$

$$- \big(e^{i(K'-K)} - 1\big)S(\kappa')\big(\text{Ai}(-\kappa)\text{Bi}'(\mu - \kappa) - \text{Ai}'(\mu - \kappa)\text{Bi}(-\kappa)\big)$$

$$+ \frac{1}{\pi}S(\kappa)\big(2\cos(K') - (e^{iK'} + e^{-iK})\big) - \frac{1}{\pi}S(\kappa')\big(2\cos(K) - (e^{iK'} + e^{-iK})\big)\Big].$$

$$(7.4.35)$$

Since the original normalization integral (7.4.28) is the product of this expression and expression (7.4.29) of the delta function, we obtain

$$\int_{-\infty}^{\infty} \overline{\Psi(x, E)}\Psi(x, E')dx$$

$$= \frac{2\pi}{\mu(\kappa - \kappa')}\Big[\big(e^{i(K'-K)} - 1\big)S(\kappa)\big(\text{Ai}(-\kappa')\text{Bi}'(\mu - \kappa') - \text{Ai}'(\mu - \kappa')\text{Bi}(-\kappa')\big)$$

$$- \big(e^{i(K'-K)} - 1\big)S(\kappa')\big(\text{Ai}(-\kappa)\text{Bi}'(\mu - \kappa) - \text{Ai}'(\mu - \kappa)\text{Bi}(-\kappa)\big)$$

$$+ \frac{1}{\pi}S(\kappa)\big(2\cos(K') - (e^{iK'} + e^{-iK})\big)$$

$$- \frac{1}{\pi}S(\kappa')\big(2\cos(K) - (e^{iK'} + e^{-iK})\big)\Big]\delta(K - K'). \tag{7.4.36}$$

There is one thing to note here. Since the value of K belongs to the range $0 < K < \pi$ for each band, if the bands are different, even if $K = K'$, it will never be $\kappa = \kappa'$. Therefore, the value of this expression becomes zero. That is, the orthogonality between different bands is derived.

When E and E' belong to the same band, this expression is an indeterminate form of $0/0$. Therefore, according to l'Hôpital's rule, we partially differentiate the denominator and numerator with respect to κ, respectively, and after that, we set $\kappa' = \kappa$. Then, of course, K is treated as a function of κ. The result is a surprisingly simple form:

$$\int_{-\infty}^{\infty} \overline{\Psi(x, E)}\Psi(x, E')dx = -\frac{4}{\mu}S(\kappa)\frac{d\cos(K)}{d\kappa}\delta(K - K'). \tag{7.4.37}$$

The part of this expression excluding the delta function should be positive definite. In fact, we can derive the following relation by using the eigenvalue equation (7.4.24) and equation (7.4.10) which is satisfied by the Airy functions Ai and Bi,

$$\frac{d\cos(K)}{d\kappa} = -\frac{\pi\mu}{2}S(\kappa). \tag{7.4.38}$$

Therefore, the orthogonality relation becomes obviously positive definite:

$$\int_{-\infty}^{\infty} \overline{\Psi(x, E)}\Psi(x, E')dx = 2\pi S^2(\kappa)\delta(K - K'). \tag{7.4.39}$$

Alternatively, if $\delta(K - K')$ is represented by $\delta(E - E')$, it becomes

$$\int_{-\infty}^{\infty} \overline{\Psi(x, E)}\Psi(x, E')dx = N^2(E)\delta(E - E'). \qquad (7.4.40)$$

Here, the normalization constant $N^2(E)$ is defined as

$$N^2(E) = 4\mu |S(\kappa)| \sin(K). \qquad (7.4.41)$$

As a result, $\Psi(x, E)/N(E)$ becomes a normalized eigenfunction.

Incidentally, as in the previous section, we will investigate the stream function here. This function is defined as $j = i\overline{\Psi}(\overleftarrow{\partial_x} - \overrightarrow{\partial_x})\Psi$ as shown in (7.3.60). By using Ψ of expression (7.4.27), we find

$$j = \frac{2\mu}{\pi}S(\kappa)\sin(K). \qquad (7.4.42)$$

As seen in the numerical calculation in the next subsection, since the sign of $S(\kappa)$ should take a positive and negative value for each band, and here, we set $0 < K < \pi$, the direction of flow repeats positive and negative values for each band alternately in the increasing order of energy. Also, as mentioned earlier, because there is a degeneracy in the wave function in this model, there is another case of a sign change in K. In this case, the direction of flow repeats the negative and positive values for each band.

We expected that there would be a big difference between the right traveling wave and the left traveling wave, because the potential dealt with here has a specific direction as a serrated model. However, the result shows that there is no difference between the right and left traveling waves.

7.4.3 *Numerical calculation*

We try to analyze numerically the eigenvalue equation (7.4.24). In calculating the Airy functions numerically, we deal with them by using the series expansion formula (7.4.14) with the sum of n up to 50 when $|z| \leq 2$, and using the asymptotic form of (7.4.15) or (7.4.16) when $|z| > 2$.

First, in the eigenvalue equation (7.4.24), for the fixed value of V_0, we calculate numerically the propagation number K for the energy E of the range $0 \leq E < 425$. The result is shown in Figs. 7.9 and Fig. 7.10 which correspond to $V_0 = 100$ and $V_0 = 200$, respectively.

Since the factor $S(\kappa)$ is represented by $S(\kappa) = (2/\pi\mu)\sin(K)(dK/d\kappa)$ from expression (7.4.38), it is noted that the sign of $S(\kappa)$ coincides with the sign of the slope of the curve in this graph for $0 < K < \pi$.

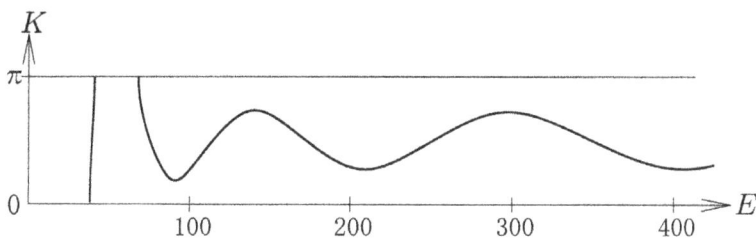

Fig. 7.9 $V_0 = 100$

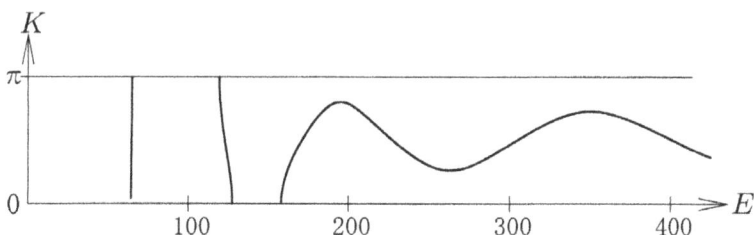

Fig. 7.10 $V_0 = 200$

In Fig. 7.9, there is an independent band near $E \cong 40$. However, from around $E \cong 70$, it becomes a continuous spectrum without forming a perfect band structure, the value of K oscillates within the range not exceeding 0 and π. Moreover, the amplitude of this vibration becomes smaller and the period becomes larger as the energy E increases. In the limit of E being infinite, since the Airy functions Ai and Bi become zero from the asymptotic form of (7.4.16), the value of K should converge to $\pi/2$. Also, in Fig. 7.10, there are two bands near $E \cong 65$ and 120, and above the neighborhood of $E \cong 160$, the spectrum is also continuous. These facts are the major features that have not been seen in the periodic potential so far. The band structure means that originally the admitted energy has a finite range and becomes discrete. The part of this continuous spectrum is considered an incomplete band due to the adhesion of the bands. The band structure which is used in the normalization of eigenfunctions should be understood as a discussion including this incomplete band structure.

It is impossible to analytically solve V_0 from the eigenvalue equation (7.4.24). Therefore, the value of V_0 is not fixed, and the value of the right-hand side of the eigenvalue equation (7.4.24) is calculated for each point

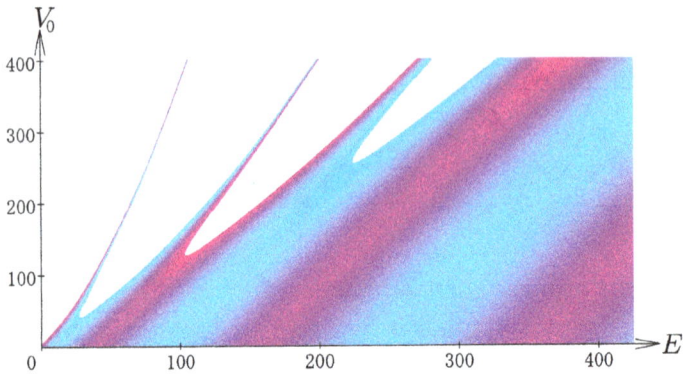

Fig. 7.11 Relation of V_0 and E

while scanning the two-dimensional space of V_0 and E. With this scanning method, Fig. 7.11 is plotted in the following way: when the absolute value of the right-hand side is more than 1, nothing is drawn; when it is 1, a magenta dot is plotted; when it is -1, a cyan dot is plotted; and when the value is in the middle of 1 and -1, the point is plotted in a color that changes continuously from magenta to cyan.

Also from this figure, the energy band structure appears in the region of $V_0 > E$ approximately. However, it can be seen that the spectrum is completely continuous without forming the band structure in the region of $V_0 < E$.

On the Appropriateness of the Name "Propagation Number" K

I analyzed here the so-called serrated shape potential problem which is not known what it is useful for. This is because I was interested in how the orthogonality of eigenfunctions can be derived through concrete examples which are solvable analytically. Regarding periodic potential problems, many technical books were written in a very ambiguous way. However, I resolved this problem through the analysis of concrete examples.

In the following, I will mention the name of the variable K which is called the "propagation number". Some books call K the "wave number" indiscriminately, and exploit the fact that e^{iK}, $\cos(K)$ have a period of 2π, and add 2π to K every time the energy band goes up by one. As mentioned in Floquet's theorem (7.3.26) and (7.3.27), the wave function in the periodic potential problem is written in the form

$$\Psi(x, E) = e^{iKx} \times [\text{ periodic function }].$$

From this expression, it might be understandable that the K is called the wave number. However, the correct definition of a wave number is the integral variable k, when the wave function is represented by Fourier transform as

$$\Psi(x, E) = \int_{-\infty}^{\infty} e^{ikx} \hat{\Psi}(k, E) dk.$$

Also, the name "propagation number" is completely unrelated to the name imagined from the "propagator" used in quantum field theory.

I have always been doubtful about the appropriateness of the name "propagation number" or "wave number". When a band and a continuous spectrum are mixed, as dealt with here, there is no meaning of propagation or wave number. Moreover, the stream function j is not proportional to the K as shown in expression (7.4.42). It should only mean that the K is simply the number that determines the phase when the period goes up by one in Floquet's theorem. Therefore, K could be called the "phase constant", or better still "Floquet's phase constant", named after the discoverer. What do readers think?

Formulas of Special Functions

Here is a summary of the special functions used in Part II of this book. These formulas are cited from *Mathematical Formula II, III*[1] and *Higher Transcendental Functions, Vol. 1, Vol. 2.*[2]

Gamma Function

The original gamma function is defined by

$$\Gamma(z) = \int_0^\infty e^{-t} t^{z-1} dt, \quad \Re(z) > 0.$$

The gamma function commonly used is the analytic continuation of this definition for z:

$$\Gamma(z+1) = z\Gamma(z), \quad \Gamma(n+1) = n!, \quad n = 0, 1, 2, \ldots.$$

When $z = -n$, n being a non-negative integer, $\Gamma(z)$ becomes a pole of order 1 of which residue is $(-1)^n/n!$.

Gauss–Legendre multiplication formula, n being a positive integer:

$$\Gamma(nz) = \frac{n^{nz}}{(2\pi)^{(n-1)/2}\sqrt{n}} \prod_{k=0}^{n-1} \Gamma\left(z + \frac{k}{n}\right).$$

[1] S. Moriguchi, K. Udagawa and S. Hitosumatsu (Eds), *Mathmatical Formula II, III*, (1956) Iwanami shoten (in Japanese).
[2] A. Erdélyi (Ed), *Higher Transcendental Functions, Vol. 1, Vol. 2*, Bateman Manuscript Project, California Institute of Technology, (1953–1955) McGraw-Hill, New York.

In the particular case of $n = 2$, double product formula:

$$\Gamma(2z) = \frac{2^{2z}}{2\sqrt{\pi}}\Gamma(z)\Gamma\left(z + \frac{1}{2}\right),$$

and in the case of $n = 3$, triple product formula:

$$\Gamma(3z) = \frac{3^{3z}}{2\pi\sqrt{3}}\Gamma(z)\Gamma\left(z + \frac{1}{3}\right)\Gamma\left(z + \frac{2}{3}\right).$$

Reflection formula:

$$\Gamma(z)\Gamma(1-z) = \frac{\pi}{\sin(\pi z)}, \quad \Gamma\left(\frac{1}{2}+z\right)\Gamma\left(\frac{1}{2}-z\right) = \frac{\pi}{\cos(\pi z)}, \quad \Gamma(1/2) = \sqrt{\pi}.$$

Hypergeometric Function

Gauss' hypergeometric function $F(\alpha, \beta; \gamma; z)$ is defined by using a series expansion

$$F(\alpha, \beta; \gamma; z) = \sum_{k=0}^{\infty} \frac{\Gamma(\alpha + k)\Gamma(\beta + k)\Gamma(\gamma)}{\Gamma(\alpha)\Gamma(\beta)\Gamma(\gamma + k)} \frac{z^k}{k!}.$$

In particular, if α or β is $-n$, n being a non-negative integer, $F(\alpha, \beta; \gamma; z)$ becomes a polynomial of order n.

This function $u = F(\alpha, \beta; \gamma; z)$ satisfies the hypergeometric differential equation

$$\left[z(1-z)\frac{d^2}{dz^2} + \left(\gamma - (\alpha + \beta + 1)z\right)\frac{d}{dz} - \alpha\beta\right]u = 0.$$

There is another independent solution in addition to $F(\alpha, \beta; \gamma; z)$:

$$u = z^{1-\gamma}F(\alpha - \gamma + 1, \beta - \gamma + 1; 2 - \gamma; z).$$

Kummer's transformation formula:

$$F(\alpha, \beta; \gamma; z) = (1 - z)^{\gamma - \alpha - \beta}F(\gamma - \alpha, \gamma - \beta; \gamma; z)$$
$$= (1 - z)^{-\alpha}F(\alpha, \gamma - \beta; \gamma; z/(z - 1)).$$

Gauss' transformation formula:

$$F(\alpha, \beta; \gamma; z) = \frac{\Gamma(\alpha + \beta - \gamma)\Gamma(\gamma)}{\Gamma(\alpha)\Gamma(\beta)}(1 - z)^{\gamma - \alpha - \beta}$$
$$\times F(\gamma - \alpha, \gamma - \beta; \gamma - \alpha - \beta + 1; 1 - z)$$
$$+ \frac{\Gamma(\gamma - \alpha - \beta)\Gamma(\gamma)}{\Gamma(\gamma - \alpha)\Gamma(\gamma - \beta)}F(\alpha, \beta; \alpha + \beta - \gamma + 1; 1 - z).$$

Bessel Functions and Modified Bessel Functions

The Bessel function $J_\nu(z)$ is defined by using the series expansion

$$J_\nu(z) = \sum_{k=0}^{\infty} \frac{(-1)^k}{k!\,\Gamma(\nu+k+1)} \left(\frac{z}{2}\right)^{\nu+2k}.$$

This function $u = J_\nu(z)$ satisfies Bessel's differential equation

$$\left[\frac{d^2}{dz^2} + \frac{1}{z}\frac{d}{dz} + 1 - \frac{\nu^2}{z^2}\right] u = 0.$$

There is another independent solution called the Neumann function $N_\nu(z)$ to this equation:

$$u = N_\nu(z) \equiv \frac{\cos(\nu\pi)J_\nu(z) - J_{-\nu}(z)}{\sin(\nu\pi)}.$$

If ν is an integer, this definition becomes the indeterminate form of $0/0$, so it is defined by taking the limit value. In addition, the first and second kind Hankel functions, which are linear combinations of $J_\nu(z)$ and $N_\nu(z)$, are defined by

$$H_\nu^{(1)}(z) = J_\nu(z) + iN_\nu(z), \quad H_\nu^{(2)}(z) = J_\nu(z) - iN_\nu(z).$$

The modified Bessel function $I_\nu(z)$ is defined by

$$I_\nu(z) = \sum_{k=0}^{\infty} \frac{1}{k!\,\Gamma(\nu+k+1)} \left(\frac{z}{2}\right)^{\nu+2k}.$$

This function $w = I_\nu(z)$ satisfies the modified Bessel's differential equation

$$\left[\frac{d^2}{dz^2} + \frac{1}{z}\frac{d}{dz} - 1 - \frac{\nu^2}{z^2}\right] w = 0.$$

There is another modified Bessel function which satisfies the equation

$$w = K_\nu(z) \equiv \frac{\pi}{2} \cdot \frac{I_{-\nu}(z) - I_\nu(z)}{\sin(\nu\pi)}.$$

Summation formulas:

$$J_0(z) + 2\sum_{r=1}^{\infty} J_{2r}(z) = 1, \quad J_0(z) + 2\sum_{r=1}^{\infty}(-1)^r J_{2r}(z) = \cos z.$$

$$\sum_{r=-\infty}^{\infty} J_r^2(z) = 1, \quad \sum_{r=-\infty}^{\infty} r^\ell J_r(z)J_{r+m}(z) = 0, \quad \ell = 0, 1, \ldots, m-1,$$

m is a positive integer.

Recurrence formulas:

$$J_{\nu-1}(z) + J_{\nu+1}(z) = \frac{2\nu}{z}J_\nu(z), \quad N_{\nu-1}(z) + N_{\nu+1}(z) = \frac{2\nu}{z}N_\nu(z),$$

$$I_{\nu-1}(z) - I_{\nu+1}(z) = \frac{2\nu}{z}I_\nu(z), \quad K_{\nu-1}(z) - K_{\nu+1}(z) = -\frac{2\nu}{z}K_\nu(z).$$

Differential formulas:

$$\frac{d}{dz}J_\nu(z) = \frac{1}{2}\left[J_{\nu-1}(z) - J_{\nu+1}(z)\right], \quad \frac{d}{dz}N_\nu(z) = \frac{1}{2}\left[N_{\nu-1}(z) - N_{\nu+1}(z)\right],$$

$$\frac{d}{dz}I_\nu(z) = \frac{1}{2}\left[I_{\nu-1}(z) + I_{\nu+1}(z)\right], \quad \frac{d}{dz}K_\nu(z) = -\frac{1}{2}\left[K_{\nu-1}(z) + K_{\nu+1}(z)\right].$$

$$\left(\frac{1}{z}\frac{d}{dz}\right)^n\left(\frac{J_\nu(z)}{z^\nu}\right) = (-1)^n\frac{J_{\nu+n}(z)}{z^{\nu+n}}, \quad \left(\frac{1}{z}\frac{d}{dz}\right)^n\left(\frac{N_\nu(z)}{z^\nu}\right) = (-1)^n\frac{N_{\nu+n}(z)}{z^{\nu+n}},$$

$$\left(\frac{1}{z}\frac{d}{dz}\right)^n\left(\frac{I_\nu(z)}{z^\nu}\right) = \frac{I_{\nu+n}(z)}{z^{\nu+n}}, \quad \left(\frac{1}{z}\frac{d}{dz}\right)^n\left(\frac{K_\nu(z)}{z^\nu}\right) = (-1)^n\frac{K_{\nu+n}(z)}{z^{\nu+n}}.$$

$$\left(\frac{1}{z}\frac{d}{dz}\right)^n\left(z^\nu J_\nu(z)\right) = z^{\nu-n}J_{\nu-n}(z), \quad \left(\frac{1}{z}\frac{d}{dz}\right)^n\left(z^\nu N_\nu(z)\right) = z^{\nu-n}N_{\nu-n}(z),$$

$$\left(\frac{1}{z}\frac{d}{dz}\right)^n\left(z^\nu I_\nu(z)\right) = z^{\nu-n}I_{\nu-n}(z),$$

$$\left(\frac{1}{z}\frac{d}{dz}\right)^n\left(z^\nu K_\nu(z)\right) = (-1)^n z^{\nu-n}K_{\nu-n}(z).$$

Integral formulas:

$$\int z^{\nu+1}J_\nu(\alpha z)dz = \frac{z^{\nu+1}}{\alpha}J_{\nu+1}(\alpha z),$$

$$\int z^{-\nu+1}J_\nu(\alpha z)dz = -\frac{z^{-\nu+1}}{\alpha}J_{\nu-1}(\alpha z).$$

These formulas can also be used for $N_\nu(\alpha z)$, $H_\nu^{(1)}(\alpha z)$ and $H_\nu^{(2)}(\alpha z)$ as it is:

$$\int J_\nu(z)dz = 2\sum_{r=0}^{\infty} J_{\nu+2r+1}(z).$$

Convolution integral:

$$\int_0^x J_\mu(x-t)J_\nu(t)dt = 2\sum_{r=0}^{\infty}(-1)^r J_{\mu+\nu+2r+1}(x).$$

Integral representation:

$$J_n(z) = \frac{1}{\pi} \int_0^\pi \cos(nt - z \sin t) dt = \frac{1}{\pi} \int_0^\pi e^{i(nt - z \sin t)} dt,$$

$$J_{2n}(z) = \frac{2}{\pi} \int_0^{\pi/2} \cos(2nt) \cos(z \sin t) dt, \quad n = 0, 1, 2, \dots,$$

$$J_{2n+1}(z) = \frac{2}{\pi} \int_0^{\pi/2} \sin((2n+1)t) \sin(z \sin t) dt, \quad n = 0, 1, 2, \dots,$$

$$J_\nu(z) = \frac{e^{i\nu\pi/2}}{\pi} \left[\int_0^\pi e^{-iz \cos t} \cos(\nu t) dt - \sin(\nu\pi) \int_0^\infty e^{-\nu t + iz \cosh t} dt \right],$$

$$0 < \arg(z) < \pi,$$

$$K_\nu(z) = \frac{1}{2} \int_{-\infty}^\infty e^{-\nu t - z \cosh t} dt = \int_0^\infty e^{-z \cosh t} \cosh(\nu t) dt.$$

Asymptotic forms, x is positive and large enough:

$$J_\nu(x) \sim \sqrt{\frac{2}{\pi x}} \cos\left(x - \frac{2\nu+1}{4}\pi\right), \quad N_\nu(x) \sim \sqrt{\frac{2}{\pi x}} \sin\left(x - \frac{2\nu+1}{4}\pi\right),$$

$$H_\nu^{(1)}(x) \sim \sqrt{\frac{2}{\pi x}} e^{i(x - (2\nu+1)\pi/4)}, \quad H_\nu^{(2)}(x) \sim \sqrt{\frac{2}{\pi x}} e^{-i(x - (2\nu+1)\pi/4)},$$

$$I_\nu(x) \sim \frac{1}{\sqrt{2\pi x}} e^x, \quad K_\nu(x) \sim \sqrt{\frac{\pi}{2x}} e^{-x}.$$

Orthoganality relation: let the zero points of the Bessel function $J_n(z)$, ($n = 0, 1, 2, \dots$) be ξ_1, ξ_2, $\dots$ in increasing order, the following orthogonality relation holds,

$$\int_0^1 J_n(\xi_k x) J_n(\xi_{k'} x)\, x dx = \frac{1}{2} \left[J_{n+1}(\xi_k) \right]^2 \delta_{k,k'}.$$

Airy Function

Two independent Airy functions $\mathrm{Ai}(z)$ and $\mathrm{Bi}(z)$ are defined by using the Bessel functions

$$\mathrm{Ai}(z) = \begin{cases} \dfrac{1}{\pi} \sqrt{\dfrac{z}{3}} K_{1/3}\left(\tfrac{2}{3} z^{3/2}\right), & z \geq 0, \\[2ex] \dfrac{1}{3} \sqrt{-z} \left[J_{-1/3}\left(\tfrac{2}{3}(-z)^{3/2}\right) + J_{1/3}\left(\tfrac{2}{3}(-z)^{3/2}\right) \right], & z < 0. \end{cases}$$

$$\mathrm{Bi}(z) = \begin{cases} \sqrt{\dfrac{z}{3}}\left[I_{-1/3}\left(\tfrac{2}{3}z^{3/2}\right) + I_{1/3}\left(\tfrac{2}{3}z^{3/2}\right)\right], & z \geq 0, \\[2mm] \sqrt{-\dfrac{z}{3}}\left[J_{-1/3}\left(\tfrac{2}{3}(-z)^{3/2}\right) - J_{1/3}\left(\tfrac{2}{3}(-z)^{3/2}\right)\right], & z < 0. \end{cases}$$

These functions $u = \mathrm{Ai}(z)$ and $u = \mathrm{Bi}(z)$ satisfy the differential equation

$$\left(\frac{d^2}{dz^2} - z\right)u = 0,$$

and they are the entire functions expressed by the series expansions

$$\mathrm{Ai}(z) = \frac{1}{3^{2/3}}\sum_{n=0}^{\infty}\frac{z^{3n}}{n!\,\Gamma(n+\tfrac{2}{3})3^{2n}} - \frac{1}{3^{4/3}}\sum_{n=0}^{\infty}\frac{z^{3n+1}}{n!\,\Gamma(n+\tfrac{4}{3})3^{2n}},$$

$$\mathrm{Bi}(z) = \frac{1}{3^{1/6}}\sum_{n=0}^{\infty}\frac{z^{3n}}{n!\,\Gamma(n+\tfrac{2}{3})3^{2n}} + \frac{1}{3^{5/6}}\sum_{n=0}^{\infty}\frac{z^{3n+1}}{n!\,\Gamma(n+\tfrac{4}{3})3^{2n}}.$$

Hereafter, the variable z is limited to a real number and written as x.

The asymptotic forms of these functions are obtained from that of the Bessel functions. When the variable x is positive and large enough, they become

$$\mathrm{Ai}(x) \sim \frac{e^{-\frac{2}{3}x^{3/2}}}{2\sqrt{\pi}\,x^{1/4}}, \qquad \mathrm{Bi}(x) \sim \frac{e^{\frac{2}{3}x^{3/2}}}{\sqrt{\pi}\,x^{1/4}}.$$

Also, the asymptotic forms when x is negative and large enough are obtained as

$$\mathrm{Ai}(x) \sim \frac{\sin\left(\tfrac{2}{3}(-x)^{3/2} + \tfrac{\pi}{4}\right)}{\sqrt{\pi}\,(-x)^{1/4}}, \qquad \mathrm{Bi}(x) \sim \frac{\cos\left(\tfrac{2}{3}(-x)^{3/2} + \tfrac{\pi}{4}\right)}{\sqrt{\pi}\,(-x)^{1/4}}.$$

The Airy integral representation:

$$\mathrm{Ai}(x) = \frac{1}{\pi}\int_0^{\infty}\cos\left(\frac{t^3}{3} + xt\right)dt,$$

$$\mathrm{Bi}(x) = \frac{1}{\pi}\int_0^{\infty}\left[\exp\left(-\frac{t^3}{3} + xt\right) + \sin\left(\frac{t^3}{3} + xt\right)\right]dt.$$

Delta Function

We cite here the formulas of the Dirac delta function used in this book, although these are not special functions:

$$\lim_{M \to \infty} \frac{\sin(Mx)}{x} = \pi\delta(x), \quad \lim_{M \to \infty} \frac{\cos(Mx)}{x} = 0, \quad \lim_{M \to \infty} \frac{e^{\pm iMx}}{x} = \pm i\pi\delta(x),$$

and

$$\lim_{M \to \infty} \sum_{n=-M}^{M} e^{inx} = \lim_{M \to \infty} \frac{\sin\left((M + \frac{1}{2})x\right)}{\sin(x/2)} = 2\pi \sum_{n=-\infty}^{\infty} \delta(x - 2\pi n).$$

If we restrict the range of x within one period $-\pi \le x \le \pi$, this formula becomes

$$\lim_{M \to \infty} \sum_{n=-M}^{M} e^{inx} = 2\pi\delta(x).$$

Epilogue to PART II

The natural world is designed so that mathematical logic can be applied. Moreover, this logic has a prophetic power to create something new.

Newton's equation predicted the existence of universal gravitation.

Maxwell's equation predicted the existence of the electromagnetic wave.

Einstein's equation predicted the existence of the gravitational wave.

Dirac's equation predicted the existence of the antiparticle.

****

I wrote in Chapter 6 about the significance of finding an exact solution.

Certainly, current simulators are very advanced and they can analyze numerically in almost every situation. For example, they can analyze the way a building or a highway-bridge with complex structure vibrates in the event of an earthquake. We input the building's structural data and earthquake data into a computer, and then we do nothing but wait for the computer to produce the results. We are amazed at the excellence of today's simulators. However, we should not be satisfied with this, because simulators have no prophetic power. We feel that the expansion of mathematical logic is more necessary for the development of future science.

Index

www.ingramcontent.com/pod-product-compliance
Lightning Source LLC
Chambersburg PA
CBHW050536190326
41458CB00007B/1800